Biochemistry in the Space of the Highest Dimension

Gennadiy Vladimirovich Zhizhin
Russian Academy of Natural Sciences, Russia

A volume in the Advances in Chemical and Materials Engineering (ACME) Book Series

IGI Global
PUBLISHER of TIMELY KNOWLEDGE

Published in the United States of America by
IGI Global
Engineering Science Reference (an imprint of IGI Global)
701 E. Chocolate Avenue
Hershey PA, USA 17033
Tel: 717-533-8845
Fax: 717-533-8661
E-mail: cust@igi-global.com
Web site: http://www.igi-global.com

Library of Congress Cataloging-in-Publication Data

Names: Zhizhin, G. V. (Gennadiĭ Vladimirovich), author.
Title: Biochemistry in the space of the highest dimension / by: Gennadiy Zhizhin.
Description: Hershey PA : Engineering Science Reference, [2024] | Includes bibliographical references. | Summary: "In general, the monograph considers a number of new fundamental problems concerning the conditions of life on the planet Earth, the health of its inhabitants, and new fundamental solutions are obtained that contribute to the implementation of these decisions"-- Provided by publisher.
Identifiers: LCCN 2023041790 (print) | LCCN 2023041791 (ebook) | ISBN 9798369305881 (hardcover) | ISBN 9798369305928 (paperback) | ISBN 9798369305898 (ebook)
Subjects: LCSH: Biochemistry--Mathematics. | Dimension theory (Topology)
Classification: LCC QP517.M3 Z94 2024 (print) | LCC QP517.M3 (ebook) | DDC 572--dc23/eng/20231113
LC record available at https://lccn.loc.gov/2023041790
LC ebook record available at https://lccn.loc.gov/2023041791

This book is published in the IGI Global book series Advances in Chemical and Materials Engineering (ACME) (ISSN: 2327-5448; eISSN: 2327-5456)

British Cataloguing in Publication Data
A Cataloguing in Publication record for this book is available from the British Library.

Advances in Chemical and Materials Engineering (ACME) Book Series

J. Paulo Davim
University of Aveiro, Portugal

ISSN:2327-5448
EISSN:2327-5456

MISSION

The cross disciplinary approach of chemical and materials engineering is rapidly growing as it applies to the study of educational, scientific and industrial research activities by solving complex chemical problems using computational techniques and statistical methods.

The **Advances in Chemical and Materials Engineering (ACME) Book Series** provides research on the recent advances throughout computational and statistical methods of analysis and modeling. This series brings together collaboration between chemists, engineers, statisticians, and computer scientists and offers a wealth of knowledge and useful tools to academics, practitioners, and professionals through high quality publications.

COVERAGE

- Polymers Engineering
- Numerical Techniques
- Statistical Methods
- Electrochemical and Corrosion
- Biomaterials
- Heat Treatments
- Toxicity of Materials
- Multifuncional and Smart Materials
- Wear of Materials
- Fracture Mechanics

IGI Global is currently accepting manuscripts for publication within this series. To submit a proposal for a volume in this series, please contact our Acquisition Editors at Acquisitions@igi-global.com or visit: http://www.igi-global.com/publish/.

The Advances in Chemical and Materials Engineering (ACME) Book Series (ISSN 2327-5448) is published by IGI Global, 701 E. Chocolate Avenue, Hershey, PA 17033-1240, USA, www.igi-global.com. This series is composed of titles available for purchase individually; each title is edited to be contextually exclusive from any other title within the series. For pricing and ordering information please visit http://www.igi-global.com/book-series/advances-chemical-materials-engineering/73687. Postmaster: Send all address changes to above address.

Titles in this Series

For a list of additional titles in this series, please visit:
http://www.igi-global.com/book-series/advances-chemical-materials-engineering/73687

FlexPDE and Finite Element Method Applications in Thermal Energy Storage and Cavities
Khashayar Hosseinzadeh (Sharif University of Technology, Iran) Mehdi Mahboobtosi (Babol Noshirvani University of Technology, Iran) and Davood Domairry Ganji (Babol Noshirvani University of Technology, Iran)
Engineering Science Reference • © 2024 • 380pp • H/C (ISBN: 9798369309322) • US $225.00

Next Generation Materials for Sustainable Engineering
Shrikaant Kulkarni (Padmashri Dr. V.B. Kolte College of Engineering, India) Boris Kharisov (Autonomous University of Nuevo Leon, Mexico) A.K. Haghi (University of Coimbra, Portugal) and Vipul Srivastava (Lovely Professional University, India)
Engineering Science Reference • © 2024 • 442pp • H/C (ISBN: 9798369313060) • US $295.00

Advancements in Powder Metallurgy Processing, Applications, and Properties
Shashanka Rajendrachari (Bartin University, Turkey) and Rayappa Shrinivas Mahale (Jain College of Engineering and Research, India)
Engineering Science Reference • © 2024 • 313pp • H/C (ISBN: 9781668493854) • US $275.00

Sustainable Approach to Protective Nanocoatings
Rajni Garg (Galgotias College of Engineering and Technology, India) and Ansar Anjum (Galgotias College of Engineering and Technology, India)
Engineering Science Reference • © 2024 • 443pp • H/C (ISBN: 9798369331361) • US $345.00

Innovations and Applications of Hybrid Nanomaterials
Virat Khanna (Maharaja Agrasen University, India) Prianka Sharma (Maharaja Agrasen University, India) and Priyanka Mahajan (Maharaja Agrasen University, India)
Engineering Science Reference • © 2024 • 336pp • H/C (ISBN: 9798369332689) • US $315.00

IGI Global
PUBLISHER of TIMELY KNOWLEDGE

701 East Chocolate Avenue, Hershey, PA 17033, USA
Tel: 717-533-8845 x100 • Fax: 717-533-8661
E-Mail: cust@igi-global.com • www.igi-global.com

Table of Contents

Detailed Table of Contents

The possibility of creating vacant electron orbitals in a space of higher dimension in the vicinity of the nucleus of an atom, leading to an increase in the electrophilicity of atoms and the chemical activity of elements due to the arrangement of electron pairs in these orbitals, is investigated. It is shown that the stationary Schrödinger equation describing the distribution of electrons in the vicinity of the atomic nucleus has a solution, in principle, for any dimensionality of the space around the nucleus. As an example, a solution of the Schrödinger equation in a five - dimensional space is obtained. It is shown that the solution of the Schrödinger equation in p - dimensional space has p quantum numbers: the principal quantum number, the orbital quantum number and p - 2 magnetic quantum numbers. Taking into account the spin quantum number, the total number of quantum numbers in p - dimensional space is p + 1. This leads to the possibility of increasing the number of quantum cells of orbitals and, consequently, to the possibility of increasing the valence of the elements.

This chapter establishes the existence of integral equality relatively, to the issue of the transmission of information by elements of lower and higher dimensions in the polytopes of the higher dimension that describe natural objects. This integral equality was established for the first time when studying the nanoworld in a higher-dimensional space (Zhizhin, 2019 a). As we will see in later chapters, this equality is of great importance in higher dimensional biochemistry. Therefore, this equality (the law of conservation of incidence) can be considered one of the concepts of biochemistry of the world of higher dimension. This integral equality is called the law of conservation of incidents. There is incidence interpreted as the transfer of information from one material body to another.

five-carbon model of a sugar molecule. When building a model of a double-stranded model of a nucleic acid molecule, a special type of polytope is used: a cross-polytope. The features of the interaction of nucleic acids in ribosomes responsible for the synthesis of protein molecules according to a given program are considered.

Chapter 6

On the basis of images of higher dimension of nucleic acids, the properties of the interaction of nucleic acids with each other have been studied. It was found that the region of interaction of nucleic acids is a cross - polytope of dimension 13 (the polytope of hereditary information). Its structure indicates that the interaction of nucleic acids has a hidden order, according to which the number of standard amino acids exactly corresponds to the number of possible locations of nitrogenous bases in the polytope of hereditary information. In accordance with the law of conservation of incidence (chapter two), the total flow of information transmitted by the polytope of hereditary information at the local location of the polytope of hereditary information on the chain of double-stranded nucleic acids is calculated. The possibility of creating polytopes of hereditary information in three and four helical nucleic acids is considered.

Chapter 7

Variants of damage to DNA and the polytope of hereditary information under the influence of radiation, chemical reactions, and the possibility of restoring DNA functions, including with the help of methyl groups, are considered. The geometric features of nucleic acids during methylation of their nitrogenous bases, which is typical for viral nucleic acids in infectious diseases, were studied. It was found that methylation of viral nucleic acids leads to the transition of their nitrogenous bases into the space of higher dimension and sharply increases (more than 2000 times) the intensity of information processes in the field of interaction of nucleic acids (in the polytopic of hereditary information). The activity of methylated viral nucleic acids can be countered by activation, i.e. a fairly high degree of methylation, nucleic acids of the organism.

Chapter 8

The established ideas about the genetic code are considered, taking into account the achievements of scientists of the previous period. The shortcomings and contradictions of these ideas are noted, which are mainly related to the assumption of the three-dimensionality of biomolecules and the linearity of nucleic acids. Since in recent years it has been clearly shown that these assumptions are far from reality, a new concept of the genetic code has been put forward, considering the higher dimension of sugar molecules, phosphoric acid residues and nucleic acids. It is shown that nitrogenous bases are located in two-dimensional coordinate planes of the polytope of hereditary information, and the number of these coordinate planes is exactly equal to the number of possible types of amino acids, i.e., each coordinate plane corresponds to some amino acid.

Chapter 9

Spatial models of the β - structures of protein molecules, forming layers of amino acids, in principle, of unlimited length for different conformations have been constructed. It is shown that the simplified flat Pauling models do not reflect the spatial structure of these layers. Using the recently developed theory of higher - dimensional polytopic prismahedrons, models of the volumetric filling of space with amino acid molecules are constructed. The constructed models for the first time mathematically describe the native structures of globular proteins. A model that simultaneously contains β - structures in parallel and anti - parallel positions and α - spirals. The turn of the polypeptide chain, which is often found in globular proteins, has been studied in higher dimensional space.

Chapter 10

The structure of the adamantane molecule, which coincides with the unit cell of diamond, is constructed. It is proved that the dimension of the adamantane molecule is nine. This is more than twice the dimension of the adamantane molecule previously considered. The higher dimension of adamantane molecules indicates the higher dimension of the molecules of substances containing adamantane molecules in their composition, in particular drugs based on adamantane. Examples of calculating the dimensions of known medicines are given. In this work, it is shown that the main part of adamantane itself, containing ten active carbon atoms, can be a means of impeding the formation of beta-amyloids that play the main role in various diseases of nerve cells.

Chapter 11

Peculiarities of the conditions of correspondence of biomolecules of higher dimension, necessary for the passage of chemical reactions between them, are considered. As specific examples, the interactions of amino acids with each other and nucleic acids with each other, and monosaccharide molecules are considered. It is shown

Preface

In recent years, it has been proved (Zhizhin, 2016, 2018, 2019 a, b, 2020, 2021, 2022) that molecules of chemical compounds have a higher dimension (i. e. their dimension, as a rule, is more than three). The dimension was determined based on the geometric shape of the molecules represented as a convex body, in the vertices of which atoms or groups of atoms (functional groups) are located. In this case, the dimension of the figure was calculated by the Poincare formula (Poincare`, 1895) using the calculation of the number of vertices, flat (two - dimensional) faces and faces of other dimensions. Moreover, chemical compounds can belong to both inanimate and living substances. It is shown that the highest dimension gives substances new properties. This is certainly important for obtaining substances with new useful properties (Zhizhin, 2021). This is extremely important for living objects of nature, since the properties of objects of living nature are inextricably linked with their health (Zhizhin, 2019). It was found that polytopes of the highest dimension corresponding to chemical compounds, only due to the highest dimension, have non - Euclidean properties (Zhizhin, 2023). This non - Euclidean of the geometry of chemical compounds is not associated with the curvature space, as took place in the geometry of Lobachevsky (Lobachevsky, 1829, 1835) and Rimann (Rimann, 1854). It is a consequence of the highest dimension of polytopes, since the axioms of Euclid's geometry (Euclid, 2012) in polytopes of highest dimensions are not performed (Zhizhin, 2023), even in the case of n - measured geometry of Euclid (Hilbert, 1901; Efimov, 1945). The higher dimension of space and its non-Euclidean properties require a revision of the patterns of motion of elementary particles within chemical compounds. As shown (Zhizhin, 2023), the principles of classical mechanics may not hold for these motions. This is especially important when analyzing possible biochemical reactions. Currently, in biochemistry (Nelson, Cox, 2023), five types of covalent interaction of components are distinguished: 1) reactions proceeding with the formation or breaking of a carbon-carbon bond; 2) intramolecular rearrangements; 3) reactions involving free radicals; 4) reactions with the transfer of functional groups; 5) redox reactions. However, in biochemistry, non-covalent interactions of components are also of great importance (Editors, Maharramov, Makhmudov, Kopylovich, Pombeiro, 2016). It should be noted that the dimension of space inside chemical compounds in all these works is traditionally taken equal to three, i.e. the highest dimension within polytopes corresponding to chemical compounds is not taken into account. The question of the dimension of space inside atoms also requires a new consideration. It was previously believed that the Schrödinger equation in a space of dimensions greater than three has no solutions (Buchel, 1963; Freeman, 1969; Gurevich, Mostepanenko, 1971). Based on the solutions of the Schrödinger equation in three-dimensional space, a classification of elements in the table of chemical elements was built (Gray, 1967). However, the solution of the Schrödinger equation in a space of higher dimension was obtained in 2018 (Zhizhin, 2018, 2019). From this it follows that in addition to the known electron orbitals in the three-dimensional space s, p, d, f, there can be other electron orbitals in the space of higher dimension. This leads to the possibility of a significant increase in the chemical activity of the elements and can serve as an explanation for the chemical reactions observed in experiments that cannot be explained using the standard table of chemical elements.

Accounting for the higher dimension of biomolecules is very important in the analysis of various processes in living organisms. First of all, it is important when considering the structure of nucleic acids and their constituents. As established, the transmission of hereditary information occurs in the polytope of hereditary information of dimension 13 (Zhizhin, 2020). This fundamentally changes ideas about the genetic code, about the possibility of resisting viral infections, about the possibility of treating nervous and other diseases. Examples of consideration of such specific issues, taking into account the highest dimension of the biomolecule, are given in this monograph.

In general, the monograph considers a number of new fundamental problems concerning the conditions of life on the planet Earth, the health of its inhabitants, new fundamental solutions are obtained that contribute to the implementation of these decisions. Combining these studies with previous studies by many other eminent researchers can give an additional impetus to the development of natural science. This monograph represents the first exposition of the basic principles of biochemistry of the world of higher dimension.

Gennadiy Vladimirovich Zhizhin
Russian Academy of Natural Sciences, Russia

REFERENCES

Büchel, W. (1963). Warum hat der Raum drei Dimensionen? *Physikalische Blätter*, *1912*(12), 547–549. doi:10.1002/phbl.19630191204

Efimov, N. V. (1945). *Higher geometry*. OGIZ.

Euclid. (2012). *Beginning*. URSS.

Freeman, I. M. (1969). Why is Space Three-Dimensional? *American Journal of Physics*, *37*(12), 1222–1224. doi:10.1119/1.1975283

Gray, H. B. (1965). *Electrons and Chemical Bonding. W.A. BENJAMIN*. INC.

Gurevich, L., & Mostepanenko, V. (1971). On the Existence of Atoms in n – Dimensional Space. *Physics Letters. [Part A]*, *35*(3), 201–202. doi:10.1016/0375-9601(71)90148-4

Hilbert, D. (1901). Gesammelte Abhandlagen. *Archivf. Mathematics and Physics*, *2*, 263–272.

Lobachevsky, N. I. (1829). *About the beginnings of geometry*. Kazan University Bulletin.

Lobachevsky, N.I. (1835). *Imaginary geometry*. Scientific notes of Kazan University.

Magerramov, A. M., Makhmudov, K. T., Kopylovich, M. N., & Pombeiro, A. J. L. (Eds.). (2016). *Non-covalent interactions in the synthesis and design of new compounds*. Wiley. doi:10.1002/9781119113874

Nelson, D. L., & Cox, M. M. (2022). *Lehninger Principles of Biochemistry*. W.H. Freeman.

Poincare, A. (1895). Analysis situs. *J. de é Ecole Polytechnique*, *1*, 1 – 121.

Riemann, B. (1854). *Ueber die Hypothesen, welche der Geometrie zu Grunde liegen*. University of Göttingen.

Zhizhin, G. V. (2016). The structure, topological and functional dimension of biomolecules. *J. International Journal of Chemoinformatics and Chemical Engineering*, *5*(2), 44–58. doi:10.4018/IJCCE.2016070104

Zhizhin, G. V. (2019 a). *The Geometry of Higher - Dimensional Polytopes.* IGI Global. doi:10.4018/978-1-5225-6968-8

Zhizhin, G. V. (2019 b). *Attractors and higher dimensions in population and molecular biology: emerging research and opportunities.* IGI Global. doi:10.4018/978-1-5225-9651-6

Zhizhin, G. V. (2020). Hidden Nucleic Acids Bond Order? *Acta Scientific Biotechnology, 1*(4), 34–36.

Zhizhin, G. V. (2021). *Nanotechnology and Clusters in the Space of Higher Dimension: Emerging Research and Opportunities.* IGI Global. doi:10.4018/978-1-7998-3784-8

Zhizhin, G. V. (2022). *The Classes of Polytopes Higher Dimensions in Physics, Chemistry and Biology.* IGI Global.

Zhizhin, G. V. (2023). *Non – Euclidean Geometry in Materials of Living and Non-Living Matter in the Space of the Highest Dimension.* Nova Science Publisher.

Chapter 1
Electrophilicity of Molecules in Higher Dimensional Space

ABSTRACT

The possibility of creating vacant electron orbitals in a space of higher dimension in the vicinity of the nucleus of an atom, leading to an increase in the electrophilicity of atoms and the chemical activity of elements due to the arrangement of electron pairs in these orbitals, is investigated. It is shown that the stationary Schrödinger equation describing the distribution of electrons in the vicinity of the atomic nucleus has a solution, in principle, for any dimensionality of the space around the nucleus. As an example, a solution of the Schrödinger equation in a five - dimensional space is obtained. It is shown that the solution of the Schrödinger equation in p - dimensional space has p quantum numbers: the principal quantum number, the orbital quantum number and p - 2 magnetic quantum numbers. Taking into account the spin quantum number, the total number of quantum numbers in p - dimensional space is p + 1. This leads to the possibility of increasing the number of quantum cells of orbitals and, consequently, to the possibility of increasing the valence of the elements.

INTRODUCTION

Many biochemical reactions occur during the interaction of nucleophiles (functional groups that are rich in electrons and are able to donate them to other atoms and groups of atoms) and electrophiles (electrophilic groups that can accept electrons from other groups). Nucleophiles, interacting with electrophiles, donate electrons to them (Metzler, 1977; Nelson and Cox, 2023). Most often, carbon atoms act as nucleophiles and electrophiles, depending on the environment. The studies carried out in the monograph of G. Zhizhin "The Geometry of Higher-Dimensional

DOI: 10.4018/979-8-3693-0588-1.ch001

Polypopes" (Zhizhin, 2019) convincingly showed that many elements of the periodic system of Mendeleev elements form molecules whose spatial structure can be geometrically represented as a convex polytope. Moreover, the dimension of this polytope, determined by the Euler-Poincaré relation between the geometric elements forming the polytope (Poincare, 1895), is usually higher than three. In addition, in this monograph it is noted that most elements of the periodic system exhibit valence in compounds higher than the valence determined by the location of the element in the periodic table. It is known that the calculation of the valence of a chemical element based on its location in the periodic table is based on the electronic structure of atoms (*s* -, *p* -, *d* -, *f* – orbitals), determined by solving the Schrodinger equation (Schrodinger, 1927) under the assumption of the three - dimensionality of the space surrounding the nucleus of the atom. To explain the observed differences in the valence of atoms from valence, determined from the location of the element in the periodic table, researchers resorted to the construction of various models of the interaction of electrons in the vicinity of atomic nuclei. These models are quite diverse and still remain at the level of hypotheses. In this connection, this chapter analyzes the possibility of obtaining a solution of the Schrödinger equation in a space of higher dimension. Although before it was believed that the solutions of the Schrödinger equation in a space with a dimension greater than three do not exist (Büchel, 1963; Freeman, 1969; Gurevich & Mostepanenko, 1971). The existence of vacant electron orbitals in the space of higher dimension in the vicinity of the nucleus of an atom can be the cause of an increase in the electrophilicity of atoms, i.e. an increase in the chemical activity of elements.

THE STATIONARY SCHRÖDINGER EQUATION IN A *P*-DIMENSIONAL METRIC SPACE

The stationary Schrödinger equation in a *p*-dimensional metric space has view

$$-\frac{\hbar}{2m}\nabla_p^2\Psi + U\Psi = E\Psi \tag{1.1}$$

Where Ψ - the wave function, U - the potential energy of an electron (the function of radius *r*), E – the kinetic energy of electron, ∇_p^2 - Laplacian in *p*-dimensional metric space, m – the mass of electron, $\hbar$ - Planck constant.

The potential energy $U(r)$ – this potential energy of the Coulomb attraction of an electron to the nucleus and can be written in explicit form,

$$U(r) = -\frac{Ze^2}{r} \tag{1.2}$$

Where eZ - ion's charge, e – electron's charge.

It is essential that the form of the potential energy (1.2) does not depend on the dimensionality of space and it is defined as a function inversely proportional to the radius in the first degree. Since in a space of any dimension, the potential energy is a function of the distance between the charges (of the electron and the nucleus). The incorrectness in previous studies of the solution of the Schrödinger equation in spaces of higher dimension is that the authors of these works (Büchel, 1963; Freeman, 1969; Gurevich & Mostepanenko, 1971) recorded potential energy without a physical basis as a function of the radius to the power of p -2. Such an entry goes to formula (2) only in three-dimensional space, but in a space of the higher dimension this record is not true and leads to an absurd conclusion about the impossibility of the existence of a discrete atomic structure in a space of higher dimension.

Equation (1.1), having transferred to the atomic system of units, taking into account (1.2), in the Cartesian coordinate system has the form,

$$\frac{\partial^2\Psi}{\partial x_1^2} + \frac{\partial^2\Psi}{\partial x_2^2} + \ldots + \frac{\partial^2\Psi}{\partial x_p^2} + \left(\frac{c}{r} + 2E\right)\Psi = 0 \tag{1.3}$$

Where c – constant, $r = \sqrt{x_1^2 + \ldots + x_p^2}$

For further analysis, it is convenient to go over into a spherical coordinate system, where the radius of one of the p variables. The remaining p - 1 variables are the angles between the vector of length r to the point of location of the particle and the unit vectors of the Cartesian system of coordinates.

It can be noted that Gurevich & Mostepanenko (1971) were confused with coordinate systems. For example, the velocity vector in the n-dimensional space has $n + 1$ components in their work. Babenko (2015), considering n-dimensional space, describes the one-dimensional Schrödinger equation, forgetting the remaining variables.

In an n - dimensional space, the analytic connection between the coordinates is expressed by equalities (Hobson, 1931)

The Laplacian in the Schrodinger equation in a spherical coordinate system in p-dimensional space can be written in the form,

$$\nabla_p^2 = \frac{1}{L_1 L_2 L_3 \ldots L_p}\left[\frac{\partial}{\partial r}\left(\frac{L_2 L_3 \ldots L_p}{L_1}\frac{\partial \Psi}{\partial r}\right) + \frac{\partial}{\partial \alpha_1}\left(\frac{L_1 L_3 \ldots L_p}{L_2}\frac{\partial \Psi}{\partial \alpha_1}\right) + \frac{\partial}{\partial \alpha_2}\left(\frac{L_1 L_2 L_4 \ldots L_p}{L_3}\frac{\partial \Psi}{\partial \alpha_2}\right) + \ldots + \frac{\partial}{\partial \alpha_{p-1}}\left(\frac{L_1 L_2 \ldots L_{p-1}}{L_p}\frac{\partial \Psi}{\partial \alpha_{p-1}}\right)\right] \quad (1.5)$$

Where L_i - the Lame coefficients, $i = 1, 2, \ldots, p$,

$$L_1 = \sqrt{\left(\frac{\partial x_1}{\partial r}\right)^2 + \left(\frac{\partial x_2}{\partial r}\right)^2 + \ldots + \left(\frac{x_p}{\partial r}\right)^2},$$

$$L_2 = \sqrt{\left(\frac{\partial x_1}{\partial \alpha_1}\right)^2 + \left(\frac{\partial x_2}{\partial \alpha_1}\right)^2 + \ldots + \left(\frac{x_p}{\partial \alpha_1}\right)^2}, \quad (1.6)$$

$$L_p = \sqrt{\left(\frac{\partial x_1}{\partial \alpha_{p-1}}\right)^2 + \left(\frac{\partial x_2}{\partial \alpha_{p-1}}\right)^2 + \ldots + \left(\frac{x_p}{\partial \alpha_{p-1}}\right)^2}$$

Differentiating the equality (1.4) with respect to the independent variables, substituting the derivatives in the Lame coefficients (1.6), we obtain after the transformations,

$$L_1 = 1, L_2 = r\sin\alpha_2 \ldots \sin\alpha_{p-1}, L_3 = r\sin\alpha_3 \ldots \sin\alpha_{p-1},$$
$$L_4 = r\sin\alpha_4 \ldots \sin\alpha_{p-1}, \ldots, L_p = r. \quad (1.7)$$

The Schrödinger equation in a spherical coordinate system in p-dimensional space has view

$$\nabla_p^2\Psi + \left(\frac{2}{r} + 2E\right)\Psi = 0 \quad (1.8)$$

Where $\nabla_p^2\Psi$ is expressed using equations (5) – (7).

A solution of equation (8) can be obtained by separating the variables assuming that

$$\Psi\left(r,\alpha_1,\ldots,\alpha_{p-1}\right)=R\left(r\right)\Phi_1\left(\alpha_1\right)\Phi_2\left(\alpha_2\right)\ldots\Phi_{p-1}\left(\alpha_{p-1}\right) \tag{1.9}$$

THE DECISION OF THE STATIONARY SCHRÖDINGER EQUATION IN A 4-DIMENSISION METRIC SPACE

We confine ourselves to the case of four-dimensional space, $p = 4$. Then Schrödinger equation (1.8) has view

$$\frac{1}{r^3}\frac{\partial}{\partial r}\left(r^3\frac{\partial\Psi}{\partial r}\right)+\frac{1}{r^2\sin{\alpha_2}^2\sin{\alpha_3}^2}\frac{\partial^2\Psi}{\partial{\alpha_1}^2}+\frac{1}{r^2\sin\alpha_2\sin{\alpha_3}^2}\left(\sin\alpha_2\frac{\partial\Psi}{\partial\alpha_2}\right)$$
$$+\frac{1}{r^2\sin{\alpha_3}^2}\frac{\partial}{\partial\alpha_3}\left(\sin{\alpha_3}^2\frac{\partial\Psi}{\partial\alpha_3}\right)+2\frac{\Psi}{r}+2E\Psi=0. \tag{1.10}$$

We seek the wave function in the form,

$$\Psi\left(r,\alpha_1,\alpha_2,\alpha_3\right)=R\left(r\right)\Phi_1\left(\alpha_1\right)\Phi_2\left(\alpha_2\right)\Phi_3\left(\alpha_3\right) \tag{1.11}$$

We substitute (1.11) into (1.10)

$$\frac{1}{rR(r)}\frac{d}{dr}\left(r^3\frac{dR(r)}{dr}\right)+2r^2\left(\frac{1}{r}+E\right)=-\frac{1}{\Phi_1(\alpha_1)\sin{\alpha_2}^2\sin{\alpha_3}^2}\frac{d^2\Phi_1(\alpha_1)}{d{\alpha_1}^2}$$
$$-\frac{1}{\Phi_2(\alpha_2)\sin\alpha_2\sin{\alpha_3}^2}\frac{d}{d\alpha_2}\left(\sin\alpha_2\frac{d\Phi_2(\alpha_2)}{d\alpha_2}\right)-\frac{1}{\Phi_3(\alpha_3)\sin{\alpha_3}^2}\frac{d}{d\alpha_3}\left(\sin{\alpha_3}^2\frac{d\Phi_3(\alpha_3)}{d\alpha_3}\right) \tag{1.12}$$

The left part of (1.12) is dependent at radius *r*, and rite part is dependent only at angels.

Therefore, both parts equality (1.12) equal to same constant C_1. From left part of (1.12) we have

$$\frac{1}{rR(r)}\frac{d}{dr}\left(r^3\frac{dR(r)}{dr}\right)+2r^2\left(\frac{1}{r}+E\right)=C_1 \tag{1.13}$$

From right part of (1.12) we have

$$C_1 = -\frac{1}{\Phi_1(\alpha_1)\sin\alpha_2{}^2\sin\alpha_3{}^2}\frac{d^2\Phi_1(\alpha_1)}{d\alpha_1{}^2}$$

$$-\frac{1}{\Phi_2(\alpha_2)\sin\alpha_2\sin\alpha_3{}^2}\frac{d}{d\alpha_2}\left(\sin\alpha_2\frac{d\Phi_2(\alpha_2)}{d\alpha_2}\right)-\frac{1}{\Phi_3(\alpha_3)\sin\alpha_3{}^2}\frac{d}{d\alpha_3}\left(\sin\alpha_3{}^2\frac{d\Phi_3(\alpha_3)}{d\alpha_3}\right) \tag{1.14}$$

The equation (1.14) we rewrite as follows,

$$C_1\sin\alpha_3{}^2+\frac{1}{\Phi_3(\alpha_3)}\frac{d}{d\alpha_3}\left(\sin\alpha_3{}^2\frac{d\Phi_3(\alpha_3)}{d\alpha_3}\right)=$$
$$-\frac{1}{\Phi_1(\alpha_1)\sin\alpha_2{}^2}\frac{d^2\Phi_1(\alpha_1)}{d\alpha_1{}^2}-\frac{1}{\Phi_2(\alpha_2)\sin\alpha_2}\frac{d}{d\alpha_2}\left(\sin\alpha_2\frac{d\Phi_2(\alpha_2)}{d\alpha_2}\right) \tag{1.15}$$

The left part of (1.15) is depended at angles only, and the right part of (1.15) is depended at angels only. Therefore, both parts of (1.15) equal to same constant C_2. From the left part (1.15) we have

$$C_1\sin\alpha_3{}^2+\frac{1}{\Phi_3(\alpha_3)}\frac{d}{d\alpha_3}\left(\sin\alpha_3{}^2\frac{d\Phi_3(\alpha_3)}{d\alpha_3}\right)=C_2 \tag{1.16}$$

From the right part of (1.15) we have

$$-\frac{1}{\Phi_1(\alpha_1)}\frac{d^2\Phi_1(\alpha_1)}{d\alpha_1{}^2}=C_2\sin\alpha_2{}^2+\frac{\sin\alpha_2}{\Phi_2(\alpha_2)}\frac{d}{d\alpha_2}\left(\sin\alpha_2\frac{d\Phi_2(\alpha_2)}{d\alpha_2}\right) \tag{1.17}$$

The left part of (1.17) is depended only at *α1* and right part of (1.17) is depended only at α_2.Therefore, both parts equal to same constant C_3. From the left part of (1.17) we have

$$-\frac{1}{\Phi_1(\alpha_1)}\frac{d^2\Phi_1(\alpha_1)}{d\alpha_1{}^2}=C_3 \tag{1.18}$$

From the right part of (1.17) we have

$$C_2 \sin\alpha_2{}^2 + \frac{\sin\alpha_2}{\Phi_2(\alpha_2)} \frac{d}{d\alpha_2}\left(\sin\alpha_2 \frac{d\Phi_2(\alpha_2)}{d\alpha_2} \right) = C_3 \tag{1.19}$$

The equation (1.18) has decision,

$$\Phi_1(\alpha_1) = Ae^{\pm i\sqrt{C_3}\alpha_1} \tag{1.20}$$

Since $\Phi_1(0)$-$\Phi_2(2\pi)$, then $A = Ae^{\pm i\sqrt{C_3}2\pi}$, i.e. $e^{\pm i\sqrt{C_3}2\pi} = 1$ (1.21)

Consequently, $\cos(2\pi\sqrt{C_3}) \pm i\sin(2\pi\sqrt{C_3}) = 1$
It may be only if $\sqrt{C_3} = 0, \pm1, \pm2, ...$

So we got the first quantum number $m_1 = \sqrt{C_3} = 0, \pm1, \pm2, ...$ (1.22)

The constant A is defined from condition of normalization,

$$\int_0^{2\pi} \Phi_1^*(\alpha_1)\Phi_1(\alpha_1)d\alpha_1 = 1 = A^2 \int_0^{2\pi} e^{-i\sqrt{C_3}\alpha_1} e^{i\sqrt{C_3}\alpha_1} d\alpha_1 = A^2 2\pi$$

Consequently, $A = \frac{1}{\sqrt{2\pi}}$ Thus, the function it is known,

$$\Phi_1(\alpha_1) = \frac{1}{\sqrt{2\pi}} e^{-im_1\alpha_1}, m_1 = \sqrt{C_3} \tag{1.23}$$

We write the equation (1.19) in the form,

$$\Phi_2(\alpha_2)[C_2 - \frac{m_1^2}{\sin\alpha_2{}^2}] + \frac{1}{\sin\alpha_2} \frac{d}{d\alpha_2}\left(\sin\alpha_2 \frac{d\Phi_2(\alpha_2)}{d\alpha_2} \right) = 0 \tag{1.24}$$

Equation (1.24) is the adjoined Legendre equation. In the case $C_2 = l(l+1)$, l are integers, its solutions are the adjoined Legendre polynomials (Landau & Lifshitz, 1963)

$$P_l^{m_1}(\cos\alpha_2) = \frac{1}{2^l l!}\sin\alpha_2^{\ m_1}\frac{d^{m_1+l}}{(d\cos\alpha_2)^{m_1+l}}(\cos\alpha_2^{\ 2} - 1)^l \quad (1.25)$$

So we got second quantum number *l*. Equation (1.16) can be rewritten in the form

$$\Phi_3(\alpha_3)\left(C_1 - l\frac{l+1}{\sin\alpha_3^{\ 2}}\right) + \frac{1}{\sin\alpha_3^{\ 2}}\frac{d}{d\alpha_3}\left(\sin\alpha_3^{\ 2}\frac{d\Phi_3(\alpha_3)}{d\alpha_3}\right) = 0 \quad (1.26)$$

Since $\sin\alpha_3^{\ 2}$ is a positive function varying from 0 to 1, then we introduce the average value of this function τ in the range α_3 from 0 to 2π,

$$\tau = \frac{1}{2\pi}\int_0^{2\pi}\sin\alpha_3^{\ 2}d\alpha_3 = \frac{1}{2}$$

Then the equation (1.26) has the form

$$C_1 - 2l(l+1) = -\frac{1}{\Phi_3(\alpha_3)}\frac{d^2\Phi_3(\alpha_3)}{d\alpha_3^{\ 2}} \quad (1.27)$$

The decision of this equation is

$$\Phi_3(\alpha_3) = Be^{\pm i\alpha_3\sqrt{C_1 - 2l(l+1)}2\pi} \quad (1.28)$$

Since $\Phi_3(0)=\Phi_3(2\pi)$, so from (1.28) we have

$$B = Be^{\pm i\sqrt{C_1 - 2l(l+1)}2\pi}\text{, i.e. } e^{\pm i\sqrt{C_1 - 2l(l+1)}2\pi} = 1$$

Consequently, $\cos(2\pi\sqrt{C_1 - 2l(l+1)}) \pm i\sin(2\pi\sqrt{C_1 - 2l(l+1)}) = 1$

It may be only if $\sqrt{C_1 - 2l(l+1)} = 0, \pm1, \pm2, \ldots$

So we got the second quantum number

$$m_2 = \sqrt{C_1 - 2l(l+1)} = 0, \pm1, \pm2, \ldots \quad (1.29)$$

The constant *B* is defined from condition of normalization (see equation (1.20))

$$\Phi_3(\alpha_3) = \frac{1}{\sqrt{2\pi}} e^{-im_2\alpha_3} \tag{1.30}$$

Now consider the equation for the radius (1.13). We perform differentiation in it

$$r^2 \frac{d^2R}{dr^2} + 3r\frac{dR}{dr} + R(2r + 2r^2E - 2l(l+1) - m_2^2) = 0 \tag{1.31}$$

We seek the solution of the equation (1.31) in the form of a series in powers of r

$$R(r) = e^{-\mu r} r^l \sum_j b_j r^j, \tag{1.32}$$

μ=(-E)$1^{/2,}$ E<0, bj are numerical coefficients.

Substituting (1.32) into (1.31), performing differentiation and combining summands with the same powers r, we obtain

$$\sum_j b_j r^{l+j}\left((l+j)^2 + 2(l+j) - 2l(l+1) - m_2^2\right) = \sum_j b_j r^{l+j+1}\left(\mu(2l+2j+3) - 2\right) \tag{1.33}$$

From equation (1.33) follows the recurrence formula for the coefficients

$$b_{j+1} = \frac{\mu(2l+2j+3) - 2}{(l+j+1)^2 + 2(l+j+1) - 2l(l+1) - m_2^2} b_j \tag{1.34}$$

In view of the need to obtain a limited physical meaningful solution, the values of the coefficients, starting from a certain value of j must vanish. This is possible if the numerator on the right-hand side of (1.34) is zero,

$$\mu(2l+2j+3) - 2 = 0 \tag{1.35}$$

From equation (1.35) follows that

$$\mu = \sqrt{-E} = \frac{2}{2l+2j+3} \tag{1.36}$$

We introduce the notation $n = l + j + 1$. From (1.36) equation we have

$$E = -\frac{2}{(2n+1)^2} \tag{1.37}$$

For big n formula (1.37) go to the formula of Bore at energy of electron.

The radial function is determined by the equality (1.32) of setting different values j. For example, with $j = 0$

$$R(r) = \mathrm{e}^{-\mu r} r^l b_0, b_0 = \frac{1}{\int_0^\infty e^{-\mu r} r^l dr} \tag{1.38}$$

With $j = 1$

$$R(r) = e^{-\mu r} r^l b_0 \left(1 + r \frac{\mu(2l+3)-2}{(l+1)^2 + 2(l+1) - 2l(l+1) - m_2^2} \right),$$

$$b_0 = \frac{1}{\int_0^\infty e^{-\mu r} r^l \left(1 + r \frac{\mu(2l+3)-2}{(l+1)^2 + 2(l+1) - 2l(l+1) - \mathrm{m}_2^2} \right) dr} \tag{1.39}$$

Thus, one can obtain various approximations for the function $R(r)$.

THE DECISION OF THE STATIONARY SCHRÖDINGER EQUATION IN A *P* - DIMENSISION METRIC SPACE

In the case of an n-dimensional metric space in a spherical coordinate system, the Schrödinger equation (1.8) with the wave function (1.9) has the form,

$$\frac{1}{r^{p-1}}\frac{\partial}{\partial r}\left(r^{p-1}\frac{\partial\Psi}{\partial r}\right)+\frac{1}{r^2\sin\alpha_2{}^2...\sin\alpha_{p-1}{}^2}\frac{\partial^2\Psi}{\partial\alpha_1{}^2}+\frac{1}{r^2\sin\alpha_2\sin\alpha_3{}^2...\sin^2\alpha_{p-1}}\left(\sin\alpha_2\frac{\partial\Psi}{\partial\alpha_2}\right)$$
$$+\frac{1}{r^2\sin\alpha_3{}^2\sin^2\alpha_4...\sin^2\alpha_{p-1}}\frac{\partial}{\partial\alpha_3}\left(\sin\alpha_3{}^2\frac{\partial\Psi}{\partial\alpha_3}\right)$$
$$+...+\frac{1}{r^2\sin^2\alpha_{p-1}}\frac{\partial}{\partial\alpha_{p-1}}\left(\sin^2\alpha_{p-1}\frac{\partial\Psi}{\partial\alpha_{p-1}}\right)+2\frac{\Psi}{r}+2E\Psi=0. \tag{1.40}$$

We substitute (1.9) into (1.40)

$$\frac{1}{r^{p-3}R(r)}\frac{d}{dr}\left(r^{p-1}\frac{dR(r)}{dr}\right)+2r^2\left(\frac{1}{r}+E\right)=-\frac{1}{\Phi_1(\alpha_1)\sin\alpha_2{}^2...\sin\alpha_{p-1}{}^2}\frac{d^2\Phi_1(\alpha_1)}{d\alpha_1{}^2}$$
$$-\frac{1}{\Phi_2(\alpha_2)\sin\alpha_2\sin\alpha_3{}^2...\sin^2\alpha_{p-1}}\frac{d}{d\alpha_2}\left(\sin\alpha_2\frac{d\Phi_2(\alpha_2)}{d\alpha_2}\right)$$
$$-\frac{1}{\Phi_3(\alpha_3)\sin\alpha_3{}^2\sin^2\alpha_4...\sin^2\alpha_{p-1}}\frac{d}{d\alpha_3}\left(\sin\alpha_3{}^2\frac{d\Phi_3(\alpha_3)}{d\alpha_3}\right)$$
$$-...-\frac{1}{\Phi_{p-1}\sin^2\alpha_{p-1}}\frac{d}{d\alpha_{p-1}}\left(\sin^2\alpha_{p-1}\frac{d\Phi_{p-1}}{d\alpha_{p-1}}\right). \tag{1.41}$$

The left part of (1.41) is dependent at radius r, and rite part is dependent only at angels.

Therefore, both parts equality (1.41) equal to same constant C_4. From left part of (1.41) there is

$$\frac{1}{r^{p-3}R(r)}\frac{d}{dr}\left(r^{p-1}\frac{dR(r)}{dr}\right)+2r^2\left(\frac{1}{r}+E\right)=C_4. \tag{12}$$

From right part of (1.41) there is

$$C_4 = -\frac{1}{\Phi_1(\alpha_1)\sin\alpha_2{}^2...\sin\alpha_{p-1}{}^2}\frac{d^2\Phi_1(\alpha_1)}{d\alpha_1{}^2}$$
$$-\frac{1}{\Phi_2(\alpha_2)\sin\alpha_2\sin\alpha_3{}^2...\sin^2\alpha_{p-1}}\frac{d}{d\alpha_2}\left(\sin\alpha_2\frac{d\Phi_2(\alpha_2)}{d\alpha_2}\right)$$
$$-\frac{1}{\Phi_3(\alpha_3)\sin\alpha_3{}^2...\sin^2{}_{p-1}}\frac{d}{d\alpha_3}\left(\sin\alpha_3{}^2\frac{d\Phi_3(\alpha_3)}{d\alpha_3}\right) \quad (1.43)$$
$$-...-\frac{1}{\Phi_{p-1}\sin^2\alpha_{p-1}}\frac{d}{d\alpha_{p-1}}\left(\sin^2\alpha_{p-1}\frac{d\Phi_{p-1}}{d\alpha_{p-1}}\right).$$

The equation (1.43) can rewrite as follows,

$$C_4\sin\alpha_{p-1}{}^2+\frac{1}{\Phi_{p-1}(\alpha_{p-1})}\frac{d}{d\alpha_{p-1}}\left(\sin\alpha_{p-1}{}^2\frac{d\Phi_{p-1}(\alpha_{p-1})}{d\alpha_{p-1}}\right)=$$
$$-\frac{1}{\Phi_1(\alpha_1)\sin\alpha_2{}^2...\sin^2{}_{p-2}}\frac{d^2\Phi_1(\alpha_1)}{d\alpha_1{}^2}-\frac{1}{\Phi_2(\alpha_2)\sin\alpha_2\sin^2\alpha_3...\sin^2\alpha_{p-2}}\frac{d}{d\alpha_2}\left(\sin\alpha_2\frac{d\Phi_2(\alpha_2)}{d\alpha_2}\right)$$
$$-\frac{1}{\Phi_3(\alpha_3)\sin^2\alpha_3...\sin^2\alpha_{p-2}}\frac{d}{d\alpha_3}\left(\sin^2\alpha_3\frac{d\Phi_3(\alpha_3)}{d\alpha_3}\right)$$
$$-...-\frac{1}{\Phi_{p-2}(\alpha_{p-2})\sin^2\alpha_{p-2}}\frac{d}{d\alpha_{p-2}}\left(\sin^2\alpha_{p-2}\frac{d\Phi_{p-2}(\alpha_{p-2})}{d\alpha_{p-2}}\right).$$

(1.44)

The left part of (1.44) is depended at angle α_{p-1} only, and the right part (1.44) depended at angles $\alpha_2,...,\alpha_{p-2}$. Therefore, both parts of (1.44) equal to same constant C_5. From the left part (1.44) there is

$$C_4\sin\alpha_{p-1}{}^2+\frac{1}{\Phi_{p-1}(\alpha_{p-1})}\frac{d}{d\alpha_{p-1}}\left(\sin\alpha_{p-1}{}^2\frac{d\Phi_{p-1}(\alpha_{p-1})}{d\alpha_{p-1}}\right)=C_5 \quad (1.45)$$

From the right part (1.44) there is

$$-\frac{1}{\Phi_1(\alpha_1)}\frac{d^2\Phi_1(\alpha_1)}{d\alpha_1^{\ 2}} = C_5 \sin^2\alpha_2 ... \sin^2\alpha_{p-2}$$
$$+\frac{\sin\alpha_2}{\Phi_2(\alpha_2)}\frac{d}{d\alpha_2}\left(\sin\alpha_2\frac{d\Phi_2(\alpha_2)}{d\alpha_2}\right)+\frac{\sin^2\alpha_2}{\Phi_3(\alpha_3)}\frac{d}{d\alpha_3}\left(\sin^2\alpha_3\frac{d\Phi_3(\alpha_3)}{d\alpha_3}\right) \quad (1.46)$$
$$+...+\frac{\sin^2\alpha_2...\sin^2\alpha_{p-3}}{\Phi_{p-2}(\alpha_{p-2})}\frac{d}{d\alpha_{p-2}}\left(\sin^2\alpha_{p-2}\frac{d\Phi_{p-2}(\alpha_{p-2})}{d\alpha_{p-2}}\right)$$

The left part of (1.46) is depended at angle *α1* only, and the right part (1.46) depended at angles $\alpha_2,...,\alpha_{p-2}$. Therefore, both parts of (1.46) equal to same constant C_6. From the left part (1.46) there is

$$-\frac{1}{\Phi_1(\alpha_1)}\frac{d^2\Phi_1(\alpha_1)}{d\alpha_1^{\ 2}} = C_6 . \quad (1.47)$$

From the right part (1.46) there is

$$C_6 = C_5 \sin^2\alpha_2 ... \sin^2\alpha_{p-2}$$
$$+\frac{\sin\alpha_2}{\Phi_2(\alpha_2)}\frac{d}{d\alpha_2}\left(\sin\alpha_2\frac{d\Phi_2(\alpha_2)}{d\alpha_2}\right)+\frac{\sin^2\alpha_2}{\Phi_3(\alpha_3)}\frac{d}{d\alpha_3}\left(\sin^2\alpha_3\frac{d\Phi_3(\alpha_3)}{d\alpha_3}\right) \quad (1.48)$$
$$+...+\frac{\sin^2\alpha_2...\sin^2\alpha_{p-3}}{\Phi_{p-2}(\alpha_{p-2})}\frac{d}{d\alpha_{p-2}}\left(\sin^2\alpha_{p-2}\frac{d\Phi_{p-2}(\alpha_{p-2})}{d\alpha_{p-2}}\right)$$

The equation (1.48) can rewrite as follows

$$\frac{C_6}{\sin^2\alpha_2}-\frac{1}{\Phi_2(\alpha_2)\sin\alpha_2}\frac{d}{d\alpha_2}\left(\sin\alpha_2\frac{d\Phi_2(\alpha_2)}{d\alpha_2}\right) = C_5 \sin^2\alpha_3 ... \sin^2\alpha_{p-2}$$
$$+\frac{1}{\Phi_3(\alpha_3)}\frac{d}{d\alpha_3}\left(\sin^2\alpha_3\frac{d\Phi_3(\alpha_3)}{d\alpha_3}\right)+...+\frac{\sin^2\alpha_3...\sin^2\alpha_{p-3}}{\Phi_{p-2}(\alpha_{p-2})}\frac{d}{d\alpha_{p-2}}\left(\sin^2\alpha_{p-2}\frac{d\Phi_{p-2}(\alpha_{p-2})}{d\alpha_{p-2}}\right)$$
(1.49)

The left part of (1.49) is depended at angle α_2 only, and the right part (1.49) depended at angles $\alpha_3,...,\alpha_{p-2}$. Therefore, both parts of (1.49) equal to same constant C_7. From the left part (1.49) there is

$$\frac{C_6}{\sin^2\alpha_2}-\frac{1}{\Phi_2(\alpha_2)\sin\alpha_2}\frac{d}{d\alpha_2}\left(\sin\alpha_2\frac{d\Phi_2(\alpha_2)}{d\alpha_2}\right)=C_7 \tag{1.50}$$

From the right part (1.49) there is

$$C_7=C_5\sin^2\alpha_3...\sin^2\alpha_{p-2}$$
$$+\frac{1}{\Phi_3(\alpha_3)}\frac{d}{d\alpha_3}\left(\sin^2\alpha_3\frac{d\Phi_3(\alpha_3)}{d\alpha_3}\right)+...+\frac{\sin^2\alpha_3...\sin^2\alpha_{p-3}}{\Phi_{p-2}(\alpha_{p-2})}\frac{d}{d\alpha_{p-2}}\left(\sin^2\alpha_{p-2}\frac{d\Phi_{p-2}(\alpha_{p-2})}{d\alpha_{p-2}}\right) \tag{1.51}$$

The equation (1.51) can rewrite as follows

$$\frac{C_7}{\sin^2\alpha_3}-\frac{1}{\Phi_3(\alpha_3)\sin^2\alpha_3}\frac{d}{d\alpha_3}\left(\sin^2\alpha_3\frac{d\Phi_3(\alpha_3)}{d\alpha_3}\right)=C_5\sin^2\alpha_4...\sin^2\alpha_{p-2}$$
$$+\frac{1}{\Phi_4(\alpha_4)}\frac{d}{d\alpha_4}\left(\sin^2\alpha_4\frac{d\Phi_4(\alpha_4)}{d\alpha_4}\right)+...+\frac{\sin^2\alpha_4...\sin^2\alpha_{p-3}}{\Phi_{p-2}(\alpha_{p-2})}\frac{d}{d\alpha_{p-2}}\left(\sin^2\alpha_{p-2}\frac{d\Phi_{p-2}(\alpha_{p-2})}{d\alpha_{p-2}}\right) \tag{1.52}$$

From the preceding it is clear that the process of separating equations for unknown angles can be continued to the end, i.e. until a complete system of equations for these unknowns is obtained. This, in principle, allows us to solve a system of equations for any finite value of p; for any finite dimension of the space, in principle, obtain a solution of the Schrödinger equation. Let, for example, the dimension of the space be five ($p = 5$). Then the equation (1.44) takes the form

$$C_4\sin\alpha_4{}^2+\frac{1}{\Phi_4(\alpha_4)}\frac{d}{d\alpha_4}\left(\sin\alpha_4{}^2\frac{d\Phi_4(\alpha_4)}{d\alpha_4}\right)=-\frac{1}{\Phi_1(\alpha_1)\sin\alpha_2{}^2\sin^2{}_3}\frac{d^2\Phi_1(\alpha_1)}{d\alpha_1{}^2}$$
$$-\frac{1}{\Phi_2(\alpha_2)\sin\alpha_2\sin^2\alpha_3}\frac{d}{d\alpha_2}\left(\sin\alpha_2\frac{d\Phi_2(\alpha_2)}{d\alpha_2}\right)-\frac{1}{\Phi_3(\alpha_3)\sin^2\alpha_3}\frac{d}{d\alpha_3}\left(\sin^2\alpha_3\frac{d\Phi_3(\alpha_3)}{d\alpha_3}\right) \tag{1.53}$$

Since both parts of the equation (1.53) depend on different variables, they are equal to some constant. From left part equation (1.53) there is

$$C_4\sin\alpha_4{}^2+\frac{1}{\Phi_4(\alpha_4)}\frac{d}{d\alpha_4}\left(\sin\alpha_4{}^2\frac{d\Phi_4(\alpha_4)}{d\alpha_4}\right)=C_5 \tag{1.54}$$

From right part of equation (1.53) there is

$$\frac{1}{\Phi_1(\alpha_1)}\frac{d^2\Phi_1(\alpha_1)}{d\alpha_1{}^2} = C_5 \sin^2\alpha_2 \sin^2\alpha_3$$
$$+\frac{\sin\alpha_2}{\Phi_2(\alpha_2)}\frac{d}{d\alpha_2}\left(\sin\alpha_2\frac{d\Phi_2(\alpha_2)}{d\alpha_2}\right)+\frac{\sin^2\alpha_2}{\Phi_3(\alpha_3)}\frac{d}{d\alpha_3}\left(\sin^2\alpha_3\frac{d\Phi_3(\alpha_3)}{d\alpha_3}\right). \quad (1.55)$$

Since both parts of the equation (1.55) depend on different variables, they are equal to some constant. From left part equation (1.55) can obtain equation (1.47).

The equation (1.47) for any dimension *p* have decision,

$$\Phi_1(\alpha_1) = Ae^{\pm i\sqrt{C_6}\alpha_1} \quad (1.56)$$

Since $\Phi_1(0)=\Phi_1(2\pi)$, then $A = Ae^{\pm i\sqrt{C_6}2\pi}$, (1.57)

i.e. $e^{\pm i\sqrt{C_6}2\pi} = 1$. Consequently, $\cos\left(2\pi\sqrt{C_6}\right) \pm i\sin\left(2\pi\sqrt{C_6}\right) = 1$.

It may be only if $\sqrt{C_6} = 0, \pm1, \pm2, \ldots$.

So we got the first quantum number $m_1 = \sqrt{C_6} - 0, \pm1, \pm2, \ldots$. (1.58)

The constant *A* is defined from condition of normalization.

$$\int_0^{2\pi} \Phi_1^*(\alpha_1)\Phi_1(\alpha_1)d\alpha_1 = 1 = A^2\int_0^{2\pi} e^{-i\sqrt{C_6}\alpha_1}e^{i\sqrt{C_6}\alpha_1}d\alpha_1 - A^2 2\pi.$$

Consequently, $A = \frac{1}{\sqrt{2\pi}}$ Thus, the function $\Phi_1(\alpha_1)$ it is known for any p,

$$\Phi_1(\alpha_1) = \frac{1}{\sqrt{2\pi}}e^{-im_1\alpha_1}, m_1 = \sqrt{C_6} \quad (1.59)$$

From left part equation (1.55) there is

$$\frac{C_6}{\sin^2\alpha_2}-\frac{1}{\Phi_2(\alpha_2)\sin\alpha_2}\frac{d}{d\alpha_2}\left(\sin\alpha_2\frac{d\Phi_2(\alpha_2)}{d\alpha_2}\right)$$
$$=C_5\sin^2\alpha_3+\frac{1}{\Phi_3(\alpha_3)}\frac{d}{d\alpha_3}\left(\sin^2\alpha_3\frac{d\Phi_3(\alpha_3)}{d\alpha_3}\right). \qquad (1.60)$$

Since both parts of the equation (1.60) depend on different variables, they are equal to some constant. From the left part of equation (1.60) we have equation (1.50). The equation (1.50) can rewrite with account (1.58) as follows,

$$\Phi_2(\alpha_2)\left(C_7-\frac{m_1^2}{\sin^2\alpha_2}\right)+\frac{1}{\sin\alpha_2}\frac{d}{d\alpha_2}\left(\sin\alpha_2\frac{d\Phi_2(\alpha_2)}{d\alpha_2}\right)=0 \qquad (1.61)$$

The equation (1.61) is the adjoined Legendre equation. In case $C_7 = l(l+1)$, where l are integers, its solutions are the adjoined Legendre polynomials (Landau & Lifshitz, 1963)

$$P_l^{m_1}(\cos\alpha_2)=\frac{1}{2^l l!}\sin^{m_1}\alpha_2\frac{d^{m_1+l}}{(d\cos\alpha_2)^{m_1+l}}\left(\cos^2\alpha_2-1\right)^l \qquad (1.62)$$

So can get second quantum number l – orbital quantum number.

From right part equation (1.60) there is

$$C_5\sin^2\alpha_3+\frac{1}{\Phi_3(\alpha_3)}\frac{d}{d\alpha_3}\left(\sin^2\alpha_3\frac{d\Phi_3(\alpha_3)}{d\alpha_3}\right)=C_7 \qquad (1.63)$$

One rewrite the equation (1.63) in the form

$$\left(C_5-l\frac{l+1}{\sin^2\alpha_3}\right)\Phi_3(\alpha_3)+\frac{1}{\sin^2\alpha_3}\frac{d}{d\alpha_3}\left(\sin^2\alpha_3\frac{d\Phi_3(\alpha_3)}{d\alpha_3}\right)=0 \qquad (1.64)$$

Since $\sin^2\alpha^3$ is a positive function varying from 0 to 1, then one introduces the average value of this function τ in the range α_3 from 0 to 2π,

$$\tau = \frac{1}{2\pi}\int_0^{2\pi} \sin^2 \alpha_3 d\alpha_3 = \frac{1}{2}$$

Then the equation has form

$$C_5 - 2l(l+1) = -\frac{1}{\Phi_3(\alpha_3)}\frac{d^2\Phi_3(\alpha_3)}{d\alpha_3{}^2} \tag{1.65}$$

The decision of this equation is

$$\Phi_3(\alpha_3) = Be^{\pm i\sqrt{C_5 - 2l(l+1)}\alpha_3} \tag{1.66}$$

Since $\Phi_3(0)=\Phi_3(2\pi)$, then $B = Be^{\pm i\sqrt{C_5-2l(l+1)}2\pi}$,
i.e. $e^{\pm i\sqrt{C_5-2l(l+1)}2\pi} = 1$ Consequently,

$$\cos\left(2\pi\sqrt{C_5 - 2l(l+1)}\right) \pm i\sin\left(2\pi\sqrt{C_5 - 2l(l+1)}\right) = 1$$

It may be only if $\sqrt{C_5 - 2l(l+1)} = 0, \pm 1, \pm 2, \ldots$

So can get the third quantum number $m_2 = \sqrt{C_5 - 2l(l+1)} = 0, \pm 1, \pm 2, \ldots$ (1.67)

The constant *B* is defined from condition of normalization,

$$\int_0^{2\pi} \Phi_3^*(\alpha_3)\Phi_3(\alpha_3)d\alpha_3 = 1 = B^2\int_0^{2\pi} e^{-i\sqrt{C_5-2l(l+1)}\alpha_3}e^{i\sqrt{C_5-2l(l+1)}\alpha_3}d\alpha_3 = B^2 2\pi$$

Consequently, $B = \frac{1}{\sqrt{2\pi}}$ Thus, the function $\Phi_3(\alpha_3)$ it is known,

$$\Phi_3(\alpha_3) = \frac{1}{\sqrt{2\pi}}e^{-im_2\alpha_3}, m_2 = \sqrt{C_5 - 2l(l+1)} \tag{1.68}$$

One now turns to equation (1.54) with respect to the angle $\alpha 4$.

The equation (1.64) can rewrite as follows,

$$\left(C_4 - 2l\frac{l+1}{\sin^2\alpha_4}\right)\Phi_4(\alpha_4) + \frac{1}{\sin^2\alpha_4}\frac{d}{d\alpha_4}\left(\sin^2\alpha_4\frac{d\Phi_4(\alpha_4)}{d\alpha_4}\right) = 0 \tag{1.69}$$

Performing the transformations in equation (1.69) for the angle *α4* are the same as in equation (1.64) for the angle *α3,* one transform equation (1.69) to the form,

$$!_4 - 4l(l+1) = -\frac{1}{\Phi_4(\alpha_4)}\frac{d^2\Phi_4(\alpha_4)}{d\alpha_4^{\ 2}} \tag{1.70}$$

The equation (1.70), so equation (1.65), have decision,

$$\Phi_4(\alpha_4) = \frac{1}{\sqrt{2\pi}}e^{-im_3\alpha_4},$$

where $m_3 = \sqrt{C_4 - 4l(l+1)} = 0, \pm 1, \pm 2, \ldots$ is fourth quantum number. (1.71)

We now consider the equation for the radius (1.42). For $p=5$ and $C_{41} = m_3^2 + 4l(l)$ this equation has the form

$$\frac{1}{Rr^2}\frac{d}{dr}\left(r^4\frac{dR}{dr}\right) + 2r^2\left(\frac{1}{r} + E\right) = m_3^2 + 4l(l+1)$$

Performing differentiation in this equation, can obtain,

$$r^2\frac{d^2R}{dr^2} + 4r\frac{dR}{dr} + R\left(2r + 2Er^2 - 4l(l+1) - m_3^2\right) = 0 \tag{1.72}$$

One seeks the solution of the equation (1.72) in the form of a series in powers of *r,*

$$R(r) = e^{-\mu r}r^l\sum_j b_j r^j \tag{1.73}$$

In (1.73) $\mu = \sqrt{-2E}, E < 0, b_j$ are numerical coefficients. One substitute (1.73) into (1.72)

$$\sum_j b_j r^{l+j}\left((l+j)^2+2(l+j)-4l(l+1)-m_3^2\right)=\sum_j b_j r^{l+j+1}\left(2\mu(l+j+2)-2\right) \quad (1.74)$$

From equation (1.74) follows the recurrence formula for the coefficients

$$b_{j+1}=\frac{2\mu(l+j+2)-2}{(l+j)^2+2(l+j)-4l(l+1)-m_3^2}b_j \quad (1.75)$$

In view of the need to obtain a limited physical meaningful solution, the values of the coefficients, starting from a certain value of *j* must vanish. This is possible if the numerator on the right-hand side of (1.75) is zero,

$$\mu(1+j+2)-1=0 \quad (1.76)$$

From equation (1.76) follows that

$$\mu=\sqrt{-E}=\frac{1}{l+j+2} \quad (1.77)$$

One introduces the notation $n = l + j + 1$ is principal quantum number. From (1.77) equation can obtain,

$$E=-\frac{1}{2(n+1)^2} \quad (1.78)$$

The radial function is determined by the equality (1.73) of setting different values *j*. For example, with $j = 0$

$$R(r)=\mathrm{e}^{-\mu r}r^l b_0, b_0=\frac{1}{\int_0^\infty e^{-\mu r}r^l dr} \quad (1.79)$$

With $j = 1$

$$R(r) = e^{-\mu r} r^l b_0 \left(1 - r\frac{2\mu(l+2)-2}{3l^2+2l+m_3^2}\right), \quad b_0 = \frac{1}{\int_0^\infty e^{-\mu r} r^l \left(1 + \frac{2-2\mu(l+2)}{3l^2+2l+\mathrm{m}_3^2}\right) dr} \tag{1.80}$$

Thus, one can obtain various approximations for the function $R(r)$.

CONCLUSION

Thus, it is proved that the Schrödinger equation has a solution in a space of higher dimension. Consequently, the discrete nature of matter exists in a space of higher dimension. The incorrectness of authors in previous studies of the solution of the Schrodinger equation in space of higher dimension was that the potential energy of electron was written in form different at law of Coulomb. The study showed that the motion of an electron around a nucleus in a space of higher dimension leads to the appearance of quantum numbers with a total number equal to the dimension of space. On a level with the principal quantum number, characterizing the average distance of the electron from the nucleus, and the orbital quantum number, characterizing the shape of the electron orbitals, a series of magnetic quantum numbers appear in a space of higher dimension. It corresponds to the possible values of the projections of the angular momentum of an electron on the coordinate axes of space. The number of these magnetic quantum numbers is equal to the dimension of the space minus two. Taking into account the principal quartic number and orbital quantum number, the total number of quantum numbers characterizing the motion of an electron around the nucleus is equal to the dimension of space. Thus, the total number of quantum numbers of an electron with allowance for the spin quantum number characterizing the internal rotation of an electron is equal to the dimension of space plus one. In this case, the electron energy in the first approximation, as well as in the model of the Bohr atom, is inversely proportional to the square of the principal quantum number (equation (1.78)). For example, in three - dimensional space the atom has besides a spin quantum number yet of three more quantum numbers. The main quantum number n, which determines the average distance of the electron from the nucleus. Orbital quantum number l, characterizing the momentum of an electron. The magnetic quantum number m_l, characterizing the possible value of the projection of the angular momentum of the electron on the axis z in a magnetic field. The main quantum number can take the values 1, 2, 3, ... The orbital quantum number can take the values 0, 1, 2, ...

Magnetic quantum numbers can take the values 0, ±1, ±2, ±3,... .

The values of the orbital quantum number *l* correspond to form of electron orbital. Taking into account the large dimensionality of the neighborhood of the nucleus in the atom, one should expect an increase in the number of quantum cells in orbitals *p, d, f* while maintaining their shape. For example, not 6 but 8 electrons (for each quantum cell with a corresponding pair of electrons per axis) can be on the orbital *p*. When analyzing chemical elements and molecules formed by them, it was shown (Zhizhin, 2018) that many elements exhibit a valence in chemical reactions much higher than the valence determined by their place in the periodic table (group number). To explain this, can resort to assumptions about various possible mechanisms of electron interaction: atomic and molecular hybridization, repulsion of divided and unpaired electron pairs, and the involvement of electrons inside the electron shells in the chemical interactions. In addition, physical experiments at high pressures reveal a sharp change in the structure of compounds that are also not explainable from the point of view of standard valence.

On the other hand, the analysis of real forms of molecules shows that their dimensionality, determined by the shape of the corresponding convex body, whose vertices are atoms, is often higher than three.

Undoubtedly, the possible loss of the number of quantum cells in orbitals *p, d, f* should not lead to a change in the number of electrons in the atomic structure in neutral atoms. The number of electrons under these conditions is of course equal to the charge of the nucleus. But the possible increase in the number of vacant quantum cells contributes to an increase in the chemical activity of atoms, allowing the creation of complex chemical compounds. This can be seen, for example, in biology.

The creation of vacant orbitals in higher-dimensional space can just serve as the basis for the electrophilicity of elements, which is characteristic of biochemical reactions (Metzler, 1977; Nelson, Cox, 2022). It should be noted that the electrophilicity of elements can also be created as a result of non-covalent interaction of elements (Magerramov, Makhmudov, Kopylovich, Pombeiro, 2016).

REFERENCES

Babenko, Yu. I. (2015). Antispace. In I.N. Toganov, Yu.I. Babenko. Anti-time and anti - space. S.-Petersburg: TIN.

Büchel, W. (1963). Warum hat ger Raume drei Dimensionen? *Physikalische Blätter*, *19*(12), 547–549. doi:10.1002/phbl.19630191204

Freeman, I. M. (n.d.). Why is Space Three-Dimensional? *American Journal of Physics*, *37*, 1222–1224. doi:10.1119/1.1975283

Gurevich, L., & Mostepanenko, V. (1971). On the Existence of Atoms in n - Dimensional Space. *Physics Letters. [Part A]*, *35*(3), 201–202. doi:10.1016/0375-9601(71)90148-4

Hobson, E. W. (1931). *The theory of spherical and ellipsoidal harmonics*. At the University Press.

Landau, L. D., & Lifshitz, E. M. (1963). Quantum mechanics. Nonrelativistic theory. Moscow: Gos. Ed. Phys.- Math. Lit.

Magerramov, A. M., Makhmudov, K. T., Kopylovich, M. N., & Pombeiro, A. J. L. (Eds.). (2016). *Non-covalent interactions in the synthesis and design of new compounds*. Wiley. doi:10.1002/9781119113874

Metzler, D. E. (1977). *BIOCHEMISTRY. The Chemical Reactions of Living Cells*. Academic Press.

Nelson, D. L., & Cox, M. M. (2022). *Lehninger Principles of Biochemistry*. W.H. Freeman.

Poincare, A. (1895). Analysis Situs. *Journal de Ecole Polytechnique*, *1*, 1–121.

Schrodinger, (1928). *Abhandlungen zur Wellenmechanik*. Leipzig.

Zhizhin, G. V. (2018). *Chemical Compound Structures and the Higher Dimension of Molecules: Emerging Research and Opportunities*. IGI Global. doi:10.4018/978-1-5225-4108-0

Zhizhin, G. V. (2019). *The Geometry of Higher–Dimensional Polytopes*. IGI Global. doi:10.4018/978-1-5225-6968-8

Chapter 2
The Law of Conservation of Incidents

ABSTRACT

This chapter establishes the existence of integral equality relatively, to the issue of the transmission of information by elements of lower and higher dimensions in the polytopes of the higher dimension that describe natural objects. This integral equality was established for the first time when studying the nanoworld in a higher-dimensional space (Zhizhin, 2019 a). As we will see in later chapters, this equality is of great importance in higher dimensional biochemistry. Therefore, this equality (the law of conservation of incidence) can be considered one of the concepts of biochemistry of the world of higher dimension. This integral equality is called the law of conservation of incidents. There is incidence interpreted as the transfer of information from one material body to another.

INTRODUCTION

In previous works of the author (Zhizhin, 2012; 2013; 2014 a, b, c; 2015; 2016; 2018; 2019 b, c, d, e; Zhizhin, & Diudea, 2016; Zhizhin, Khalaj, & Diudea, 2016), it was convincingly shown that various objects of the nanoworld have a higher dimension. In particular, many molecules of chemical compounds geometrically have the form of the polytopes of the higher dimension. Moreover, many of them are constructions with the main part in the form of an n - simplex or n - cross - polytope. Strictly speaking, any polytope can be reduced to an n - simplex by adding a sufficient number of edges to it without changing the number of vertices. As it turned out, an important place among polytopes describing chemical compounds are occupied by the cross - polytopes of higher dimension. In nanoclusters of chemical compounds, multi - shell structures of the highest dimension are widespread (Zhizhin, 2019 d). In

DOI: 10.4018/979-8-3693-0588-1.ch002

the analysis of the continuously distributed nanoworld, it was found (Zhizhin, 2015, 2019 b) that the fundamental areas of their structures are the polytopic prismahedrons of the higher dimension. A polytopic prismahedron is a multidimensional prism with bases in the form of polytopes of higher dimension (Zhizhin, 2019 b). The simplest polytopic prismahedron is the golden hyper - rhombohedron of dimension 4, which was found in the analysis of intermetallic diffraction patterns (Zhizhin, 2013; Shevchenko, Zhizhin, Mackay, 2013; Zhizhin, 2014 c). It is interesting to note that our children's ideas about elementary bricks (cubes and parallelepipeds) filling the space, based on the assumption of the three - dimensionality space, have no continuation in the space of the nanoworld. A logical continuation of our children`s ideas in the space of higher dimension could be a cube of the higher dimension, but the n - cube is not a fundamental area in the space of the nanoworld, since the study of numerous chemical compounds did not reveal elementary cells in the form of an n – cube (Zhizhin, 2018).

This work is devoted to the presentation and proof of the law found in the space of the nanoworld, characterized by the higher dimension of the components. It is connected with the concept of incidence in polytopes of the higher dimension.

In the monograph (Zhizhin, 2019 b) has studied in detail the geometry of higher - dimensional polytopes, based on an analysis of the structures of chemical compounds (Zhizhin, 2018). The monograph (Zhizhin, 2019 b) introduced the concept of the incidence coefficients of elements of lower dimension with respect to elements of the higher dimension and elements of higher dimension with respect to elements of the lower dimension. The first characterizes the number of elements of a certain higher dimension to which the given element of a lower dimension belongs. The second characterizes the number of elements of a given lower dimension that are included in a particular element of a higher dimension. Here we must remember that the vertices of geometric elements of various dimensions are atoms, molecules or functional groups. Therefore, the incidence of geometric elements to a friend means contacts between particles of the matter, including living matter. The contact between particles of matter can be interpreted as the transfer of information on material structures, including biological structures. The monograph (Zhizhin, 2019 e) showed that the elements of nanostructures of living matter are objects of higher dimension. Information sharing is an inherent property of living beings, without which even primitive organisms would not be able to maintain an extremely delicate balance, which depends on their survival (Mancuso, Viola, 2013; Mancuso, 2017). The monograph (Zhizhin, 2019 e) found that DNA, RNA molecules are polytopes of higher dimension, and nitrogenous bases connecting the double helix in the DNA molecule form a cross - polytopes of dimension 13. In this connection, the phenomenon of living organisms associated with DNA modification due to the binding of the methyl group CH_3 to the nitrogenous bases of DNA (methylation). This provides the memory of living organisms, i. e. transfer of inherited traits without changing the sequence of genes in DNA (Mancuso, 2017; Lindquist et al., 2016; Hawkes, Hennelly, Novikova et al., 2016). Therefore, it is of interest to study in detail the incidence of the entire polygamy of elements of different dimensions in higher dimension polytopes, considering it as a study of the possibility of transmitting information in biological molecules. This study is devoted to the article. It considers all possible incidents in convex bodies, starting with the simplest bodies and gradually complicating them.

Thus, the article for the first time posed the question of the possibility of transmitting information as a result of the contact of material objects having different dimensions in the space of the nanoworld. It is in this world that molecular formations have geometric shapes of higher dimensionality. This issue is of fundamental importance in the transmission of hereditary information.

Incident Conservation Law for N - Simplex

In a n - simplex (Zhizhin, 2014 c, 2019 b) the number of elements with dimension d is:

$$f(n) = C^{d+1} \tag{2.1}$$

For example, if $d = 0$, so $f_0(n) = C^1_{n+1} = \frac{(n+1)!}{n!} = n+1$. (2.2)

This is the number of vertices in n – simplexes. If $d = 1$, so

$$f_1(n) = C^2_{n+1} = \frac{(n+1)!}{(n-1)!2!} - \frac{(n+1)n}{2}. \tag{2.3}$$

This is the number of edges in n - simplexes. We introduce the notation: $k_{d^i_j d_u}$ is the number of elements of dimension u, which include an element of dimension j ($u > j$) with number i. Thus, $k_{d^i_j d_u}$ is the incidence factor of element i with dimension j relative to elements with dimension u. We introduce the notation also: $k_{d_j d^i_u}$ is the number of elements of dimension j, which included in element i with a dimension u ($u > j$). Thus, $k_{d_j d^i_u}$ is the incidence factor of element i with dimension u relative to elements with dimension j.

If $n = 2$, so $f_0(2)=3$. $f_1(2)=3$, see (2.2), (2.3). The factors of incidents (from smaller dimension to larger) are $k_{d^i_0 d_1} = 2, i = 1, 2, 3$; $k_{d^i_1 d_2} = 1, i = 1, 2, 3$.

Sum up the incidence coefficients for all vertices and edges of the triangle

$$\sum_{i=1}^{3} k_{d_0^i d_1} = 6, \sum_{i=1}^{3} k_{d_1^i d_2} = 3, \sum_{i=1}^{3} k_{d_0^i d_1} + \sum_{i=1}^{3} k_{d_1^i d_2} = 9. \tag{2.4}$$

The factors of incidents (from larger dimension to smaller) are $k_{d_0 d_1^i} = 2, i = 1, 2, 3; k_{d_1 d_2} = 3.$

Sum up the incidence coefficients for all elements of the triangle with dimension larger of zero

$$\sum_{i=1}^{3} k_{d_0 d_1^i} + k_{d_1 d_2} = 9. \tag{2.5}$$

Comparing (2.4) and (2.5) you can see that the sum of incidents in a triangle from elements with a lower dimension to elements with a higher dimension is equal to the sum of incidents from elements with a higher dimension to elements with a lower dimension. Thus, the sum of incidents retains its value when the direction of the relationship between the elements it is change.

If $n = 3$, so $f_0(3)=4$, $f_1(3)=6$, $f_2(3)=4$. The factors of incidents (from smaller dimension to larger) are

$$k_{d_0^i d_1} = 3, i = 1 \div 4; \ k_{d_0^i d_2} = 3, i = 1 \div 4; \ k_{d_0^i d_3} = 1, i = 1 \div 4;$$
$$k_{d_1^i d_2} = 2, i = 1 \div 6; \ k_{d_1^i d_3} = 1, i = 1 \div 6; \ k_{d_2^i d_3} = 1, i = 1 \div 4.$$

Sum up the incidence coefficients for all elements of the tetrahedron

$$\sum_{i=1}^{4} k_{d_0^i d_1} = 12, \sum_{i=1}^{4} k_{d_0^i d_2} = 12, \sum_{i=1}^{4} k_{d_0^i d_3} = 4, \sum_{i=1}^{6} k_{d_1^i d_2} = 12, \sum_{i=1}^{6} k_{d_1^i d_3} = 6, \sum_{i=1}^{4} k_{d_2^i d_3} = 4,$$
$$\sum_{i=1}^{4} k_{d_0^i d_1} + \sum_{i=1}^{4} k_{d_0^i d_2} + \sum_{i=1}^{4} k_{d_0^i d_3} + \sum_{i=1}^{6} k_{d_1^i d_2} + \sum_{i=1}^{6} k_{d_1^i d_3} + \sum_{i=1}^{4} k_{d_2^i d_3} = 50. \tag{2.6}$$

The factors of incidents (from larger dimension to smaller) are

$$k_{d_0 d_1^i} = 2, i = 1 \div 6; k_{d_0 d_2^i} = 3, i = 1 \div 4; k_{d_1 d_2^i} = 3, i = 1 \div 4; k_{d_0 d_3} = 4, k_{d_1 d_3} = 6, k_{d_2 d_3} = 4.$$

Sum up the incidence coefficients for all elements of the tetrahedron

$$\begin{array}{l}\sum_{i=1}^{6} k_{d_0 d_1^i} = 12, \sum_{i=1}^{4} k_{d_0 d_2^i} = 12, \sum_{i=1}^{4} k_{d_1 d_2^i} = 12, \\ \sum_{i=1}^{6} k_{d_0 d_1^i} + \sum_{i=1}^{4} k_{d_0 d_2^i} + \sum_{i=1}^{4} k_{d_1 d_2^i} + k_{d_0 d_3} + k_{d_1 d_3} + k_{d_2 d_3} = 50\end{array}. \tag{2.7}$$

Comparing (2.6) and (2.7) you can see that the sum of incidents in a tetrahedron from elements with a lower dimension to elements with a higher dimension is equal to the sum of incidents from elements with a higher dimension to elements with a lower dimension. Thus, the sum of incidents retains its value when the direction of the relationship between the elements it is change.

Let us prove the general statement for a simplex of arbitrary finite dimension *n*.

Theorem 2.1. In any simplex of dimension *n*, the sum of all incidents of elements of a lower dimension with respect to all elements of a higher dimension is equal to the sum of all incidents of elements of a higher dimension with respect to all elements of a lower dimension and equals the sum of the series,

$$f_0(n)(2^n - 1) + f_1(n)(2^{n-1} - 1) + f_2(n)(2^{n-2} - 1) + ... + f_{n-1}(n), f_d(n) = C_{n+1}^{d+1}, d = 0 \div (n-1).$$

Proof.

The number of edges (elements of dimension 1) that have a certain vertex d_0^i its end point (given that in a simplex each vertex is connected by edges with all other vertices) is determined by the incidence factor,

$$k_{d_0^i d_1} = C_n^1 = f_0(n) - 1 = n, i = 1 \div f_0(n). \tag{2.8}$$

The number of elements of dimension 2 that have its vertex a certain vertex d_0^i is determined by the incident factor,

$$k_{d_o^i d_2} = C_n^2, i = 1 \div f_0(n). \tag{2.9}$$

The number of elements of dimension 3 that have its vertex a certain vertex d_0^i is determined by the incident factor $k_{d_o^i d_3} = C_n^3, i = 1 \div f_0(n)$. And so on.

The number of elements of dimension (*n* - 1) that have its vertex a certain vertex d_0^i is determined by the incident factor $k_{d_o^i d_{n-1}} = C_n^{n-1}, i = 1 \div f_0(n)$. The number of elements of dimension n that have its vertex a certain vertex d_0^i is determined by the incident factor $k_{d_0^i d_n} = C_n^n = 1, i = 1 \div f_0(n)$.

The total number of incidents of vertices to elements of higher dimensions in a n - simplex will be,

$$f_0(n)\sum_{j=1}^{n} k_{d_0^i d_j} = f_0(n)(C_n^1 + C_n^2 + ... + C_n^n) = f_0(n)(2^n - 1).$$

The number of elements of dimension 2 that have a certain edge d_1^i its edge is determined by the incidence factor,

$$k_{d_1^i d_2} = C_{n-1}^1 = f_0(n) - 2 = n - 1, i = 1 \div f_1(n).$$

The number of elements of dimension 3 that have its edge an element d_1^i is determined by the incident factor $k_{d_1^i d_3} = C_{n-1}^2, i = 1 \div f_1(n)$. And so on.

The number of elements of dimension (n - 1) that have its edge a certain edge d_1^i is determined by the incident factor $k_{d_1^i d_{n-1}} = C_{n-1}^{n-2}, i = 1 \div f_1(n)$. The number of elements of dimension n that have its edge a certain edge d_1^i is determined by the incident factor $k_{d_1^i d_n} = C_{n-1}^{n-1} = 1, i = 1 \div f_1(n)$.

The total number of incidents of edges to elements of higher dimensions in n - simplex will be,

$$f_1(n)\sum_{j=2}^{n} k_{d_1^i d_j} = f_1(n)(C_{n-1}^1 + C_{n-1}^2 + ... + C_{n-1}^{n-1}) = f_1(n)(2^{n-1} - 1). \quad (2.10)$$

In the same way, it is proved that the total number of incidences of elements of dimension 2 to elements of higher dimension in a n - simplex is determined by the expression,

$$f_2(n)\sum_{j=3}^{n} k_{d_2^i d_j} = f_2(n)(C_{n-2}^1 + C_{n-2}^2 + ... + C_{n-2}^{n-2}) = f_2(n)(2^{n-2} - 1).$$

The total number of incidences of elements of dimension 3 to elements of higher dimension in a n - simplex is determined by the expression,

$$f_3(n)\sum_{j=4}^{n} k_{d_3^i d_j} = f_3(n)(C_{n-3}^1 + C_{n-3}^2 + ... + C_{n-3}^{n-3}) = f_3(n)(2^{n-3} - 1).$$

And so on. Thus, the sum of all incidents of elements of a lower dimension with respect to all elements of a higher dimension in n - simplexes is equal,

$$f_0(n)(2^n - 1) + f_1(n)(2^{n-1} - 1) + f_2(n)(2^{n-2} - 1) + ... + f_{n-1}(n). \tag{2.11}$$

To complete the proof of Theorem 2.1, it is necessary to show that the same expression is equal to the sum of all incidents of elements of a higher dimension with respect to all elements of a lower dimension in n - simplexes.

We pose the question: how many elements of a lower dimension belong to one element of a higher dimension? For example, how many vertices belong to a single edge $d_1^i, i = 1 \div f_1(n)$? Obviously, this number is 2. This is the value of the coefficient of incidence $k_{d_0 d_1^i} = C_2^1 = 2$. The product $C_2^1 f_1(n)$ gives the number of vertices belonging to all edges of an n - simplex. This number $2C_{n+1}^2 = 2\frac{(n+1)!}{(n-1)!2!} = (n+1)n$ is equal to the product $k_{d_0^i d_1} = C_n^1 C_{n+1}^1 = n(n+1)$, i.e. the number of edges with a peak point $d_0^i, i = 1 \div f_0(n)$ (2.8). The number of vertices belonging to a single element of dimension 2 (triangle) is given by the incidence coefficient $k_{d_0 d_2^i}, i = 1 \div f_2(n)$. Obviously, this number is $3 = C_3^1$. The product $C_3^1 f_2(n)$ corresponds to the number of vertices belonging to all elements of dimension 2 in the n - simplex. This number is

$$C_3^1 f_2(n) = \frac{3!}{2!} C_{n+1}^1 = \frac{3!}{2!} \frac{(n+1)!}{(n-2)!3!} = \frac{(n+1)!}{(n-2)!2!}.$$

Let us compare this number with the number of elements of dimension 2 with vertices (2.9), i.e. with the product

$$C_n^2 f_0(n) = \frac{n!}{(n-2)!2!} C_{n+1}^1 = \frac{n!}{(n-2)!2!} \frac{(n+1)!}{n!} = \frac{(n+1)!}{(n-2)!2!}.$$

It is clear that $C_n^2 f_0(n) = C_3^1 f_2(n)$. This proves that the number of vertices belonging to all elements of dimension 2 in an *n* - simplex is equal to the number of elements of dimension 2 that have their vertices d_0^i.

Let $k_{d_0 d_j^i}$ be the number of vertices belonging to some one element of dimension *j*, that is d_j^i. Obviously, this number is equal to $C_{j+1}^1 = j+1, i=1 \div f_j(n)$. The product $C_{j+1}^1 f_j(n)$ corresponds to the number of vertices belonging to all elements of dimension *j* in the *n* - simplex. Compare this number with the number of elements of dimension *j* having vertices d_0^i at $i=f_0(n)$, i.e. with the product

$$C_n^j f_0(n) = (n+1)\frac{n!}{(n-j)!\,j!}.$$

On the other hand, we have

$$C_{j+1}^1 f_0(n) = \frac{(j+1)!}{j!}\frac{(n+1)!}{(n-j)!(j+1)!} = \frac{n!(n+1)}{j!(n-j)!}.$$

It can be seen that these values coincide. This proves that the number of vertices belonging to all elements of dimension *j* in the *n* - simplex is equal to the number of elements of dimension *j* that have their vertices d_0^i.

Let $k_{d_1 d_j^i}$ be the number of edges belonging to some one element of dimension *j*, that is d_j^i, $j > 1, i = 1 \div f_j(n)$. Obviously, this number edges belonging all elements with dimension *j* is

$$f_1(j) f_j(n) = C_{j+1}^2 C_{n+1}^{J+1} = \frac{(j+1)!}{(j-1)!2!}\frac{(n+1)!}{(n-j)!(j+1)!} = \frac{(n+1)!}{(j-1)!2!(n-j)!}.$$

Compare this number with the number of elements dimension *j* having element d_1 (see (2.10))

$$f_1(n) k_{d_1^i d_j} = f_1(n) C_{n-1}^{j-1} = C_{n+1}^2 C_{n-1}^{j-1} = \frac{(n+1)n}{2}\frac{(n-1)!}{(n-j)!(j-1)!}.$$

It can be seen that these numbers are the same.

Let $k_{d_h d_j^i}$ be the number of elements with dimension h belonging to some one element of dimension j ($h < j$), that is $d_j^i, i = 1 \div f_j(n)$. Obviously, this number equal $f_h(j)$. The product $f_h(j)f_j(n)$ corresponds to the number of elements with dimension h belonging to all elements of dimension j in the n - simplex. This number is

$$f_h(j)f_j(n) = C_{j+1}^{h+1}C_{n+1}^{j+1} = \frac{(j+1)!}{(j-h)!(h+1)!}\frac{(n+1)!}{(n-j)!(j+1)!} = \frac{(n+1)!}{(j-h)!(h+1)!(n-j)!}.$$

Compare this number with the number of elements dimension j having elements with dimension h ($d_h^i, i = f_h(n)$)

$$f_h(n)k_{d_h^i d_j} = f_h(n)C_{n-h}^{j-h} = C_{n+1}^{h+1}C_{n-h}^{j-h} = \frac{(n+1)!}{(n-h)!(h+1)!}\frac{(n-h)!}{(n-j)!(j-h)!}.$$

It can be seen that these numbers are the same.

This proves that the number of elements of dimension h, belonging to all elements of dimension j in n - simplex, is equal to the number of elements of dimension j, which have elements of dimension h. Thus, the total number of incidents of elements of a smaller dimension to elements of a larger dimension is equal to the total number of incidents of elements of a higher dimension to elements of a smaller dimension. This exists in spite of the difference in the incidence coefficients of elements of a smaller dimension to elements of a higher dimension and the incidence coefficients of elements of a higher dimension to elements of a smaller dimension.

This fact can be considered the law of conservation of incidences in the n - simplex. The total number of incidents in the n - simplex is described by equality (2.11).

Q.E.D.

It can be seen that the total number of incidents in the n - simplex depends only on the dimension n and increases sharply with increasing n. For example, it is 12 for $n = 2$, 50 for $n = 3$, 180 for $n = 4$, 602 for $n = 5$.

$$C_3^1 f_2(n) = \frac{3!}{2!}C_{n+1}^1 = \frac{3!}{2!}\frac{(n+1)!}{(n-2)!3!} = \frac{(n+1)!}{(n-2)!2!}.$$

Let us compare this number with the number of elements of dimension 2 with vertices (2.9), i.e. with the product

$$C_n^2 f_0(n) = \frac{n!}{(n-2)!2!} C_{n+1}^1 = \frac{n!}{(n-2)!2!} \frac{(n+1)!}{n!} = \frac{(n+1)!}{(n-2)!2!}.$$

It is clear that $C_n^2 f_0(n) = C_3^1 f_2(n)$. This proves that the number of vertices belonging to all elements of dimension 2 in an n - simplex is equal to the number of elements of dimension 2 that have their vertices d_0^i.

Let $k_{d_0 d_j^i}$ be the number of vertices belonging to some one element of dimension j, that is d_j^i. Obviously, this number is equal to $C_{j+1}^1 = j+1$, i=1÷$f_j(n)$. The product $C_{j+1}^1 f_j(n)$ corresponds to the number of vertices belonging to all elements of dimension j in the n - simplex. Compare this number with the number of elements of dimension j having vertices d_0^i at i=$f_0(n)$, i.e. with the product

$$C_n^j f_0(n) = (n+1)\frac{n!}{(n-j)!j!}.$$

On the other hand, we have

$$C_{j+1}^1 f_0(n) = \frac{(j+1)!}{j!} \frac{(n+1)!}{(n-j)!(j+1)!} = \frac{n!(n+1)}{j!(n-j)!}.$$

It can be seen that these values coincide. This proves that the number of vertices belonging to all elements of dimension j in the n - simplex is equal to the number of elements of dimension j that have their vertices d_0^i.

Let $k_{d_1 d_j^i}$ be the number of edges belonging to some one element of dimension j, that is d_j^i, $j > 1, i = 1 \div f_j(n)$. Obviously, this number edges belonging all elements with dimension j is

$$f_1(j) f_j(n) = C_{j+1}^2 C_{n+1}^{J+1} = \frac{(j+1)!}{(j-1)!2!} \frac{(n+1)!}{(n-j)!(j+1)!} = \frac{(n+1)!}{(j-1)!2!(n-j)!}.$$

Compare this number with the number of elements dimension j having element d_1 (see (2.10))

$$f_1(n)k_{d_1^i d_j} = f_1(n)C_{n-1}^{j-1} = C_{n+1}^{2}C_{n-1}^{j-1} = \frac{(n+1)n}{2}\frac{(n-1)!}{(n-j)!(j-1)!}.$$

It can be seen that these numbers are the same.

Let $k_{d_h d_j^i}$ be the number of elements with dimension h belonging to some one element of dimension j ($h < j$), that is $d_j^i, i = 1 \div f_j(n)$. Obviously, this number equal $f_h(j)$. The product $f_h(j)f_j(n)$ corresponds to the number of elements with dimension h belonging to all elements of dimension j in the n - simplex. This number is

$$f_h(j)f_j(n) = C_{j+1}^{h+1}C_{n+1}^{j+1} = \frac{(j+1)!}{(j-h)!(h+1)!}\frac{(n+1)!}{(n-j)!(j+1)!} = \frac{(n+1)!}{(j-h)!(h+1)!(n-j)!}.$$

Compare this number with the number of elements dimension j having elements with dimension h ($d_h^i, i = f_h(n)$)

$$f_h(n)k_{d_h^i d_j} = f_h(n)C_{n-h}^{j-h} = C_{n+1}^{h+1}C_{n-h}^{j-h} = \frac{(n+1)!}{(n-h)!(h+1)!}\frac{(n-h)!}{(n-j)!(j-h)!}.$$

It can be seen that these numbers are the same.

This proves that the number of elements of dimension h, belonging to all elements of dimension j in n - simplex, is equal to the number of elements of dimension j, which have elements of dimension h. Thus, the total number of incidents of elements of a smaller dimension to elements of a larger dimension is equal to the total number of incidents of elements of a higher dimension to elements of a smaller dimension. This exists in spite of the difference in the incidence coefficients of elements of a smaller dimension to elements of a higher dimension and the incidence coefficients of elements of a higher dimension to elements of a smaller dimension.

This fact can be considered the law of conservation of incidences in the n - simplex. The total number of incidents in the n - simplex is described by equality (2.11).

Q.E.D.

It can be seen that the total number of incidents in the n - simplex depends only on the dimension n and increases sharply with increasing n. For example, it is 12 for $n = 2$, 50 for $n = 3$, 180 for $n = 4$, 602 for $n = 5$.

INCIDENT CONSERVATION LAW FOR N - GOLDEN - HYPER - RHOMBOHEDRON

The smallest dimension of the golden hyper - rhombohedron is 4 (Figure 1) (Zhizhin, 2014 c, 2019 b).

Figure 1. The four - golden - hyper – rhombohedron

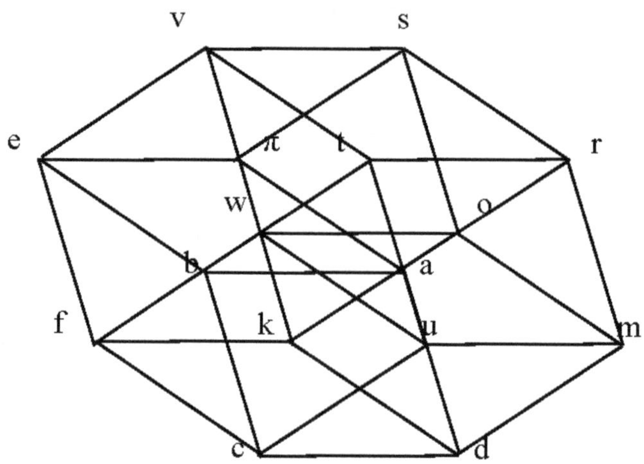

From Figure 1 it follows that in this case

$f_0(4)=16, f_1(4)=32, f_2(4)=24, f_3(4)=8.$

The factors of incidents (from smaller dimension to larger) are

$$k_{d_0^i d_1} = 4, i = 1 \div 16; k_{d_0^i d_2} = 6, i = 1 \div 16; k_{d_0^i d_3} = 4, i = 1 \div 16; k_{d_0^i d_4} = 1, i = 1 \div 16; k_{d_1^i d_2} = 3, i = 1 \div 32;$$
$$k_{d_1^i d_3} = 3, i = 1 \div 32; k_{d_1^i d_4} = 1, i = 1 \div 32; k_{d_2^i d_3} = 2, i = 1 \div 24; k_{d_2^i d_4} = 1, i = 1 \div 24; k_{d_3^i d_4} = 1, i = 1 \div 8.$$

Sum up the incidence coefficients for all vertices, edges, two - dimension faces and three - dimension faces of the 4 - golden - hyper - rhombohedron

$$\sum_{i=1}^{16} k_{d_0^i d_1} = 4 \cdot 16 = 64, \sum_{i=1}^{16} k_{d_0^i d_2} = 6 \cdot 16 = 96, \sum_{i=1}^{16} k_{d_0^i d_3} = 4 \cdot 16 = 64, \sum_{i=1}^{16} k_{d_0^i d_4} = 16, \sum_{i=1}^{32} k_{d_1^i d_2} = 3 \cdot 32 = 96,$$
$$\sum_{i=1}^{32} k_{d_1^i d_3} = 3 \cdot 32 = 96, \sum_{i=1}^{32} k_{d_1^i d_4} = 32, \sum_{i=1}^{24} k_{d_2^i d_3} = 2 \cdot 24 = 48, \sum_{i=1}^{24} k_{d_2^i d_4} = 24, \sum_{i=1}^{8} k_{d_3^i d_4} = 1 \cdot 8 = 8. \qquad (2.12)$$
$$\sum_{i=1}^{16} k_{d_0^i d_1} + \sum_{i=1}^{16} k_{d_0^i d_2} + \sum_{i=1}^{16} k_{d_0^i d_3} + \sum_{i=1}^{16} k_{d_0^i d_4} + \sum_{i=1}^{32} k_{d_1^i d_2} + \sum_{i=1}^{32} k_{d_1^i d_3} + \sum_{i=1}^{32} k_{d_1^i d_4} + \sum_{i=1}^{24} k_{d_2^i d_3} + \sum_{i=1}^{24} k_{d_2^i d_4} + \sum_{i=1}^{8} k_{d_3^i d_4} = 544.$$

The factors of incidents (from larger dimension to smaller, Figure 2.1) are

$$k_{d_0 d_1^i} = 2, i = 1 \div 32; k_{d_0 d_2^i} = 4, i = 1 \div 24; k_{d_0 d_3^i} = 8, i = 1 \div 8; k_{d_0 d_4} = 16; k_{d_1 d_2^i} = 4, i = 1 \div 24;$$
$$k_{d_1 d_3^i} = 12, i = 1 \div 8; k_{d_1 d_4} = 32; k_{d_2 d_3^i} = 6, \mathrm{i} = 1 \div 8; k_{d_2 d_4} = 24; k_{d_3 d_4} = 8.$$

Sum up the incidence coefficients for all elements of the 4 - golden - hyper - rhombohedron with dimension larger of zero

$$\sum_{i=1}^{32} k_{d_0 d_1^i} {}_0 \sum_{i=1}^{24} k_{d_0 d_2^i} + \sum_{i=1}^{24} k_{d_1 d_2^i} + \sum_{i=1}^{8} k_{d_0 d_3^i} + \sum_{i=1}^{8} k_{d_1 d_3^i} + \sum_{i=1}^{8} k_{d_2 d_3^i} + k_{d_0 d_4} + k_{d_1 d_4} + k_{d_2 d_4} + k_{d_3 d_4} = 544. \qquad (2.13)$$

Comparing (2.12) and (2.13) you can see that the sum of incidents in a 4 - golden - hyper - rhombohedron from elements with a lower dimension to elements with a higher dimension is equal to the sum of incidents from elements with a higher dimension to elements with a lower dimension. Thus, the sum of incidents retains its value when the direction of the relationship between the elements it is changes.

It is easy to see that each incidence coefficient from a smaller dimension to a higher dimension corresponds to its inverse incidence factor from a larger dimension to a smaller dimension. For example, $k_{d_0^i d_2} = 6, k_{d_0 d_2^i} = 4$. Moreover, the number of such incidents from a smaller dimension to a higher dimension is 16, the number of corresponding inverse incidents is 24. So, the product of the incidence coefficient and the corresponding number retains its value, 6•16=4•24=96. It can be seen that this situation exists for all incidence rates. Therefore, the sum of incidents in the forward and reverse direction is preserved.

In order to write down the sum of incidents for a golden - hyper - rhombohedron of any dimension in general, it is necessary to write expressions for the incidence coefficients and numbers of incidences in the forward and reverse directions, as a function of the golden - hyper - rhombohedron dimension. This will also allow us to prove the next statement:

Theorem 2.2. In any golden - hyper - rhombohedron of dimension n, the sum of all incidents of elements of a lower dimension with respect to all elements of a higher dimension is equal to the sum of all incidents of elements of a higher dimension with respect to all elements of a lower dimension and equals the sum of the series

$$f_0(n)(2^n - 1) + f_1(n)(2^{n-1} - 1) + f_2(n)(2^{n-2} - 1) + ... + f_{n-1}(n), f_d(n) = 2^{n-d} C_n^d, d = 0 \div n - 1.$$

P r o o f.

It should be borne in mind that the numbers of elements of different dimensions in the 4 - golden - hyper - rhombohedron and in the 4 - cube are equal (Zhizhin, 2014 c, 2019 b). Moreover, the increase in the dimension of these figures occurs according to one mechanism (sequential multiplication of figures by a one - dimensional segment). Therefore, to calculate the numbers of elements of different dimensions for the n - golden - hyper - rhombohedron, one can use the formulas obtained to calculate the numbers of elements of different dimensions in the n - cube (Zhizhin, 2014 c, 2019 b). It should be remembered that the three - dimensional faces of a 4 - golden - hyper - rhombohedron are rhombohedrons, the flat faces of which are rhombuses with angles connected by a golden section. Thus, according to (Zhizhin, 2014 c; 2019 b) the number of elements with dimension d of the n - golden - hyper - rhombohedron is

$$f_d(n) = 2^{n-d} C_n^d, d = 0 \div n. \tag{2.14}$$

The number of vertices in n - golden - hyper - rhombohedron is equal $f_0 = 2^n$ (see (2.14)). The number of edges in n - golden - hyper - rhombohedron incident to any of the vertices of a golden - hyper - rhombohedron, by its definition, is equal to $k_{d_0^i d_1} = C_n^1, i = 1 \div f_0$.

The number of elements of two - dimension incident to any of the vertices of a golden - hyper - rhombohedron is equal $k_{d_0^i d_2} = C_n^2, i = 1 \div f_0$. We can also record the number of elements of any dimension incident to any of the vertices of the golden - hyper - rhombohedron up to the golden - hyper - rhombohedron itself $k_{d_0^i d_n} = C_n^n, i = 1 \div f_0$.

Multiplying each of these incidence coefficients by the number of vertices in the n - golden - hyper - rhombohedron and summing up the obtained products, we find that the total number of incidences of the vertices to the elements of higher dimensionality in the n - golden - hyper - rhombohedron is determined by the expression,

$$f_0(C_n^1 + C_n^2 + ... + C_n^n) = f_0(2^n - 1).$$

This expression coincides with the expression of the total number of incidences of vertices to elements of higher dimension in the n - simplex. The numerical difference between these expressions consists in a different value of the vertices in the n - golden - hyper – rhombohedron, and in the n - simplex, i.e. in the meaning of $f_0(n)$.

The number of edges in n - golden - hyper - rhombohedron is equal $f_1(n) = 2^{n-1}C_n^1$ (see (2.14)). The number of two - dimension elements in n - golden - hyper - rhombohedron incident to any of the edges of a golden - hyper - rhombohedron, by its definition, is equal to $k_{d_1^i d_2} = C_{n-1}^1, i = 1 \div f_1$.

The number of three - dimension elements incident to any of the edges of a golden - hyper - rhombohedron is equal $k_{d_1^i d_3} = C_{n-1}^2, i = 1 \div f_1$. We can also record the number of elements of any dimension incident to any of the edges of the golden - hyper - rhombohedron up to the golden - hyper - rhombohedron itself $k_{d_1^i d_n} = C_{n-1}^{n-1}, i = 1 \div f_1$.

Multiplying each of these incidence coefficients by the number of edges in the n - golden - hyper - rhombohedron and summing up the obtained products, we find that the total number of incidences of the edges to the elements of higher dimensionality in the n - golden - hyper - rhombohedron is determined by the expression,

$$f_1(C_{n-1}^1 + C_{n-1}^2 + ... + C_{n-1}^{n-1}) = f_1(2^{n-1} - 1).$$

This expression coincides with the expression of the total number of incidences of edges to elements of higher dimension in the n - simplex. The numerical difference between these expressions consists in a different value of the edges in the n - golden - hyper - rhombohedron and in the n - simplex, i.e. in the meaning of $f_1(n)$.

Continuing to write expressions for the total numbers of incidents of elements of different dimensions to elements of a higher dimension in a n - golden - hyper - rhombohedron, summing up these numbers, we obtain a general expression for the number of incidences of elements of a lower dimension in relation to elements of a higher dimension in n - golden - hyper – rhombohedron,

$$f_0(n)(2^n - 1) + f_1(n)(2^{n-1} - 1) + f_2(n)(2^{n-2} - 1) + ... + f_{n-1}(n), f_d(n) = 2^{n-d}C_n^d, d = 0 \div (n-1). \quad (2.15)$$

It is easy to make sure that the calculations of the incidence flow in the n - golden - hyper - rhombohedron according to expression (2.15) at values n=4 coincide with the values of these quantities obtained earlier by analyzing the images of n - golden - hyper - rhombohedron at these value n.

Let $k_{d_h d_j^i}$ be the number of elements of dimension h in a n - golden - hyper - rhombohedron belonging to some one element of dimension j ($h<j$), that is d_j^i. Obviously, this number is equal to $f_h(j)$, and $i=1\div f_j(n)$. The product $f_h(j)f_j(n)$ corresponds to the number of elements of dimension h belonging to all elements of dimension j. For a n - golden - hyper - rhombohedron, this product is equal to

$$f_h(j)f_j(n) = 2^{j-h}C_j^h 2^{n-j}C_n^j = 2^{n-h}\frac{n!}{(n-j)!(j-h)!h!}.$$

Let us compare this number with the number of elements of dimension j, which have elements of dimension h in the n - golden - hyper – rhombohedron,

$$k_{d_h^i d_j} f_h(n) = C_{n-h}^{j-h} 2^{n-h} C_n^h = 2^{n-h}\frac{n!}{(n-h)!h!}\frac{(n-h)!}{(n-j)!(j-h)!}.$$

Obviously, these numbers are equal to each other. This proves that the number of elements of dimension h in a n - golden - hyper - rhombohedron belonging to all dimensions of j ($j>h$) in a n - golden - hyper - rhombohedron is equal to the number of elements of dimension j, which have elements of dimension h as elements.

Thus, the total number of incidents of elements of a smaller dimension with respect to elements of a higher dimension is equal to the total number of incidents of elements of a higher dimension with respect to elements of a smaller dimension. This the total number define expression (2.15).

Q.E.D.

It is significant that equality of incidence (information) flows exists despite the difference in the incidence coefficients of elements of a lower dimension in relation to elements of a higher dimension and the incidence coefficients of elements of a higher dimension in relation to elements of a smaller dimension. Therefore, the incidence conservation law found in the previous section for the n - simplex is also valid for the n - golden - hyper - rhombohedron. Moreover, the analytical form of this law is preserved, expressed through functions $f_i(n)$ that determine the values of the numbers of the elements of dimension i in the polytope of dimension n.

INCIDENT CONSERVATION LAW FOR N - CROSS – POLYTOPE

In n - cross - polytope (Zhizhin, 2014 c; 2019 b) the number of elements with dimension d is

$$f_d(n) = 2^{d+1} C_n^{n-1-d}, d = 0 \div n. \qquad (2.16)$$

For example, if $d = 0$,

$$\text{so } f_0(n) = 2C_n^{n-1} = 2n. \qquad (2.17)$$

This is the number of vertices in n - cross - polytope. If $d = 1$, so

$$f_1(n) = 2^2 C_n^{n-2} = 2\frac{n!}{(n-2)!} = 2(n-1)n. \qquad (2.18)$$

This is the number of edges in n - cross - polytope. If $d = 2$, so

$$f_2(n) = 2^3 C_n^{n-3} = \frac{4}{3}(n-2)(n-1)n. \qquad (2.19)$$

This is the number of two - dimension faces in n - cross - polytope. If $d = 3$, so

$$f_3(n) = 2^4 C_n^{n-4} = \frac{2}{3}(n-3)(n-2)(n-1)n. \qquad (2.20)$$

This is the number of three - dimension faces in n - cross - polytope.

The smallest dimension of the cross - polytope is 4. From (2.17) - (2.20) it follows that in this polytope there are 8 vertices, 24 edges, 32 two - dimensional faces, 16 three - dimensional faces (Figure 2).

The factors of incidents (from smaller dimension to larger) are

$$k_{d_0^i d_1} = 6, i = 1 \div 8; k_{d_0^i d_2} = 12, i = 1 \div 8; k_{d_0^i d_3} = 8, i = 1 \div 8; k_{d_0^i d_4} = 1, i = 1 \div 8; k_{d_1^i d_2} = 4, i = 1 \div 24;$$
$$k_{d_1^i d_3} = 4, i = 1 \div 24; k_{d_1^i d_4} = 1, i = 1 \div 24; k_{d_2^i d_3} = 2, i = 1 \div 32; k_{d_2^i d_4} = 1, i = 1 \div 32; k_{d_3^i d_4} = 1, i = 1 \div 16.$$

Sum up the incidence coefficients for all vertices, edges, two - dimension faces and three -dimension faces of the 4 - cross – polytope

Figure 2. The four - cross – polytope

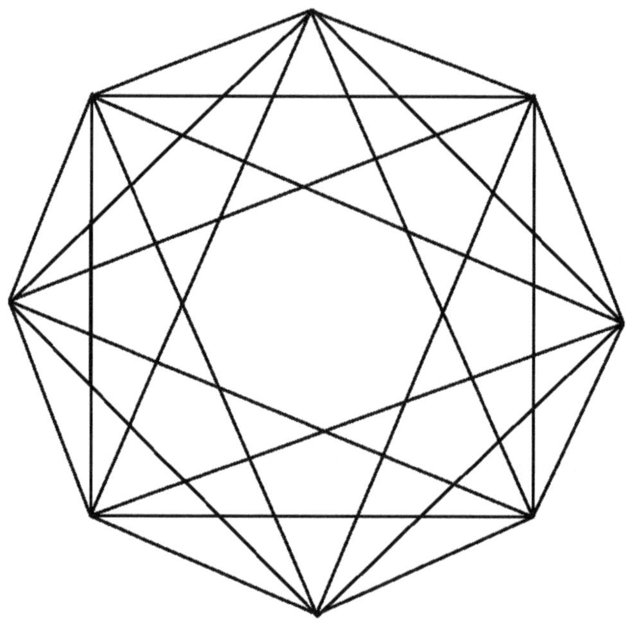

$$\sum_{i=1}^{8} k_{d_0^i d_1} = 6 \cdot 8 = 48, \sum_{i=1}^{8} k_{d_0^i d_2} = 12 \cdot 8 = 96, \sum_{i=1}^{8} k_{d_0^i d_3} = 8 \cdot 8 = 64, \sum_{i=1}^{8} k_{d_0^i d_4} = 8, \sum_{i=1}^{24} k_{d_1^i d_2} = 4 \cdot 24 = 96,$$
$$\sum_{i=1}^{24} k_{d_1^i d_3} = 4 \cdot 24 = 96, \sum_{i=1}^{24} k_{d_1^i d_4} = 24, \sum_{i=1}^{32} k_{d_2^i d_3} = 2 \cdot 32 = 64, \sum_{i=1}^{32} k_{d_2^i d_4} = 32, \sum_{i=1}^{16} k_{d_3^i d_4} = 16$$
$$\sum_{i=1}^{8} k_{d_0^i d_1} + \sum_{i=1}^{8} k_{d_0^i d_2} + \sum_{i=1}^{8} k_{d_0^i d_3} + \sum_{i=1}^{8} k_{d_0^i d_4} + \sum_{i=1}^{24} k_{d_1^i d_2} + \sum_{i=1}^{24} k_{d_1^i d_3} + \sum_{i=1}^{24} k_{d_1^i d_4} + \sum_{i=1}^{32} k_{d_2^i d_3} + \sum_{i=1}^{16} k_{d_3^i d_4} + \sum_{i=1}^{32} k_{d_2^i d_4} = 544. \quad (2.21)$$

The factors of incidents (from larger dimension to smaller, Figure 2) are

$$k_{d_0 d_1^i} = 2, i = 1 \div 24; k_{d_0 d_2^i} = 3, i = 1 \div 32; k_{d_0 d_3^i} = 4, i = 1 \div 16; k_{d_0 d_4} = 8; k_{d_1 d_2^i} = 3, i = 1 \div 32;$$
$$k_{d_1 d_3^i} = 6, i = 1 \div 16; k_{d_1 d_4} = 24; k_{d_2 d_3^i} = 4, \mathrm{i} = 1 \div 8; k_{d_2 d_4} = 32; k_{d_3 d_4} = 16.$$

Sum up the incidence coefficients for all elements of the 4 – cross - polytope with dimension larger of zero

$$\sum_{i=1}^{24} k_{d_0 d_1^i} + \sum_{i=1}^{32} k_{d_0 d_2^i} + \sum_{i=1}^{32} k_{d_1 d_2^i} + \sum_{i=1}^{16} k_{d_0 d_3^i} + \sum_{i=1}^{16} k_{d_1 d_3^i} + \sum_{i=1}^{16} k_{d_2 d_3^i} + k_{d_0 d_4} + k_{d_1 d_4} + k_{d_2 d_4} + k_{d_3 d_4} = 544. \quad (2.22)$$

Comparing (2.21) and (2.22) you can see that the sum of incidents in a 4 - cross - polytope from elements with a lower dimension to elements with a higher dimension is equal to the sum of incidents from elements with a higher dimension to elements with a lower dimension. Thus, the sum of incidents retains its value when change the direction of the relationship between the elements.

Theorem 2.3. In any cross - polytope of dimension *n*, the sum of all incidents of elements of a lower dimension with respect to all elements of a higher dimension is equal to the sum of all incidents of elements of a higher dimension with respect to all elements of a lower dimension and equals the sum of the series,

$$f_0(n)3^{n-1} + f_1(n)3^{n-2} + f_2(n)3^{n-3} + ... + f_{n-1}(n)3^0, f_d(n) = C_n^{n-1-d}2^{1+d}, d = 0 \div (n-1).$$

Proof.

According to equation (2.17), each cross - polytope of dimension *n* has $2n=f_0(n)$ vertices. The peculiarity of a cross - polytope is that each of its vertices has an opposite vertex, with which it is not connected by an edge. Moreover, there is one edge between this vertex and all other vertices. We subtract from the total number of vertices two vertices (the selected vertex and its opposite) 2n - 2. This is the possible number of edges emanating from the selected vertex. Thus, the incidence coefficient of the edges of any vertex is $k_{d_0^i d_1} = 2(n-1) = 2\,C_{n-1}^1, i = 1 \div f_0(n)$. A cross - polytope of any dimension can be depicted as a projection on a two - dimensional plane (Zhizhin, 2019 b). In this image, all its vertices are located on a circle, with the selected vertex and its opposite vertex located symmetrically relative to the center of the circle. A mentally drawn line through these two vertices halves the circle (Figure 2.2). Therefore, the number of variants the location of the edges from the selected vertex to the other vertices in one of the halves of the circle is *n* - 1, i.e. the number of combinations of *n* - 1 vertices one by one. Since there are two halves of a circle, then for the total number of vertex selection options for edge formation, this number of combinations should be multiplied by 2. This is the meaning of the expression for the incidence rate of any vertex in the *n* - cross - polytope with respect to the edge. To further prove the theorem 3 and clarify the nature of the formation of the coefficients of incidence in the *n* - cross - polytope, we will use this technique. We arrange the vertices of the 4 - cross - polytope in two lines, so that the vertices unlinked by an edge form vertical pair: each vertex in the top line is not connected by an edge to the vertex in the bottom line located strictly under this vertex in the top line,

$$\begin{matrix} d_0^1 & d_0^2 & d_0^3 & d_0^4 \\ d_0^5 & d_0^6 & d_0^7 & d_0^8 \end{matrix} \qquad (2.23)$$

We choose some vertex in the top line (2.23), for example d_0^1. It cannot be connected by an edge with a vertex d_0^5 in the bottom line, but it can be connected by an edge with other vertices of the bottom line. The number of such options is 3, i.e. with $n = 4$ is $3 = C_{n-1}^1 = C_3^1$. However, a vertex d_0^1 may be connected by an edge with any of the remaining three vertices in the top line (2.23). Therefore, the incidence coefficient of a vertex with respect to an edge in 4 - cross - polytope $k_{d_0^i d_1} = 2\,C_{n-1}^1 = 6, i = 1 \div 8$. When considering the belonging of a vertex d_0^1 to two - dimensional faces, it is necessary to determine the number of options for the participation of two vertices in the top line (2.23) (without a vertex d_0^1) and in the bottom line (2.23) (without a vertex d_0^5). This will be the number $2C_3^1$. In addition, as the vertices of the triangle, there may be vertices located in the upper and lower lines (2.23) in a cross - section way. There are 6 such variants in (2.23), i.e. more $2C_3^1$. Therefore, the incidence coefficient of a vertex with respect to a two - dimensional element in 4 - cross - polytope is $k_{d_0^i d_2} = 2^2\,C_{n-1}^2 = 12, i = 1 \div 8$. Multiplying this number by the number of vertices in 4 - cross - polytope (8) you can get the total number of incidents of vertices to two - dimensional elements in 4 - cross - polytope equal to 96. This number coincides with the corresponding number, defined earlier in Figure 2.2. Combination d_0^1 with three the different vertices from (2.23) can get the incidence coefficient of vertex to the three - dimension body (tetrahedron) for condition absent in combination opposite vertices. Can to show that this combination is 8, $k_{d_0^1 d_3} = 2^3 C_3^3 = 8, i = 1 \div 8$. Multiplying this number by the number of vertices in 4 - cross - polytope (8) you can get the total number of incidents of vertices to three - dimensional elements in 4 - cross - polytope equal to 64. This number coincides with the corresponding number, defined earlier in Figure 2.2. Combination two vertices, for example d_0^1 and d_0^2 (the edge $d_0^1\, d_0^2$) with two the different vertices from (2.23) you can get the incidence coefficient of edge to three - dimension body (tetrahedron) for condition absent in combination opposite vertices. Can to show that this combination is 4, $k_{d_1^1 d_3} = 2^2 C_2^2 = 4, i = 1 \div 8$. Multiplying this number by the number of edges in 4 - cross - polytope (8) you can get the total number of incidents of edges to three - dimensional elements in 4 - cross - polytope equal to 96. This number coincides with the corresponding number, defined earlier in Figure 2.4. Combination two vertices (the edge $d_0^1\, d_0^2$) with one the vertex from (2.23) can get the incidence coefficient of edge to triangle for condition absent in combination opposite vertices. Can show that this combination is 4, $k_{d_1^1 d_2} = 2C_2^1 = 4, i = 1 \div 8$. Multiplying this number by the number of edges in 4 - cross - polytope (8) you can get the total number of incidents of edges to triangle in 4 - cross - polytope equal to 24. This number coincides with the corresponding number, defined earlier in Figure 2.2.

In general case can write of $2n$ vertices of the n - cross - polytope in two lines

$$\begin{matrix} d_0^1 & d_0^2 & d_0^3 & \ldots & d_0^n \\ d_0^5 & d_0^6 & d_0^7 & \ldots & d_0^{2n} \end{matrix}, \tag{2.24}$$

where vertex in the top line is not connected by an edge to the vertex in the bottom line located strictly under this vertex in the top line. The incidence coefficient of a vertex with respect to an edge in n - cross - polytope is $k_{d_0^i d_1} = 2\,\mathrm{C}_{n-1}^1, i = 1 \div f_0(n)$. The incidence coefficient of a vertex with respect to a two - dimensional element in n - cross - polytope is $k_{d_0^i d_2} = 2^2\,\mathrm{C}_{n-1}^2, i = 1 \div f_0(n)$. The incidence coefficient of a vertex with respect to a three - dimensional edge in n - cross - polytope is $k_{d_0^i d_3} = 2^3\,\mathrm{C}_{n-1}^3, i = 1 \div f_0(n)$.

Go on can to say the incidence coefficient of a vertex with respect to a $(n-1)$ - dimensional element in n - cross - polytope is $k_{d_0^i d_{n-1}} = 2^{n-1}\,\mathrm{C}_{n-1}^{n-1}, i = 1 \div f_0(n)$. Obviously, that the incidence coefficient of a vertex with respect to n - cross - polytope is $k_{d_0^i d_n} = 1, i = 1 \div f_0(n)$.

Multiplying of the incidence coefficients of a vertex with respect to different dimension elements in n - cross - polytope and sum the product you can get the common express for the number of incident vertices to elements of different dimension in n - cross – polytope,

$$f_0(2^0 + 2^1 C_{n-1}^1 + 2^2 C_{n-1}^2 + \ldots + 2^{n-1} C_{n-1}^{n-1}) = f_0(n) \sum_{i=0}^{n-1} 2^i C_{n-1}^i = f_0(n) 3^{n-1}.$$

In this way you can get the common express for the number of incident edges to elements of different dimension more one in n - cross - polytope

$$f_1(2^0 + 2^1 C^1_{n-2} + 2^2 C^2_{n-2} + \dots + 2^{n-2} C^{n-2}_{n-2}) = f_1(n)\sum_{i=0}^{n-2} 2^i C^i_{n-2} = f_1(n)3^{n-2},$$

and go on.

In result, one can get the common express for the sum of all incidents of elements of a lower dimension with respect to all elements of a higher dimension in n - cross – polytope,

$$f_0(n)3^{n-1} + f_1(n)3^{n-2} + \dots + f_{n-1}3^0 = \sum_{i=0}^{n-1} f_i(n)3^{n-1-i}, f_i(n) = C^{n-1-i}_n 2^{1+i}. \quad (2.25)$$

Let $k_{d_h d_j^i}$ be the number of elements of dimension h in a n - cross - polytope belonging to some one element of dimension j ($h<j$), that is d_j^i. Obviously, this number is equal to $f_h(j)_s$ for simplex, and $i=1 \div f_j(n)_{cr}$ for n - cross - polytope. So, elements of cross - polytope are simplexes the product $f_h(j)f_j(n)_{cr}$ corresponds to the number of elements of dimension h belonging to all elements of dimension j for a simplex. This product is equal to $f_d(n) = C^{d+1}_{n+1}$

$$C^{h+1}_{j+1} 2^{1+j} C^{n-1-j}_n = 2^{1+j} \frac{(j+1)!}{(h+1)!(j-h)!} \frac{n!}{(n-1-j)!(1+j)!}.$$

Let us compare this number with the number of elements of dimension j, which have elements of dimension h in the n - cross – polytope,

$$k_{d_h^i d_j} f_h(n)_{cr} = C^{n-1-h}_n 2^{1+h} C^{j-h}_{n-h-1} 2^{j-h} = 2^{1+j} \frac{n!}{(n-1-h)!(1+h)!} \frac{(n-h-1)!}{(n-j-1)!(j-h)!}.$$

Obviously, these numbers are equal to each other. This proves that the number of elements of dimension h in a n - cross - polytope belonging to all dimensions of j ($j>h$) in a n - cross - polytope is equal to the number of elements of dimension j, which have elements of dimension h as elements.

Thus, the total number of incidents of elements of a smaller dimension with respect to elements of a higher dimension is equal to the total number of incidents of elements of a higher dimension with respect to elements of a smaller dimension. This the total number define expression (2.25).

Q.E.D.

INCIDENT CONSERVATION LAW FOR CHEMICAL COMPOUND

Incident Conservation Law for the Mackay Cluster

Consider the Mackay cluster (Mackay, 2001; Lord et al., 2006) (Figure 3).

In it, the values of the numbers of elements of different dimensions (Zhizhin, 2019 d) are determined by the equalities f_0=24, f_1=72, f_2=70, f_3=22, f_4=1.

We calculate the values of the incidence coefficients of the elements of lower dimension with respect to the elements of higher dimension.

Figure 3. The Mackay cluster

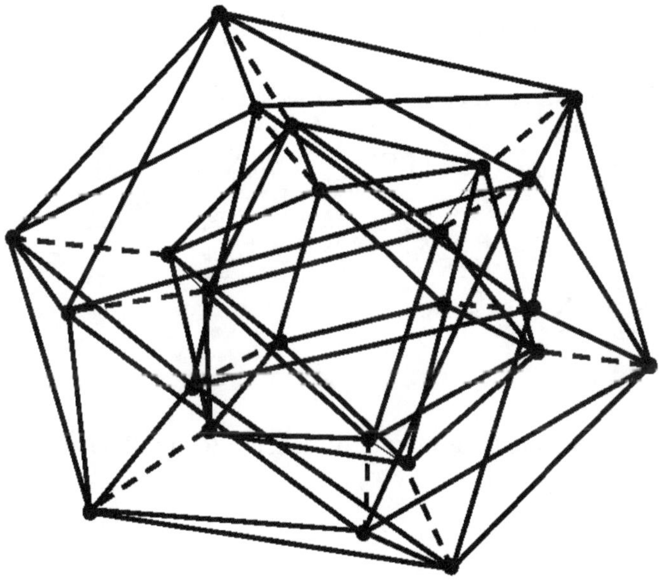

The factors of incidents (from smaller dimension to larger) are

$$k_{d_0^i d_1} = 6, i = 1 \div 24; k_{d_0^i d_2} = 10, i = 1 \div 24; k_{d_0^i d_3} = 6, i = 1 \div 24; k_{d_0^i d_4} = 1, i = 1 \div 24;$$
$$k_{d_1^i d_2} = 3, i = 1 \div 60 \text{ (edges on icosahedron)};$$
$$k_{d_1^i d_2} = 5, i = 1 \div 12 \text{ (edges in connected of icosahedrons)};$$
$$k_{d_1^i d_3} = 3, i = 1 \div 60 \text{(edges on icosahedron)};$$
$$k_{d_1^i d_2} = 5, i = 1 \div 12 \text{ (edges in connected of icosahedrons)},$$
$$k_{d_1^i d_4} = 1, i = 1 \div 72; k_{d_2^i d_3} = 2, i = 1 \div 70; k_{d_2^i d_4} = 1, i = 1 \div 70; k_{d_3^i d_4} = 1, i = 1 \div 22.$$

Sum up the incidence coefficients for all vertices, edges, two - dimension faces and three - dimension faces of the Mackay cluster

$$\sum_{i=1}^{24} k_{d_0^i d_1} = 6 \cdot 24 = 144, \sum_{i=1}^{24} k_{d_0^i d_2} = 10 \cdot 24 = 240, \sum_{i=1}^{24} k_{d_0^i d_3} = 6 \cdot 24 = 144, \sum_{i=1}^{24} k_{d_0^i d_4} = 24,$$
$$\sum_{i=1}^{60} k_{d_1^i d_2} = 3 \cdot 60 = 180, \text{(edges on icosahedron)},$$
$$\sum_{i=1}^{5} k_{d_1^i d_2} = 5 \cdot 12 = 60 \text{(edges in connected of icosahedrons)},$$
$$\sum_{i=1}^{60} k_{d_1^i d_3} = 3 \cdot 60 = 180 \text{(edges on icosahedron)},$$
$$\sum_{i=1}^{12} k_{d_1^i d_3} = 5 \cdot 12 = 60 \text{(edges in connected of icosahedrons)},$$
$$\sum_{i=1}^{72} k_{d_1^i d_4} = 72, \sum_{i=1}^{70} k_{d_2^i d_3} = 2 \cdot 70 = 140, \sum_{i=1}^{70} k_{d_2^i d_4} = 70, \sum_{i=1}^{22} k_{d_3^i d_4} = 22,$$
$$\sum_{i=1}^{24} k_{d_0^i d_1} + \sum_{i=1}^{24} k_{d_0^i d_2} + \sum_{i=1}^{24} k_{d_0^i d_3} + \sum_{i=1}^{24} k_{d_0^i d_4} + \sum_{i=1}^{72} k_{d_1^i d_2} + \sum_{i=1}^{72} k_{d_1^i d_3} + \sum_{i=1}^{72} k_{d_1^i d_4}$$
$$+ \sum_{i=1}^{70} k_{d_2^i d_3} + \sum_{i=1}^{70} k_{d_2^i d_4} + \sum_{i=1}^{22} k_{d_3^i d_4} = 1336. \tag{2.26}$$

The factors of incidents (from larger dimension to smaller) are

$k_{d_0 d_1^i} = 2, i = 1 \div 72; k_{d_0 d_2^i} = 3, i = 1 \div 40 \text{(at triangle faces)};$
$k_{d_0 d_2^i} = 4, i = 1 \div 30 \text{(at quadrangular faces)},$
$k_{d_0 d_3^i} = 6, i = 1 \div 20 \text{(at prisms)};$
$k_{d_0 d_3^i} = 12, i = 1 \div 2 \text{(at icosahedrons)},$
$k_{d_0 d_4} = 24; k_{d_1 d_2^i} = 3, i = 1 \div 40 \text{ (at triangle)};$
$k_{d_1 d_2^i} = 4, i = 1 \div 30 \text{(at quadrangular faces)},$
(edges in connected of icosahedrons);
$k_{d_1 d_3^i} = 3, i = 1 \div 60 \text{(edges on icosahedron)};$
$k_{d_1 d_2^i} = 5, i = 1 \div 12 \text{ (edges in connected of icosahedrons)},$
$k_{d_1 d_4} = 1, i = 1 \div 72; k_{d_2 d_3^i} = 2, i = 1 \div 70; k_{d_2 d_4} = 1, i = 1 \div 70; k_{d_3 d_4} = 1, i = 1 \div 22.$

Sum up the incidence coefficients for all vertices, edges, two - dimension faces and three - dimension faces of the Mackay cluster

$$\sum_{i=1}^{72} k_{d_0 d_1^i} = 2 \cdot 72 = 144, \sum_{i=1}^{70} k_{d_0 d_2^i} = 3 \cdot 40 + 4 \cdot 30 = 240,$$
$$\sum_{i=1}^{22} k_{d_0 d_3^i} = 6 \cdot 20 + 2 \cdot 12 = 144, \sum_{i=1}^{24} k_{d_0 d_4} = 24, \sum_{i=1}^{70} k_{d_1 d_2^i} = 3 \cdot 40 + 4 \cdot 30 = 240,$$
$$\sum_{i=1}^{22} k_{d_1 d_3^i} = 9 \cdot 20 + 2 \cdot 20 = 240, \sum_{i=1}^{72} k_{d_1 d_4} = 72, \sum_{i=1}^{22} k_{d_2 d_3^i} = 5 \cdot 20 + 2 \cdot 20 = 140,$$
$$\sum_{i=1}^{70} k_{d_2 d_4} = 70, \sum_{i=1}^{22} k_{d_3 d_4} = 22, \sum_{i=1}^{72} k_{d_0 d_1^i} + \sum_{i=1}^{70} k_{d_0 d_2^i} + \sum_{i=1}^{22} k_{d_0 d_3^i} + \sum_{i=1}^{24} k_{d_0 d_4} + \sum_{i=1}^{70} k_{d_1 d_2^i} + \sum_{i=1}^{22} k_{d_1 d_3^i}$$
$$+ \sum_{i=1}^{72} k_{d_1 d_4} + \sum_{i=1}^{22} k_{d_2 d_3^i} + \sum_{i=1}^{70} k_{d_2 d_4} + \sum_{i=1}^{22} k_{d_3 d_4} = 1336. \tag{2.27}$$

Comparing (2.26) and (2.27) you can see that the sum of incidents in the Mackay cluster from elements with a lower dimension to elements with a higher dimension is equal to the sum of incidents from elements with a higher dimension to elements with a lower dimension. Thus, the sum of incidents retains its value when changing the direction of the relationship between the elements.

The Law of Conservation of Incidence in Chemical Compound With a Central Atom

Let us verify the implementation of the law of conservation of incidence in chemical compounds with a central atom. Such compounds are widely distributed in intermetallic phases (Bergman et al., 1952, 1957; Samson, 1972; Gubin, 2019) and in biochemistry (Metzler, 1977). As an example, take an octahedron with a center. In this polytope (Figure 4)

Figure 4. The octahedron with a centrum

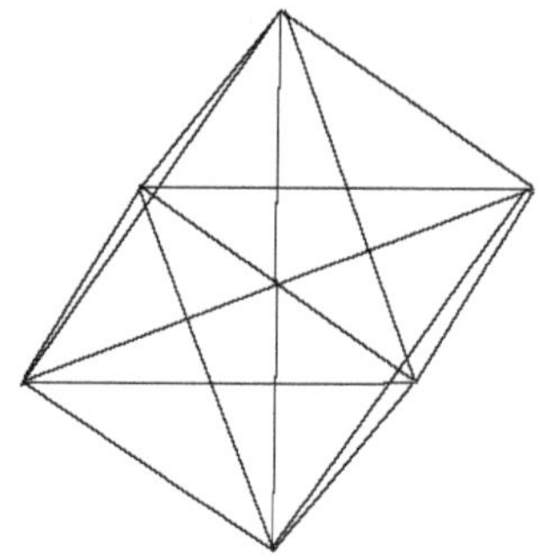

The values of the numbers of elements of different dimensions are determined by the equalities f_0=7, f_1=18, f_2=16, f_3=9, f_4=1.

We calculate the values of the incidence coefficients of the elements of lower dimension with respect to the elements of higher dimension. The factors of incidents (from smaller dimension to larger) are

$$k_{d_0^i d_1} = 5, i = 1 \div 6 (\text{for vertices of octohedron}); k_{d_0 d_1} = 6 (\text{for atom at the center});$$
$$k_{d_0^i d_2} = 8, i = 1 \div 6 (\text{for vertices of octohedron}); k_{d_0 d_2} = 12 (\text{for atom at the center});$$
$$k_{d_0^i d_3} = 5, i = 1 \div 6 (\text{for vertices of octohedron}); k_{d_0 d_3} = 8 (\text{for atom at the center});$$
$$k_{d_0^i d_4} = 1, i = 1 \div 7; k_{d_1^i d_2} = 4, i = 1 \div 6 (\text{edges from the center}); k_{d_1^i d_2} = 3 (\text{for edges of octahedron});$$
$$k_{d_1^i d_3} = 4, i = 1 \div 6 (\text{edges from the center}); k_{d_1^i d_3} = 3 (\text{for edges of octahedron});$$
$$k_{d_1^i d_4} = 1, i = 1 \div 18; k_{d_2^i d_3} = 2, i = 1 \div 16; k_{d_2^i d_4} = 1, i = 1 \div 16; k_{d_3^i d_4} = 1, i = 1 \div 9.$$

Sum up the incidence coefficients for all vertices, edges, two - dimension faces and three - dimension faces

$$\sum_{i=1}^{7} k_{d_0^i d_1} + \sum_{i=1}^{7} k_{d_0^i d_2} + \sum_{i=1}^{7} k_{d_0^i d_3} + \sum_{i=1}^{7} k_{d_0^i d_4} + \sum_{i=1}^{18} k_{d_1^i d_2} + \sum_{i=1}^{18} k_{d_1^i d_3} + \sum_{i=1}^{18} k_{d_1^i d_4} +$$
$$\sum_{i=1}^{16} k_{d_2^i d_3} + \sum_{i=1}^{16} k_{d_2^i d_4} + \sum_{i=1}^{9} k_{d_3^i d_4} = 324. \tag{2.28}$$

The factors of incidents (from larger dimension to smaller) are

$$k_{d_0 d_1^i} = 2, i = 1 \div 18; k_{d_0 d_2^i} = 3, i = 1 \div 16; k_{d_0^i d_3} = 4, i = 1 \div 9(\text{at tetrahedrons});$$
$$k_{d_0 d_3^i} = 6, i = 1(\text{at octahedron}),$$
$$k_{d_0 d_4} = 7; k_{d_1 d_2^i} = 3, i = 1 \div 16;\ k_{d_1 d_3^i} = 6, i = 1 \div 8; k_{d_1 d_3^i} = 12, i = 1(\text{for octahedron}),$$
$$k_{d_1 d_4} = 18; k_{d_2 d_3^i} = 4, i = 1 \div 8; k_{d_2 d_3} = 8(\text{for octahedron}); k_{d_2 d_4} = 16; k_{d_3 d_4} = 9.$$

Sum up the incidence coefficients for all vertices, edges, two - dimension faces and three -dimension faces

$$\sum_{i=1}^{18} k_{d_0 d_1^i} + \sum_{i=1}^{16} k_{d_0 d_2^i} + \sum_{i=1}^{10} k_{d_0 d_3^i} + k_{d_0 d_4} + \sum_{i=1}^{16} k_{d_1 d_2^i} + \sum_{i=1}^{9} k_{d_1 d_3^i} + k_{d_1 d_4} +$$
$$\sum_{i=1}^{9} k_{d_2 d_3^i} + k_{d_2 d_4} + k_{d_0 d_4} = 324. \tag{2.29}$$

Comparing (2.28) and (2.29) you can sce that the sum of incidents in the octahedron with center of elements with a lower dimension to elements with a higher dimension is equal to the sum of incidents from elements with a higher dimension to elements with a lower dimension. Thus, the sum of incidents retains its value when changing the direction of the relationship between the elements.

CONCLUSION

In the modern scientific world, information exchange is of great importance. This is characteristic for various fields of knowledge. In mathematics, there is the concept of incidence, i.e. belonging of some mathematical element to another mathematical element. If mathematical elements are models of some material bodies, then incidence can be interpreted as the transfer of information from one material body to another. In the framework of this monograph, the polytopes of higher dimension are the mathematical elements that model material bodies in nanoworld, including in the nanoworld of living matter. With their help, the structure of clusters of chemical compounds is described, i.e. bonded to each other by chemical bonds of atoms, molecules, functional groups. The geometric elements in this case are the faces of different dimensions of polytopes. Of interest is the question of transferring information from elements of lower dimension to elements of higher dimension and vice versa from elements of higher dimension to elements of lower dimension. This chapter establishes the existence of integral equality in relation to the issue of the transmission of information by elements of lower and higher dimensions that describe natural objects of nanoworld. This integral equality is called the law of conservation of incidents. At the beginning, using simple geometric examples, a direct calculation of incidents from images of geometric shapes shows that the sum of incidents of elements of lower dimension with respect to elements of higher dimension is equal to the sum of incidents of elements of higher dimension with respect to elements of lower dimension. Then, the fulfillment of the law of conservation of incidents for the *n* - simplex of the *n* - golden - hyper - rhombohedron and the *n* - cross - polytope is proved in general terms. It is significant that the flow of incidents from elements of lower dimension to elements of higher dimension and vice versa increases sharply with increasing dimension of the polytope. Interestingly, the flow of incidents (information) is of greatest importance to the *n* - cross - polytope. If we compare the values of these flows calculated from the obtained expressions, then with the same dimension value (13) for 13 - simplex this value is $4.75 \bullet 10^6$, for 13 - golden - hyper - rhombohedron this value is a thousand times greater ($1.2 \bullet 10^9$), and in the 13 - cross - polytope is 1.5 times more than the 13 - golden hyper - rhombohedron and is $1.8 \bullet 10^9$. In the monograph (Zhizhin, 2019 e), that the binding of helices in a DNA molecule with the help of nitrogen bases occurs in a cross - polytope of dimension 13. This corresponds to a significant flow of information between the elements of the DNA molecule. The latter can serve as a mathematical basis for the recently discovered epigenetic principle of the transmission of hereditary information without changing the sequence of genes in DNA and RNA molecules.

REFERENCES

Bergman, G., Waugh, J. L. T., & Pauling, L. (1952). Crystal structure of the intermetallic compound Mg_{32} (Al, Zn) $_{49}$ and related phases. *Nature*, *169*(4312), 1057–1058. doi:10.1038/1691057a0

Bergman, G., Waugh, J. L. T., & Pauling, L. (1957). The crystal structure of the metallic phase Mg_{32} (Al, Zn) $_{49}$. *Acta Crystallographica*, *10*(4), 254–259. doi:10.1107/S0365110X57000808

Gubin, S. P. (2019). *Chemistry clusters. Basics of classification and structure*. Editorial URSS.

Lindquist, S. (2016). Luminidependens (LD) is an arabidopsis protein with prion behavior. *Proceedings of the National Academy of Sciences of the United States of America*, *113*(21), 6065–6070. doi:10.1073/pnas.1604478113 PMID:27114519

Lord, E. A., Mackay, A. L., & Ranganathan, S. (2006). *New Geometry for New Materials*. Cambridge University Press.

Mackay, A. (2001). On Complexity. *Crystallography Reports*, *46*(4), 524–526. doi:10.1134/1.1387117

Mancuso, S. (2017). *PLANT REVOLUTION. Le piante hanno già inventato il nostro futuro*. GUIUTI EDITORE.

Mancuso, S., & Viola, A. (2013). *Sensibilità e intelligenza del mondo vegetale*. GUIUTI EDITORE.

Metzler, D. E. (1977). *The Chemical Reaction of Living Cells*. Academic Press.

Samson, S. (1972). Complex cubic A_6B compounds. II. The crystal structure of Mg_6Pd. *Acta Crystallographica. Section B, Structural Crystallography and Crystal Chemistry*, *28*(3), 936–945. doi:10.1107/S0567740872003437

Shevchenko, V., Zhizhin, G., & Mackay, A. (2013). On the structure of the quasicrystals in the high dimension space. In M. V. Diudea (Ed.), *Diamond and Related Nanostructures*. Springer. doi:10.1007/978-94-007-6371-5_17

Zhizhin, G. V. (2012, October). *Hierarchical filling of spaces with polytopes*. Paper presented at "St. Petersburg Scientific Forum: Science and Human Progress". 7th St.- Petersburg meeting of Nobel Prize laureates, St. – Petersburg, Russia.

Zhizhin, G. V. (2013). Relations for the numbers of faces of various dimensions in the tower of n - dimensional convex regular polytopes. *Proceedings of the 9th All - Russian School "Mathematical Studies in the Natural Sciences."* Apatity: Geological Institute, Russian Academy of Sciences.

Zhizhin, G. V. (2014a). The fractal nature of disproportionate phases. [th all-Russian Scientific School "*Mathematical research in the natural sciences*". Apatity: Geological institute KSC RAS.]. *Materials (Basel)*, 10.

Zhizhin, G. V. (2014b). Incommensurable and fluctuating structures in the terrestrial space. *Biosphere*, *3*, 211–221.

Zhizhin, G. V. (2014c). *World – 4D*. Polytechnic Service.

Zhizhin, G. V. (2015, November). *Polytopic prismahedrons – fundamental regions of the n-dimension nanostructures*. Paper presented at The International conference "Nanoscience in Chemistry, Physics, Biology and Mathematics," Cluj – Napoca, Romania.

Zhizhin, G. V. (2016). The structure, topological and functional dimension of biomolecules. *J. International Journal of Chemoinformatics and Chemical Engineering*, *5*(2), 44–58. doi:10.4018/IJCCE.2016070104

Zhizhin, G. V. (2018). *Chemical Compound Structures and the Higher Dimension of Molecules: Emerging Research and Opportunities*. IGI Global. doi:10.4018/978-1-5225-4108-0

Zhizhin, G. V. (2019a). The Law of Conservation of Incidents in the Space of Nanoworld. *International Journal of Chemoinformatics and Chemical Engineering*, *8*(1), 25–46. doi:10.4018/IJCCE.2019010103

Zhizhin, G. V. (2019 b). *The Geometry of Higher – Dimensional Polytopes*. IGI Global. doi:10.4018/978-1-5225-6968-8

Zhizhin, G. V. (2019c). The metric structure of nucleic acids and the higher dimension of their constituents. *J. Chemoinformatics and Chemical Engineering*, *7*(2), 1–15.

Zhizhin, G. V. (2019d). Higher Dimension of Clusters of Intermetallic Compounds. *J. Applied Nanotechnology Research*, *4*(1), 8–25. doi:10.4018/IJANR.2019010102

Zhizhin, G. V. (2019 e). *Attractors and Higher Dimensions in Population and Molecular Biology: Emerging Research and Opportunities*. IGI Global. doi:10.4018/978-1-5225-9651-6

Zhizhin, G. V., & Diudea, M. V. (2016). Space of Nanoworld. In M. V. Putz & M. C. Mirica (Eds.), *Sustainable Nanosystems, Development, Properties, and Applications* (pp. 214–236). IGI Global.

Zhizhin, G. V., Khalaj, Z., & Diudea, M. V. M.V. (2016). Geometrical and topological dimensions of the diamond. Distance, symmetry and topology in carbon nanomaterials. Springer.

Chapter 3
Activated Adamantane as a Higher Dimensional Drug That Prevents the Growth of Beta Amyloids in the Space of Higher Dimension

ABSTRACT

The images of saccharide and polysaccharide molecules in spaces of various dimensions are considered. A method has been developed for obtaining simplified three - dimensional images of sugar molecules and their chains based on their images in spaces of higher dimensions. It was found that three-dimensional images of furanose and pyranose molecules fundamentally differ from each other to form convex and, accordingly, non - convex bodies. This leads to fundamental differences in the structure of polysaccharides from these molecules. In the chains of - anomers, one of the projections is linearly transmitted along the chain, and the projection perpendicular to it along the chain along a curved line. In chains of α - anomers (spirals), it is observed that when moving along a chain of projections in mutually perpendicular planes, the pyranose molecules rotate in opposite directions.

INTRODUCTION

In the previous chapter, the synthesis and the structure of the protein were considered in normal condition. Synthesis of protein in living cells from individual amino acids occurs very quickly. However, the case in ribosomes does not end with the synthesis of the peptide bond - the protein should be correctly touched and packed. The process of folding a large polypeptide chain is very complicated. Not all proteins can spontaneously curtail after synthesis in the cage. To fold many proteins, chaperons

DOI: 10.4018/979-8-3693-0588-1.ch003

are needed - proteins, which interact with partially rolled or incorrectly touched by polypeptide chains and provide proper folding or form micro -infection in which such properly coming turning can occur. Despite the numerous mechanisms that ensure proper folding, in the cell, failures of this process can still occur (Nelson, Cox, 2022). Incorrect folding of proteins is a significant problem: about a quarter of all polypeptides synthesized by the cell are eliminated, since they are incorrectly reduced. In some cases, improper curtailing proteins leads to serious diseases, or contributes to their development. Many diseases, such as type 2 diabetes, Alzheimer, Huntington, Parkinson's disease, are associated with a violation of the normal process of protein folding. The soluble protein, which is normally secreted by the cell, standing out in improperly reduced form, turns into insoluble extracellular amyloid fibers. Such diseases are attributed to amyloidosis. Forming undelivered fibers have an ordered structure and contain many beta - structural areas. Many proteins can take the structure of amyloid fibers as an alternative to normal conformation. Squirrels are secreted in a not completely reduced state. The fibers are increased in extracellular space. The interaction of the beta layers of different fibers begins with each other, forming amyloid fibers. Part of amyloidosis is systemic diseases that occur with the involvement of many tissues. Primary systemic amyloidosis is caused by the deposition of fibers consisting of incorrectly touched light immunoglobulin chains. Some types of amyloidosis are associated with other types of disease. A special growly amyloidosis is associated with the deposition of an amyloid that initiates the degradation of nerve cells, mainly in the elderly (Wlodamer et al., 1989). Alzheimer's disease is accompanied by the deposition of extracellular amyloid, consisting of a beta amyloid peptide, formed from a larger transmembrane protein (predecessor of Beta Amyloid), which is expressed in most human tissues (Milton, et al., 1992). Beta amyloids - a general name for several peptides consisting of approximately 40 amino acid residues and formed from the transmembrane protein - the predecessor of the Beta amyloid. The main varieties of a peptide of 40 amino acid residues (and 40) and peptide of 42 Ami closing residues (and 42) are. Peptide A 40 does not have pathogenic properties, but peptide A 42 is considered one of the main factors that provoke Alzheimer's disease. In the brain of the patient suffering from Alzheimer's disease, this peptide can form the so -called amyloid plaques, which are from the clusters of the peptide, touched in the form of beta folds. Currently, there are no drugs that can slow down the development of Alzheimer's disease, as there is no complete understanding of the causes of the disease. To reduce level A 42, a search is carried out that could prevent its formation in the brain or can remove already formed plaques in the tissues.

Adamantane as a chemical compound was discovered in 1933 (Landa, Machacek, 1933). The adamantane molecule consists of 10 carbon atoms, repeating the arrangement of carbon atoms in the crystal lattice of diamond, and 16 hydrogen atoms, connected to carbon atoms by their unsaturated carbon atoms valence bonds. The discovery of adamantane served as an impetus for the development of the chemistry of organic polyhedranes. Adamantane derivatives have found practical application as drugs with different biological activity and for various purposes (antiviral drugs, antispasmodics, antiparkinsonian drugs, etc.): amantadine (Hubsher, Haider, Okun, 2012), memantine (Schifitto, Nava et all., 2007), rimantadine, tromantadine, etc. All these drugs repeat the structural group of carbon atoms, characteristic for adamantane, only the structural groups attached to the carbon atoms change. Among the inorganic and elemental-organic compounds, there are a number of structural analogues of the adamantane molecule, in which atoms of other elements act instead of carbon atoms. Moreover, between the atoms that are analogues of carbon atoms in the adamantane molecule, other atoms can be located that do not change the topological structure of the molecule. These compounds include Pb_4O_6, As_4O_6, Sb_4O_6, P_4O_{10}, P_4O_4, $SiCl_4$, BcF_2, HgS, Ag_2O, Cu_2O, ZnS, Cu_2Cl (Zhizhin, 2018). Despite the wide distribution in nature of adamantane molecules and their analogues, it can be assumed that the structure of this molecule has not yet been sufficiently studied. It is well known that the elementary cell of the diamond, which makes up the main part of the adamantane molecule, was identified by various authors of three - dimensional polytopes of various forms: tetrahedron, octahedron and other polyhedrons (Shafranovsky, 1964). In the future, it was found that the reason for such a variety is that the main part of the adamantane molecule of 10 atoms has a dimension of 4, i.e. the highest dimension (Zhizhin, 2014), which includes tetrahedrons, octahedrons, prisms, and pyramids as three - dimensional figures. However, due to the complexity of the image of the main part of the adamantane molecule on the plane, the composition of three - dimensional figures included in the main part of the adamantane molecule was not completely determined. This work is devoted to clarifying the result of 10 years ago, it shows that the dimension of the main part of the adamantane molecule is much higher and equal to 9.

THE STRUCTURE OF THE MAIN PART OF THE ADAMANTANE MOLECULE

Let us denote 10 atoms of the main part of the adamantane molecule by symbols $a_1,\ldots,a_{10}$. Let us take into account (Zhizhin, 2014) that 6 atoms ($a_2, a_3, a_4, a_6, a_8, a_9$) out of these ten atoms are located in the centers of the flat faces of the three - dimensional cube (thin black lines in Figure 1), and each of the four remaining atoms (a_1, a_5, a_7, a_{10}) is equally removed from the three nearest centers of the flat faces of the cube and the corresponding common to these faces vertices of the cube.

Figure 1. The structure of the main part of the adamantane molecules

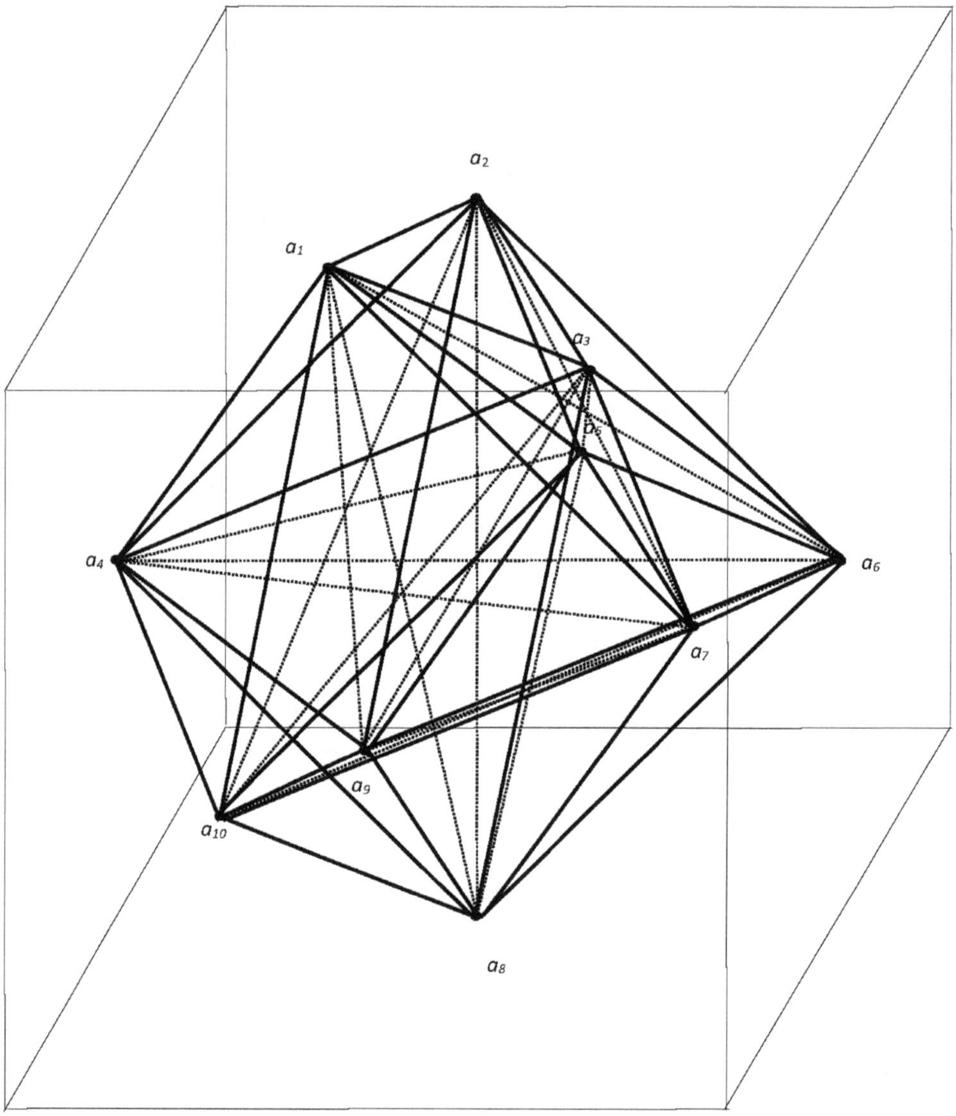

Let us connect with black lines the atoms equidistant from the three nearest flat faces of the cube with the centers of these flat faces of the cube. This creates edges

$a_1a_2, a_1a_3, a_1a_4, a_5a_2, a_5a_6, a_5a_9, a_7a_3, a_7a_6, a_7a_8, a_{10}a_4, a_{10}a_8, a_{10}a_9$.

We also connect the atoms in the centers of the flat faces of the cube closest to each other with thick black lines. This creates edges

$$a_2a_4, a_2a_6, a_2a_9, a_2a_3, a_3a_4, a_3a_6, a_3a_8, a_4a_8, a_4a_9, a_9a_8, a_9a_6, a_8a_6.$$

In addition, let's connect with thick black lines the atoms equidistant from the corresponding three nearest flat faces of the cube. This creates edges

$$a_1a_5, a_5a_{10}, a_5a_7, a_1a_{10}, a_1a_7, a_7a_{10}.$$

The total number of edges is 30. They make up the polytope shown in Figure 1.

As you can see, the polytope in Figure 1 contains a number of three - dimensional figures inside, in particular 5 tetrahedrons, 8 octahedrons, and, in addition, many prisms with triangular bases (trigonal prisms). Each octahedron forms a complex with six trigonal prisms. These prisms break down into pairs of prisms. Each pair of prisms has a rectangular face that coincides with one of the three rectangular sections of the octahedron. Therefore, 48 trigonal prisms are possible here. Each complex of an octahedron with trigonal prisms is included in the total set of such complexes, in general, forming a rather complex formation of combined three - dimensional figures. It should be noted that the combination of three - dimensional figures is a common property of higher - dimensional polytopes. To determine the dimension of polytopes of higher dimension, the Euler - Poincaré equation is usually used (Poincaré, 1895),

$$\sum_{i=0}^{n-1}(-1)^i f_i(n) = 1 + (-1)^n, \tag{3.1}$$

where f_i is the number of geometric elements with the dimension i, and n is the dimension of the polytope P.

With a large number of elements of different dimensions included in the polytope, it is rather difficult to use equation (3.1). This requires an unambiguous and accurate recalculation of all elements of different dimensions. However, another way can be used, which, in particular, was used in the previous works of the author on the study of the garnet structure (Zhizhin, 2022 a, b). It consists in bringing it closer to the polytope, the dimension of which is determined by a simple algebraic formula, without changing the number of vertices of the polytope. This can be done, for example, by increasing the number of edges of the polytope. After all, even in simpler cases, the edges of a polytope are determined from the condition of giving the figure under study a closed convex form (Zhizhin, 2019). In this case, in Figure 1, 6 edges emanate from each vertex of the polytope to other vertices. It is enough to add edges to this figure so that 9 edges come out of each vertex, i.e. each of the 10 vertices was connected by an edge to all the other vertices of the polytope. Then the polytope takes the known form of a simplex - type polytope (Zhizhin, 2019). Additional edges in Figure 3.1 are indicated by thin dotted lines. The dimension of a simplex polytope is one less than the number of polytope vertices (Zhizhin, 2019). Therefore, the dimension of the polytope in Figure 3.1, taking into account the dotted lines, is 10 - 1 = 9.

The number of faces of different dimensions *i* in a simplex polytope of dimension *n* is determined by the formula (Zhizhin, 2019 a, b)

$$f_i(n) = C_{n+1}^{i+1}, i = 0, 1, 2, ..., n. \tag{3.2}$$

In accordance with this formula, for example, there are $C_{10}^1 = 10$ vertices, $C_{10}^2 = 45$ edges, $C_{10}^3 = 120$ two - dimensional faces (triangles), $C_{10}^4 = 210$ three - dimensional faces (tetrahedrons) in the Adamantane structure, $C_{10}^5 = 252$ simplexes with dimension 4, $C_{10}^6 = 210$ simplexes with dimension 5, $C_{10}^7 = 120$ simplexes with dimension 6, $C_{10}^8 = 45$ simplexes with dimension 7, $C_{10}^9 = 10$ simplexes with dimension 8, $C_{10}^{10} = 1$ simplex with dimension 9. The Euler - Poincare equation (3.1) is valid for a simplex polytope of any dimension (Zhizhin, 2019). Therefore, it also holds for an Adamantane structure of dimension 9. Substituting the obtained values of the numbers of simplices of different dimensions into the Euler-Poincaré equation (3.1), we obtain

$$10 - 45 + 120 - 210 + 252 - 210 + 120 - 45 + 10 = 2.$$

The Euler–Poincaré equation holds for n = 9. This proves that adamantane is a simplex-type polytope of dimension 9. This proves that the constructed Adamantane structure is a polytope.

Since adamantane molecules are included in many drugs, this indicates a higher dimension of biological processes associated with the use of these drugs.

Creation of Chemically Inert Particles From Adamantane That Prevent the Formation of Beta-Amyloid

As follows from Figure 3.1 (see also Zhizhin, 2014), the main part of adamantane has carbon atoms with free valences. Atoms a_1, a_5, a_7, a_{10} each have one free valence, while atoms $a_2, a_3, a_4, a_6, a_8, a_9$ each have two free valences. Saturated valence bonds connect the a_1 atom with the a_2,a_3,a_4, atoms (one bound is unsaturated); the a_2 atom is connected to the a_1,a_5 atoms (two bounds are unsaturated) ; the a_3 atom is connected to the a_1,a_7 atoms (two bounds are unsaturated); the a_4 atom is connected to the a_1,a_{10} atoms (two bounds are unsaturated); the a_5 atom is connected to the a_2,a_9,a_6 atoms (one bound is unsaturated); the $a_6$atom is connected to a_5,a_7 (two bounds are unsaturated); the a_7 atom is connected to the a_6,a_3,a_8 atoms (one bound is unsaturated); the a_8 atom is connected to the a_7,a_{10} atoms (two bounds are unsaturated); the a_9 atom is connected to the a_5,a_{10} atoms (two bounds are unsaturated); the a_{10} atom is connected to the a_9,a_8,a_4 atoms (one bound is unsaturated). Thus, the main part of the adamantane molecule has 16 unsaturated valence bonds. These bonds allow at least 16 amino acid residues to be attached to the main part of one adamantane molecule. In the amino acid residue, there are also unsaturated valence bonds (Figure 2) at the carbon atom and at the nitrogen atom, along which amino acid residues are connected to each other in protein molecules.

Figure 2. The structure of the residuum of the molecule amino acid

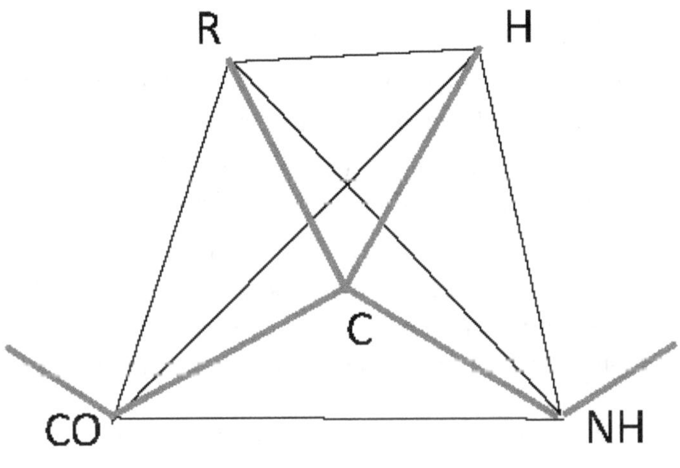

When meeting with the main part of the adamantane molecule, one of the free valence bonds of the amino acid residue can be attached to adamantane, and the other free valence bond can be used to connect amino acid residues to each other. The free end of the chain of amino acid residues can also be attached to the top of the free - valence adamantane molecule. Thus, around the main part of the adamantane molecule, a shell is formed from amino acid residues with extinguished valence bonds. Therefore, a chemically inactive particle is formed based on the main part of the adamantane molecule. This will prevent the formation of amino acid polymer chains to form beta-amyloids.

Activated adamantane gives additional activity to medical preparations in which adamantane is part of them.

Activated Amantadine

Amantadine (adamantane derivative) is an antiviral and at the same time antiparkinsonian drug.

It differs from adamantane in that an amino group NH_2 is connected to one of the vertices of adamantane with three valence bonds via the fourth valence bond. This reduces the number of free carbon valence bonds but adds additional possible nitrogen valence bonds. If we consider the amino group as an additional vertex compared to adamantane, then the polytope corresponding to amantadine will have 11 vertices. Its dimension will correspondingly increase and become equal to 10, i.e. this is a simplex-type polytope of dimension 10 if, as in adamantane, each vertex of the polytope is connected by an edge with all other vertices.

In accordance with formula (10.2), for example, there are $C_{11}^{1}=11$ vertices, $C_{11}^{2}=55$ edges, $C_{11}^{3}=165$ two - dimensional faces (triangles), $C_{11}^{4}=330$ three - dimensional faces (tetrahedrons) in the Amantadine structure, $C_{11}^{5}=462$ simplexes with dimension 4, $C_{11}^{6}=462$ simplexes with dimension 5, $C_{11}^{7}=330$ simplexes with dimension 6, $C_{11}^{8}=165$ simplexes with dimension 7, $C_{11}^{9}=55$ simplexes with dimension 8, $C_{11}^{10}=11$ simplexes with dimension 9, $C_{11}^{11}=1$ simplex with dimension 10. The Euler - Poincare equation (3.1) is valid for a simplex polytope of any dimension (Zhizhin, 2019). Therefore, it also holds for an Amantadine structure of dimension 10. Substituting the obtained values of the numbers of simplices of different dimensions into the Euler-Poincaré equation (3.1), we obtain

$$11 - 55 + 165 - 330 + 462 - 462 + 330 - 165 + 55 - 11 = 0.$$

The Euler–Poincaré equation holds for n = 10. This proves that amantadine is a simplex-type polytope of dimension 10. This proves that the constructed Amantadine structure is a polytope.

Amantadine was originally introduced and utilized as an antiviral medication. A single patient noticed relief in her Parkinson disease symptoms after taking amantadine for a flu infection, and this observation sparked an interest, and several important studies that eventually led to a new drug indication. Amantadine has over the years fallen out of favor as a drug to address influenza symptomatic treatment of Parkinson disease.

Activated Rimantadine

Rimantadine differs from adamantane by adding one more carbon atom to one of the vertices of adamantane with already existing three valence bonds, to which an amino group NH_3 and a group CH_3 are additionally attached.

Rimantadine was first obtained in 1963 by the American company DuPont & Co. a group led by William Pritchard, who investigated amantadine derivatives, and was patented in the United States in 1965 in Wilmington, Delaware (for the chemical formula U.S. Patent 3 352 912, 1965 and for the synthesis method U.S. Patent 3 592 934, 1967). The methods of synthesis proposed by W. Prichard were based on the preparation of rimantadine from ketoxime by reduction with lithium aluminum hydride.

The efficacy of rimantadine in preventing influenza was demonstrated in animals in 1965 and in clinical trials in 1968 (Zlydnikov, et al., 1969). A group led by Sergio Rabinovitch at the University of Iowa demonstrated the therapeutic effect of rimantadine in a 1969 influenza epidemic among inmates at Fort Madison (Rabinovich, et al., 1969).

It differs from adamantane in that carbon atom is connected to one of the vertices of adamantane with three valence bonds via the fourth valence bond. This reduces the number of free carbon valence bonds but adds additional possible valence bonds of added carbon atom. If we consider the amino group, added carbon atom and group CH_3 as an added vertices compared to adamantane, then the polytope corresponding to rimantadine will have 13 vertices. Its dimension will correspondingly increase and become equal to 12, i.e. this is a simplex-type polytope of dimension 12 if, as in adamantane, each vertex of the polytope is connected by an edge with all other vertices.

In accordance with formula (3.2), for example, there are $C_{13}^1 = 13$ vertices, $C_{13}^2 = 78$ edges, $C_{13}^3 = 286$ two - dimensional faces (triangles), $C_{13}^4 = 715$ three - dimensional faces (tetrahedrons) in the rimantadine structure, $C_{13}^5 = 1287$ simplexes with dimension 4, $C_{13}^6 = 1716$ simplexes with dimension 5, $C_{13}^7 = 1716$ simplexes with dimension 6, $C_{13}^8 = 1287$ simplexes with dimension 7, $C_{13}^9 = 715$ simplexes with dimension 8, $C_{13}^{10} = 286$ simplexes with dimension 9, $C_{13}^{11} = 78$ simplexes with dimension 10, $C_{13}^{12} = 13$ simplexes with dimension 12, $C_{13}^{13} = 1$ simplexes with dimension 13. The Euler - Poincare equation (3.1) is valid for a simplex polytope of any dimension (Zhizhin, 2019). Therefore, it also holds for an Amantadine structure of dimension 13. Substituting the obtained values of the numbers of simplices of different dimensions into the Euler-Poincaré equation (3.1), we obtain

$$13 - 78 + 286 - 715 + 1287 - 1716 + 1716 - 1287 + 715 - 286 + 78 - 13 = 0.$$

The Euler–Poincaré equation holds for $n = 13$. This proves that rimantadine is a simplex-type polytope of dimension 13. This proves that the constructed rimantadine structure is a polytope.

CONCLUSION

It has been established that the structure of the main part of the adamantane molecule is much more complex than previously thought. The dimension of this structure is determined by reducing it to the well-known simplex polytope form. It is determined that the dimension of this polytope is 9. Adding various functional groups to the main part of the adamantane molecule only increases its dimension. Consequently, many drugs, which include adamantane molecules with different functional groups, have a higher dimension. Examples of calculating the dimensions of known medicines are given. It has been established that 10 carbon atoms in the main part of the adamantane molecule have 16 active (non-extinguished) valence bonds. At least 16 amino acid residues can join them. In this case, in addition, the amino acid residues, connecting with each other, form a chemically inert shell around the main part of the adamantane molecule. Thus, the main part of the adamantane molecule without functional groups added in various drugs can concentrate amino acid residues around itself, forming chemically inert particles, thereby preventing the creation of beta-amyloids, leading to the development of various prion diseases of nerve cells, such as Alzheimer's disease, Parkinson's disease, and other diseases (Nussbaum, Seward, Bloom, 2013; Milton R, Milton S, Kent, 1992; Wlodamer et al., 1989; Zlydnikov, et al., 1969; Rabinovich, et. al., 1969; Hubsher, Haider, Okun, 2012; Tatarnikova, Orlov, Bobkova, 2015; Lam, Takechi, Mamo, 2021; Piller, 2022).

REFERENCES

Gerald, K. (2013). *Cell and Molecular Biology. Concepts and Experiments*. John Wiley and Sons.

Hubsher, G., Haider, M., & Okun, M. S. (2012). Amantadine the journey from fighting flu to treating Park'inson disease. *Neurology*, *78*(14), 1096–1099. doi:10.1212/WNL.0b013e31824e8f0d PMID:22474298

Lam, V., Takechi, R., & Mamo, J. C. L. (2021, September 14). Synthesis of human amyloid restricted to liver results in an alzheimer disease – like neurodegenerative phenotype. *PLoS Biology*, *19*(9), e3001358. doi:10.1371/journal.pbio.3001358 PMID:34520451

Landa, S., & Machacek, V. (1933). Sur l'adamantane, nouvel hydrocarure exait du naphte. *Collection of Czechoslovak Chemical Contributions*, *5*, 1–5. doi:10.1135/cccc19330001

Milton, R. C., Milton, S. C., & Kent, S. B. (1992). Total chemical synthesis of a D – enzyme: The enantiomers of HIV – 1 proteus show demonstration of reciprocal chiral substrate specificity. *Science*, *256*(5062), 1445–1448. doi:10.1126/science.1604320 PMID:1604320

Nussbaum, J. M., Seward, M. E., & Bloom, G. S. (2013). Alzheimer disease a tale of two prions. *Prion*, *7*(1), 14–19. doi:10.4161/pri.22118 PMID:22965142

Piller, C. (2022). Research backing experimental Alzheimer`s drug was first target of suspicion. *Science, 377*, 6604.

Poincare, A. (1895). Analysis situs. *J. de e'Ecole Polytechnique*, *1*, 1–121.

Rabinovich, S., Baldini, J. T., & Bannister, R. (1969). The therapeutic efficacy of rimantadine HCl in naturally influenza A2 outbreak. *The American Journal of the Medical Sciences*, *257*(5), 328–335. doi:10.1097/00000441-196905000-00005 PMID:4891873

Schifitto, G., Navia, B.A., & Yiannoutsos, C.T. (2007). Memantine and HIV-associated conqnitive impairmcnt o ncuropsychological and proton magnetic resonance spectroscopy study. *AIDS journal, 21*(14), 1877-1886.

Shafranovsky, I. I. (1964). *Diamonds*. Science.

Tatarnikova, O.G., Orlov, M.A., Bobkova, N.V. (2015). Beta-amyloid and tau protein: structure, interaction and primary properties. *The successes of biological chemistry*, *55*, 351-390.

Wlodamer, A., Miller, M., Jaskólski, M., Sathayanarayana, B. K., Baldin, E., Weber, I. T., Selk, L. M., Chawson, L., Schneider, J., & Kent, S. B. (1989). Conserved folding in retroviral proteases: Crystal structure of synthetic HIV – 1 protease. *Science*, *245*(4918), 616–621. doi:10.1126/science.2548279 PMID:2548279

Zhizhin, G. V. (2014). On High Dimensionalities in Nature. *Biosphere.*, *6*(4), 313–318.

Zhizhin, G. V. (2018). *The Geometry of Higher – Dimensional Polytopes*. IGI Global.

Zhizhin, G. V. (2022a). Structure of Garnet. *J. of Nature Science and Sustainable Technology.*, *16*(2), 1–9.

Zhizhin, G. V. (2022b). Structure of Minerals of the Garnet Family in the Higher-Dimensional Space. *J. of Nature Science and Sustainable Technology.*, *16*(4), 1–9.

Zhizhin, G. V. (2023). Adamantane as a Means of High Dimension, Preventing the Formation of the Beta – Amyloid. *Annals of Advanced Biomedical Sciences*, *6*(1), 1–4.

Zlydnikov, D. M., Kubar, O. I., Kovaleva, T. P., & Kamorin, L. E. (1969). Study of Rimantadine in USSR: A Review of the literature. *Clinical Infectious Diseases*, *3*(3), 408–421. doi:10.1093/clinids/3.3.408 PMID:7025146

Chapter 4 The Structure of the Sugar Molecules in Higher Dimensional Space

ABSTRACT

The geometrical cause of the formation of different forms by molecules nucleic acids (right and left spirals with different number of α - D – ribose and 2 – deoxy – α – D - ribose molecules in the period, including closed chains) has been determined. Substituting the known effective values of the lengths of chemical bonds (carbon - carbon, oxygen - oxygen, phosphorus - oxygen) into the structure of polytopes, the values of the characteristic geometric parameters of molecules nucleic acids were calculated: their effective diameter and period. It turned out that the calculated values of these parameters are in good agreement with their values, determined earlier experimentally. It is shown that the set of single - stranded nucleic acids (both DNA and RNA) is broken into two sets of chiral forms. Each form in one set contains a chiral form in another set. Moreover, in each set there are possible rotation of the spirals both in the right and in the left direction.

INTRODUCTION

Sugar molecules are widespread in nature and play an important role in all living organisms. Sugars are found in nucleic acids, in fruit sikes ("grape sugar"), as well as in human and animal blood ("blood sugar"). They are part of many glycolipids and glycoproteins. Sugars are used by plants to transport and store carbohydrates in soluble form. Sugarcane and sugar beets are rich in sugars (cane and beet sugar). Bees use sugar from flower nectar to produce honey. This class of compounds includes simple sugars (monosaccharides) and their polymers - oligosaccharides and polysaccharides. The classification of monosaccharides is based on the number of

DOI: 10.4018/979-8-3693-0588-1.ch004

carbon atoms in the molecule (pentoses, hexoses, etc.), as well as on the chemical nature of the carbonyl group (aldoses and ketoses). A small fraction of sugar molecules (less than 0.1%) in living organisms is linear. The vast majority of monosaccharide molecules are present in living organisms with the formation of rings of carbon atoms containing one oxygen atom in the ring (pyranoses are rings with five carbon atoms and one oxygen atom, furanoses are rings with four carbon atoms and one oxygen atom). The linear forms of monosaccharides are usually depicted in the form of Fischer projections, and the cyclic forms of monosaccharides are depicted in the form of Haworth projections (Metzler, 1980; Lehninger, 1982; Koolman and Roehm, 2013; Nelson, Cox, 2022). Neither projection gives a spatial image of monosaccharide molecules. Although it is clear that spatial images of molecules allow us to understand their structure, location in space and the possibility of their participation in various chemical reactions in living organisms. Speaking about spatial models of molecules, we mean the definition of which spatial forms are formed by the atoms of molecules in the form of a convex body or a set of convex bodies. The structure of these bodies determines their dimension in accordance with the Euler - Poincaré equation (Poincaré, 1895). Given the large dimensionality of molecules, it is often useful for ease of use to construct their three - dimensional images based on their images of higher dimensionality, without forgetting that their complete models have higher dimensionality (Zhizhin, 2019 c). In this chapter, these approaches are used to analyze spatial models of molecules of various sugars and their compounds. The structure of monosaccharides and polysaccharides (pyranoses and furanoses) is studied by constructing closed convex polytopes of higher dimension and their chains. For the convenience of practical use, a general method is being developed for obtaining approximate three -dimensional images of figures based on polytopes of higher dimension. This method is used to geometrically describe saccharides and their compounds. Using this method, an approximate three - dimensional image is obtained from a complete spatial model of a pyranose molecule in the form of a polytope of dimension 15 in the form of a connected pentagonal prism and a pyramid with hexagonal and pentagonal bases. An approximate three - dimensional image is obtained from the complete spatial model of furanose in the form of a polytope of dimension 12 in the form of a connected quadrangular prism and a quadrangular pyramid. The resulting three - dimensional image of the pyranose molecule allows changes taking into account possible conformations of the molecule. Three - dimensional images of saccharide molecules are used to build chains of saccharide molecules. It was found that the obtained three - dimensional image of the pyranose molecule unambiguously leads to the twisting of the polysaccharide depending on the valence angle of the oxygen bonding atom in this molecule. A three - dimensional image of the furanose molecule using the example of the D - ribose molecule was constructed in the author's previous works (Zhizhin, 2016, 2018, 2019 a, b, c, 2020 a, b). This three - dimensional image made it possible to establish the existence of a degree of freedom of arrangement of nitrogen bases in nucleic acids hidden in space of higher dimension. From which a one - correspondence of the number of possible locations of nucleotides and amino acids follows.

ABOUT OF THREE–DIMENSION IMAGES OF HIGHER DIMENSIONAL POLYTOPES

The noted possibilities of an approximate consideration of polytopes of higher dimension as polytopes of a lower dimension up to a three - dimensional dimension is a common property of polytopes of a higher dimension. It is based on the previously studied (Zhizhin, 2019 d) property of incidence of elements of different dimensions in polytopes of higher dimension. In particular, it was proved that some element of dimension i in a polytope of type simplex of dimension n simultaneously belongs to $n - i$ elements of dimension $n + 1$ of the polytope. For example, in a polytope of type simplex dimension 4 each two - dimensional face belongs simultaneously to two three - dimensional faces of this polytope. It should be noted, and this is significant, that among 10 two - dimensional faces of the polytope 4 - simplex, 5 faces belong to two three -dimensional faces located on one side of the two - dimensional face. At the same time, 5 other two - dimensional faces belong to two three - dimensional faces located on opposite sides of two - dimensional faces (Figure 1).

Figure 1. The polytope 4 – simplex

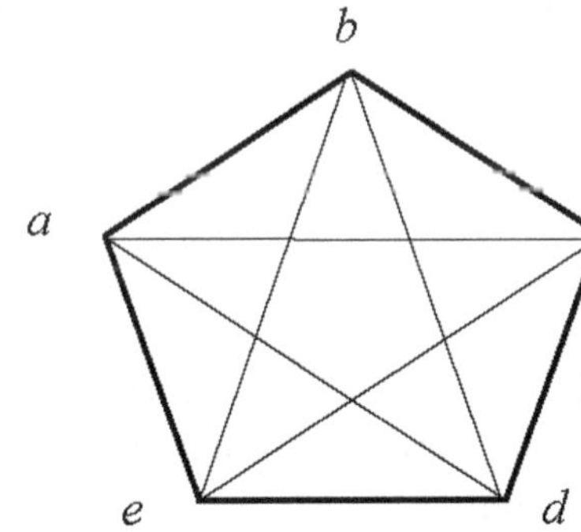

Thus, flat faces *abe, abc, bcd, ecd, ead* each belong simultaneously to two tetrahedra located on one side of the flat face: *abe* ∈ *abed, abce; abc* ∈ *abcd, abce; bcd* ∈ *abcd, bcde; ecd* ∈ *aced, bcde; ead* ∈ *abde, aced*. Flat faces *abd, acd, ebd,*

ace, ebc each belong simultaneously to two tetrahedra located on opposite sides of the flat face: *abd* ∈ *abcd, abed; acd* ∈ *abcd, aced; ebd* ∈ *abde, ebcd, ;ace* ∈ *aced, abce; ebc* ∈ *abce, ebcd.* One can assume that the planar faces to which the tetrahedra are adjacent on the other side of the face are external to the 4 - simplex. Flat faces to which tetrahedrons are adjacent on both sides of the face can be considered internal for the 4 - simplex. External flat faces can form a closed three - dimensional body if one of the edges is removed from the 4 - simplex. For example, if you remove the edge of *ac,* then the 4 - simplex turns into a three - dimensional body, if the triangle is not considered as a flat face, but only as a contour along which the tetrahedrons are in contact (Figure 2).

Figure 2. The body formed from four – simplex after missing the edge ac

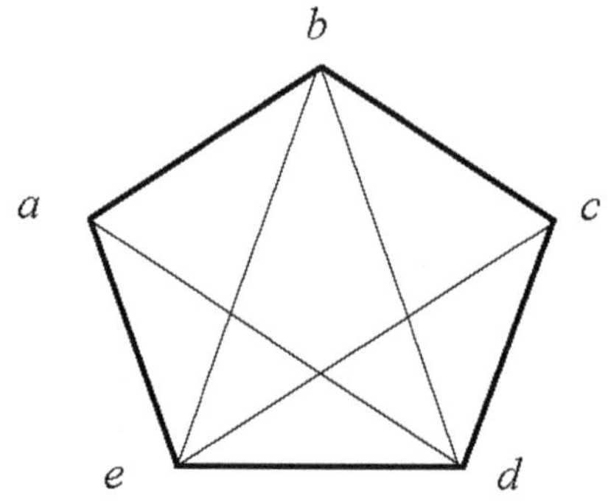

We can check whether the body in Figure 2 is three - dimensional according to the Euler -Poincaré equation (Poincaré, 1895)

$$\sum_{i=0}^{n-1} f_i(-1)^i = 1-(-1)^i, \tag{4.1}$$

where f_i is the number of elements with dimension i in polytope with dimension n.

In this case, in the body in Figure 2, the number of vertices is 5 (f_0=5), the number of edges is 9 (f_1=9), and the number of flat faces is 6 (f_2=6). Substituting these values into equation (3.1), we obtain

$5 - 9 + 6 = 2.$

Thus, in this case, the Euler – Poincaré equation holds for $n = 3$. This proves that the body in Figure 4.2 has dimension 3.

If the triangle is considered as a continuous two - dimensional face, to which two tetrahedrons adjoin from different sides of the face, then the number of two - dimensional faces increases by one. Moreover, the number of three - dimensional faces is 3 (the original body and two formed tetrahedrons). In this case, equation (4.1) has the form,

$$5 - 9 + 7 - 3 = 0,$$

i.e. the Euler – Poincaré equation holds for $n = 4$. This proves that the body in Figure 4.2 in this case has dimension 4.

A similar pattern is observed when removing any of the five edges *ac, bd, ad, eb, ec*. In all these cases, we obtain five three - dimensional topologically equivalent images of polytope 4 - simplex. Similar transformations are allowed by other polytopes of dimension 4, for example, 4 -cube, 4 – cross - polytopes.

Spatial Models of Pyranose Molecules

One of the most important monosaccharides is glucose. It makes up a significant part of all biomass (cellulose, starch), found in the blood of humans and animals. The most important aldohexose is *D* - glucose. In a linear form, the *D* - glucose molecule is represented by the Fisher formula in Figure 3.

Figure 3. Fisher formula for molecule D – glucose

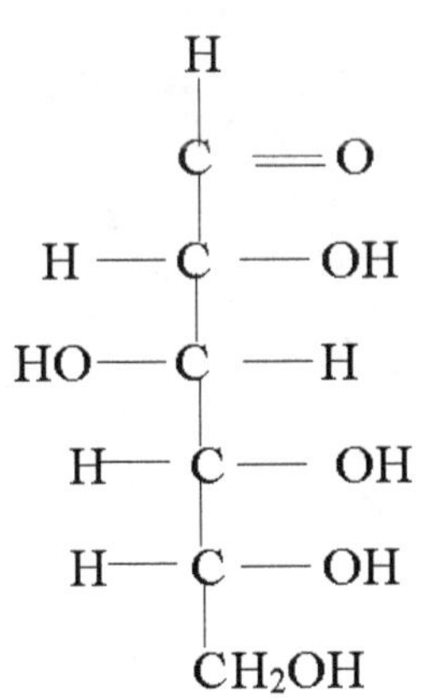

Glucose is the main source of energy for the brain. Interruption of the supply of glucose to the brain can have dire consequences: lethargy, coma, permanent brain damage and death. In animals, there is a complex hormonal mechanism that ensures that blood glucose is high enough to meet the needs of the brain, but not too high, since excess blood glucose can also lead to serious physiological consequences. In the body of people suffering from insulin-dependent diabetes mellitus, an insufficient amount of the hormone insulin is produced, which normally serves to lower the concentration of glucose in the blood. In the absence of treatment, the concentration of glucose in the blood of diabetic patients can exceed the norm several times. High blood glucose is one of the causes of such serious consequences of progressive diabetes as soil insufficiency, cardiovascular diseases, blindness, and also leads to a violation of the wound healing process.

The α - D - glucose molecule with a closed carbon cycle is formed when a double bond breaks at the oxygen atom. The image of this molecule in the form of a closed polytope (Zhizhin, 2016, 2019 c) is presented in Figure 4.

Figure 4. The molecule of α - D – glucose

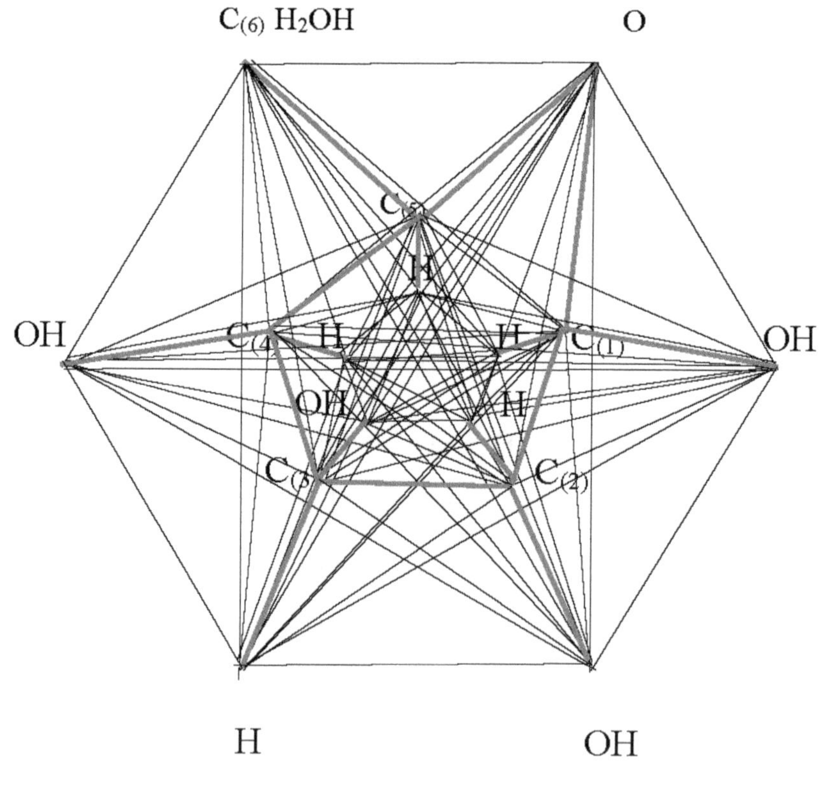

When depicting the polytope in Figure 4, the assumption that the atoms of the carbon cycle are located in the plane was not used. Therefore, this image is suitable for any conformation of molecule α -D - glucose. In the model in Figure 4, the edges corresponding to chemical covalent bonds are indicated in red. The remaining edges have only the geometric meaning of creating a convex polytope. They are marked black. In the aggregate, each vertex of a polytope (atom or functional group) is connected with all other vertices by edges. This means that the image in question is a simplex type polytopic (Zhizhin, 2019 d). To determine the dimension of this polytope, one can use the Euler - Poincaré equation (4.1).

The polytope in Figure 3.4 contains 16 vertices, i.e. f_0=16n+1; 120 edges ($f_1 = C^2_{n+1} = 120$);560 triangles ($f_2 = C^3_{n+1} = 560$); 1820 tetrahedrons ($f_3 = C^4_{n+1} = 1820$); 4368 4 – simplexes ($f_4 = C^5_{n+1} = 4368$); 8008 5 – simplexes ($f_5 = C^6_{n+1} = 8008$); 114406–simplexes ($f_6 = C^7_{n+1} = 11440$); 128707–simplexes ($f_7 = C^8_{n+1} = 12870$); 114408–simplexes ($f_8 = C^9_{n+1} = 11440$); 80089–simplexes ($f_9 = C^{10}_{n+1} = 8008$); 4368 10–simplexes ($f_{10} = C^{11}_{n+1} = 4368$); 182011–simplexes ($f_{11} = C^{12}_{n+1} = 1820$); 560 12 – simplexes ($f_{12} = C^{13}_{n+1} = 560$); 120 13 – simplexes ($f_{13} = C^{14}_{n+1} = 120$); 16 14 – simplexes ($f_{14} = C^{15}_{n+1} = 16$).

Substituting the values f_i, ($0 \leq i \leq 15$) in Euler's – Poincaré equation (4.1) can see that it holds for n = 15

$$\sum_{i=0}^{14=n-1} f_i(-1)^i = 16 - 120 + 560 - 1820 + 4368 - 8008 + 11440$$

$$-12870 + 11440 - 8008 + 4368 - 1820 + 560 - 120 + 16 = 2.$$

This confirms that the polytope in Figure 4.4 has the dimension n = 15.

To obtain a three - dimensional simplified model of the molecule α -D - glucose, in Figure 4.4 we leave only the edges corresponding to chemical covalent bonds, the edges of the external contour, and the edge connecting the $C_{(1)}$,$C_{(5)}$ atoms. We assume as a first approximation that the ring of carbon atoms $C_{(1)} \div C_{(5)}$ is a regular pentagon with a side length of 0.15 nm (the standard length of the carbon - carbon chemical bond). Since carbon atoms exhibit tetrahedral coordination, other valence bonds of carbon atoms must be located either above the plane of the ring or below the plane of the ring. It is convenient to assume for further geometric analysis that the atoms of hydrogen and the hydroxyl group located in Figure 4 inside the carbon ring are in space above the plane of the carbon ring. While the oxygen atom and the functional groups of the outer contour (larger) are under the plane of the carbon ring. A part of this spatial structure located under the plane of the carbon ring is shown in Figure 5.

Figure 5. The part of the glucose molecule located under the carbon ring

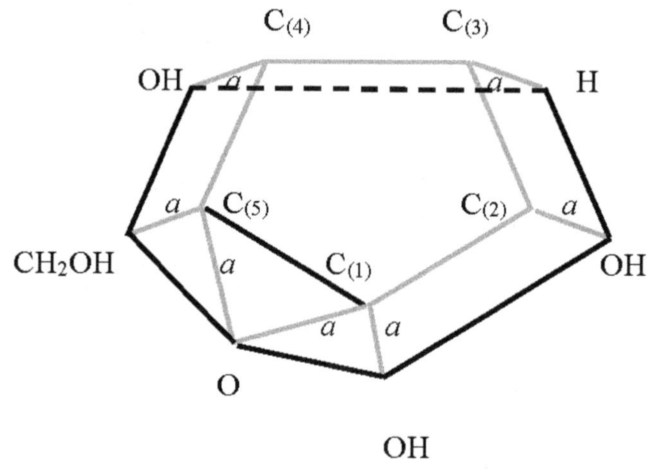

The length of the carbon – hydrogen and carbon – oxygen bonds is denoted by the letter a = 0.1 nm. Then the carbon - carbon bond length takes the form 1.5a. Thus, the part of the construction of the α -D - glucose molecule under the carbon ring is a pyramid with a lower base in the form of an irregular hexagon and an upper base in the form of a regular pentagon.

To determine the yet unknown geometric characteristics of the figure in Figure 5, it is first necessary to find the value of the angle of inclination of the side edges of the pyramid to its base. For this, we consider separately the flat lateral faces of the pyramid in the vicinity of the oxygen atom. We double these faces and designate the vertices and some characteristic points of this structure in separate letters (Figure 6).

We denote the angle CAK by the letter φ. Given that $\angle AKC = \pi / 2$ we have $CK=LS=BE=a\sin\varphi$, so AC=CD=DE=EF=a, $AK=a\cos\varphi$. Since the carbon cycle is a regular pentagon, then $\angle AKL = \pi / 2 - 54 = 126^o$. Since $CE=1.5a$ and point S is middle of segment CE then $DS = \sqrt{a^2 - (3/4)^2 a^2} = a\sqrt{7}/4$. From triangle DSL we have $DL = \sqrt{DS^2 - SL^2} = \frac{a}{4}\sqrt{7 - 16\sin^2\varphi}$. Therefore, from the triangle KLD we obtain $KD = \sqrt{DK^2 + KL^2} = a\cos\varphi = AK$. Then $\angle KDC = \varphi$. Besides, $\sin\angle DKL = \frac{DL}{KD} = \frac{\sqrt{7-16\sin^2\varphi}}{4\cos\varphi}$. Therefore,

$$\angle AKD = \angle AKL - \angle DKL = 126^o - \arcsin\frac{\sqrt{7-16\sin^2\varphi}}{4\cos\varphi}.$$

Figure 6. The enlarged part of the side surface of the pyramid (Figure 5) in the vicinity of the oxygen atom

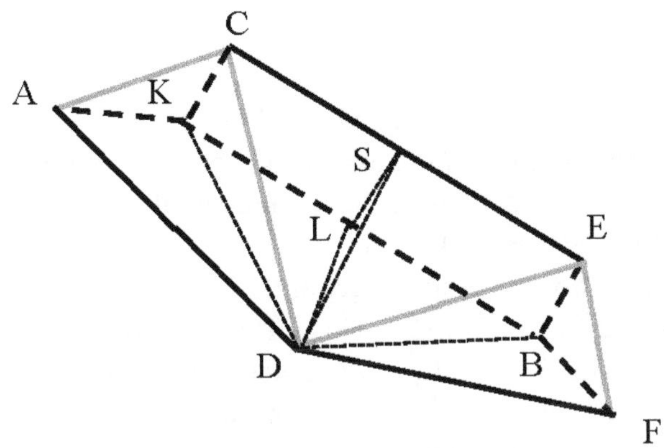

From the triangle *KAD* we obtain $\frac{AD}{2a\cos\varphi} = \sin\frac{\angle AKD}{2}$. From here we find

$$AD = 2\cos\varphi\sin(63^o - \frac{1}{2}\arcsin\frac{\sqrt{7-16\sin^2\varphi}}{4\cos\varphi}). \qquad (4.2)$$

Besides, from four-angle *AKLD* we have *KN/AK=cos (126° -90°)*. Then,

$$KN = a\cos\varphi\cos 36^o = LM.AM = AN + NM = a\cos\varphi\sin 36^o + \frac{3}{4}a.$$

Therefore, $MD = DL - LM = \frac{a}{4}\sqrt{7-16\sin^2\varphi} - a\cos\varphi\cos 36^o$. Thus, we find also one express for segment *AD*

$$AD = \sqrt{MD^2 + AM^2} = \left[(\frac{a}{4}\sqrt{7-16\sin^2\varphi} - a\cos\varphi\cos 36^o)^2 + (a\cos\varphi\sin 36^o + \frac{3}{4}a)^2 \right]^{1/2}. \qquad (4.3)$$

Comparing (4.2) and (4.3) after transformations, we obtain the equation for finding the angle φ for any a

$$\sin(63^o - \frac{1}{2}\arcsin\frac{\sqrt{7-16\sin^2\varphi}}{4\cos\varphi}) = \left[\frac{1}{2} + \frac{3}{8}\frac{\sin 36^o}{\cos\varphi} - \cos 36^o \frac{\sqrt{7-16\sin^2\varphi}}{8\cos\varphi}\right]^{1/2}. \quad (4.4)$$

Solving equation (4.4) numerically, we find that the angle $\varphi = 36o$. Thus, the height of the lower part of the α -*D* - glucose molecule is $a\sin\varphi$= 0.058778 nm, the characteristic size of the upper base is 0.243 nm, and the characteristic size of the lower base is 0.243+2a$\cos\varphi$= *0*.405 nm.

The angle of inclination of the side edges of the lower part of the model, as follows from the solution, is different from 90. This reflects in averaged form the presence of atoms in the conformations of the α -*D* - glucose molecule located quite close to the equatorial plane of the molecule (Koolman, Roehm, 2013).

The upper part of the α -*D* - glucose molecule above the carbon cycle should be depicted in the form of a pentagonal prism, directing the volatile bonds from the carbon atoms of the cycle vertically upward. This is consistent with the tetrahedral coordination of atoms around a carbon atom. Thus, the α -*D* - glucose molecule in a simplified three - dimensional form is a pyramid in contact with the carbon cycle with hexagonal and pentagonal bases and a straight pentagonal prism. However, in order to compare the obtained image with the known images of the α -*D* - glucose molecule in the Haworth projection (Koolman, Roehm, 2013), it is necessary to expand so that the pentagonal prism is at the bottom and the arrangement of other atoms and functional groups is consistent. In this case, it is necessary to take into account the possibility of changing the position of the hydroxyl group at the carbon atom $C(_{1)}$. In this way, two spatial simplified images of the glucose molecule are obtained, shown in Figure 7 (β- anomer) and Figure 8 (α- *a*nomer).

Make sure that the constructed model of the α -*D* - glucose molecule is three - dimensional, i.e. one can verify the fulfillment of the Euler - Poincaré equation (4.1) for this model. In the models in Figure 4.7 or Figure 4.8, there are 16 vertices(f_0=16), 28 edges (f_1=28), 14 flat faces (excluding the carbon cycle face), (f_2=14). Substituting the obtained values (f_i, i=0÷2) in equation (4.1), one can obtain

$$16 – 28 + 14 = 2.$$

We see that the Euler – Poincaré equation holds for n = 3. This proves that the constructed models of the α -*D* - glucose molecule are three - dimensional. It should be noted that if a face corresponding to the carbon cycle is added to equation (4.1), and this will require the addition of 3 three - dimensional regions, then the Euler – Poincaré equation takes the form

$16 - 28 + 15 - 3 = 0.$

Consequently, the Euler - Poincaré equation holds in this case for $n = 4$. Thus, in this case, the dimension of the models is 4.

Figure 7. The spatial simplified images of the α -D - glucose molecule (β - anomer)

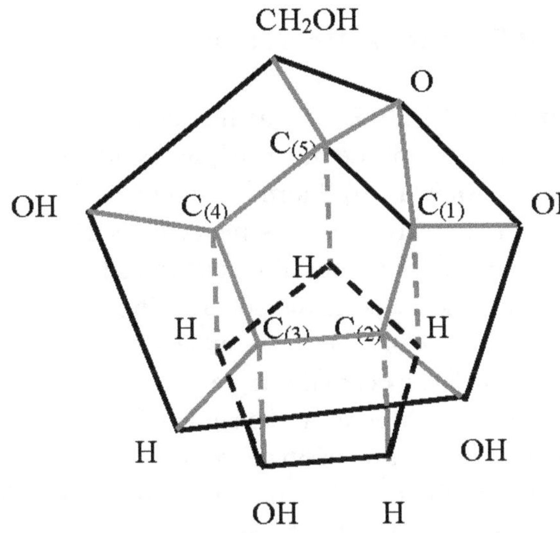

Figure 8. The spatial simplified images of the α -D - glucose molecule (α - anomer)

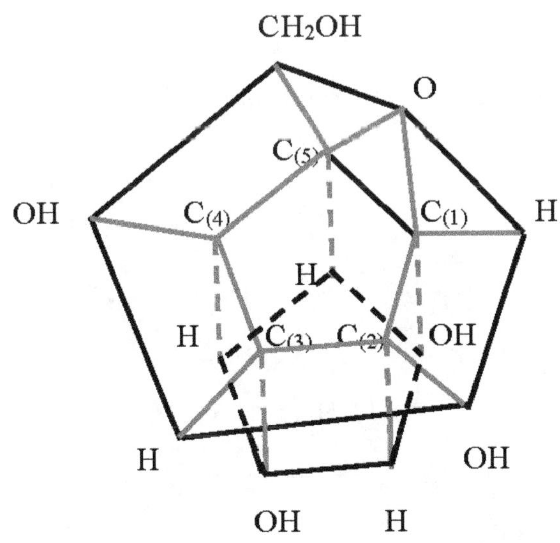

In the molecules of other pyranoses, the atoms or functional groups partially connected to the carbon atoms of the carbon cycle partially change. In this case, the structure of the molecules, up to functional groups, remains the same for glucose molecules in the pyranose cycle carbon atoms. Therefore, the dimension of molecules such as *D* - mannose, *D* - galactose, *L* - fucose is also equal to 15. Three - dimensional models of these molecules will retain their appearance, images are shown in Figure 7 and Figure 8.

Spatial Models of Furanose Molecules

In recent works (Zhizhin, 2016, 2018, 2019 a), the geometry of one of the important molecules of furanose, the *D* - ribose molecule, was studied in detail. The spatial model of this molecule is shown in full form in Figure 9.

It was shown that the dimension of this molecule is 12. A simplified three - dimensional model of this molecule was also constructed (Figure 10).

It should be noted that the shape of the polytope in Figure 4.10 is determined, as well as in the case of the α -*D* - glucose molecule as a result of solving the corresponding equation with respect to the angle of inclination of valence bonds to the plane of the furanose ring. Moreover, the angle of inclination was 90 degrees, in contrast to the angle of inclination of 36 degrees in the α -*D* - glucose molecule.

We will make sure that the model of the D - ribose molecule in Figure 4.10 is really three -dimensional, so this proof was not done in previous works. In the model in Figure 4.10 there are 13 vertices(f_0=13), 25 edges (f_1=25), 14 flat faces (excluding the carbon cycle face), (f_2=14). Substituting the obtained values (f_i, i=0÷2) in equation (4.1), one can obtain

Figure 9. The spatial model the D – ribose molecule

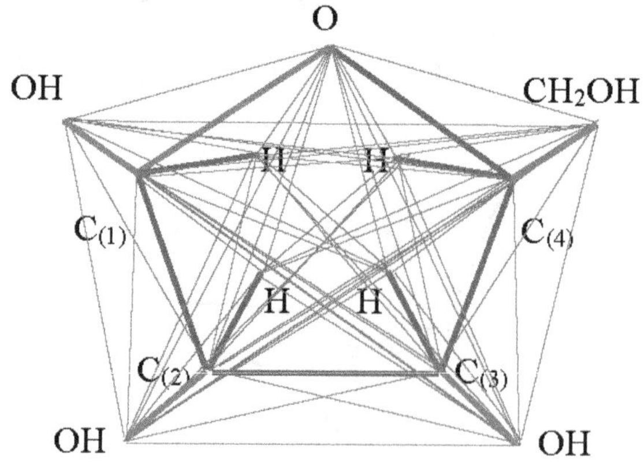

Figure 10. The simplified three – dimensional model of D – ribose molecule

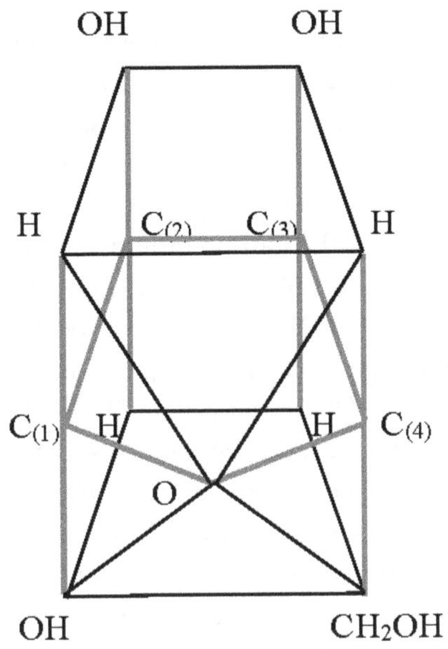

13 – 25 + 14 = 2.

We see that the Euler – Poincaré equation holds for $n = 3$. This proves that the constructed models of the *D* - ribose molecule are three - dimensional. It should be noted that if a face corresponding to the carbon cycle is added to equation (4.1), and this will require the addition of 3 three - dimensional regions, then the Euler – Poincaré equation takes the form

13 – 25 + 15 – 3 = 0.

Consequently, the Euler - Poincaré equation holds in this case for $n = 4$. Thus, in this case, the dimension of the models is 4.

In the molecules of other furanoses, the atoms or functional groups partially connected to the carbon atoms of the carbon cycle partially change. In this case, the structure of the molecules, up to functional groups, remains the same for furanose molecules as the *D* – ribose molecule. Therefore, the dimension of molecules such as *D* - xylose, *D* - arabinose, *D* - ribulose and others is also equal to 12. Three - dimensional models of these molecules will retain their appearance, image is shown in Figure 10.

Monosaccharide Chains

Monosaccharide molecules combine with each other thanks to the combination of two hydroxyl groups with the release of a water molecule. The remaining oxygen atom connects the remaining part of the monosaccharides. The most common chains of *α -D* - glucose monosaccharide residues. Here, residual *α -D* - glucose molecules can be joined through an oxygen atom, connecting carbon atoms 1 - 1, 1 - 4, 4 - 4, 4 - 1. In some cases, the carbon atom $C_{(2)}$ in the cycle will be involved in the compound. Branching chains of glucose molecules occurs at the sixth carbon atom in the functional group CH_2OH. Monosaccharide chains have different forms depending on the type of these molecules, i.e. depending on whether these molecules are *β*- anomers or α- anomers. Using the obtained simplified three - dimensional models of α *-D* - glucose molecules, one considers in more detail these compounds in a metric image, taking into account the angles between the valence bonds. Let two *β* - anomers of the α *-D* - glucose molecule join together, linking the carbon atom $C_{(1)}$ of the first α *-D* - molecule to the carbon atom $C_{(4)}$ of the second α *-D* - molecule (lactose). Based on the geometric image in Figure 7 in chapter 3, you can imagine a top view of this connection (Figure 11).

Figure 11. Projections of β - anomer disaccharides

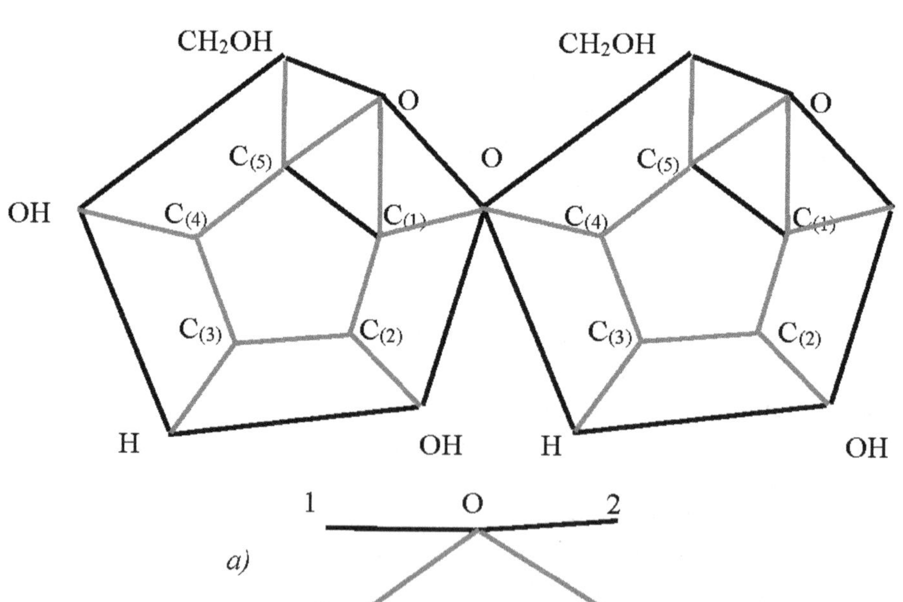

It is easy to see that the same attachment of the third α -*D* - glucose molecule leads to a linear arrangement of molecules when viewed from above on this compound. Linearity will not change with the following similar connections. However, given the three - dimensionality of the image, you should look at this connection from a different view. Figure 11a) shows an image of the junction of two α -*D* - glucose molecules when viewed from the front. The black segments in Figure 11a) are the traces of the intersection of the upper bases of the prism with a plane passing through the valence bonds of the oxygen atom and carbon atoms. According to the obtained solution of equation (4.4), the angle between the base of the prism and the valence bond is 36 degrees. both the first α -*D* - glucose molecule and the second α -*D* - glucose molecule. The valence angle at the oxygen atom in compounds with two carbon atoms in the chain of α -*D* - glucose molecules, as you know, is slightly larger than the normal valence angle and is about 112 degrees. Thus, the angle 1O2 in Figure 11a) is 176 degrees. This means that the base of the second α -*D* - glucose molecule has a slope to the base of the first α -*D* - glucose molecule other than zero. More precisely, this slope is 4 degrees, therefore the second α -*D* - glucose molecule, when viewed in full view, turns to the left relative to the first α -*D* - glucose molecule. It is clear that such an addition of a third α -*D* - glucose molecule will increase the angle of inclination of this molecule relative to the first molecule. Thus, the statement about the linearity of the chain of α -*D* - glucose molecules when connecting them in the case of anomers is not entirely accurate. One can say that this statement is somewhat one - sided.

Let two α - anomers of the α -*D* - glucose molecule join together, linking the carbon atom $C_{(1)}$ of thc first α -*D* - molecule to the carbon atom $C_{(4)}$ of the second α -*D* - molecule (maltose). Based on the geometric image in Figure 3.8, you can imagine a top view of this connection (Figure 12).

It is easy to see that the same attachment of the third α -*D* - glucose molecule leads to a nonlinear arrangement of molecules when viewed from above on this compound. Nonlinearity will save with the following similar connections. Thus, the chain of molecules turns to the right. However, given the three - dimensionality of the image, you should look at this connection from a different view. Figure 12a) shows an image of the junction of two α -*D* - glucose molecules when viewed from the front. The black segments in Figure 4.12a) are the traces of the intersection of the upper bases of the prism with a plane passing through the valence bonds of the oxygen atom and carbon atoms. According to the obtained solution of equation (4.4), the angle between the base of the prism and the valence bond is 36 degrees both the first α -*D* - glucose molecule and the second α -*D* - glucose molecule. Thus, on the Figure 4.12a the angle 1H $C_{(1)}$ is 36 degrees and the angle $C_{(1)}OC_{(4)}$ is 112 degrees. Since the angle $C_{(4)}O2$ is 36 degrees too, so the angle between bases 1H and O2 is 14 degrees. This means that the base of the second α -*D* - glucose molecule has a consequently slope to the base of the first α -*D* - glucose molecule to the left. It is clear that such an addition of a third α -*D* - glucose molecule will increase the angle of inclination of this molecule relative to the first molecule. Thus, chains of α - anomers of α -*D* - glucose molecules rotate simultaneously in opposite directions in perpendicular planes.

The study of furanose chains using simplified three - dimensional models was carried out using nucleic acids as an example, in which furanose models were connected to each other using phosphoric acid residues. These studies showed that, when nucleic acids bound by nitrogen bases interact, there is a latent degree of freedom of arrangement of nitrogen bases in higher dimensional space, leading to an unambiguous corresponding number of nucleotides taking into account their possible location in space and the number of amino acids (Zhizhin, 2020, a, b).

Figure 12. Projections of α - anomer disaccharides

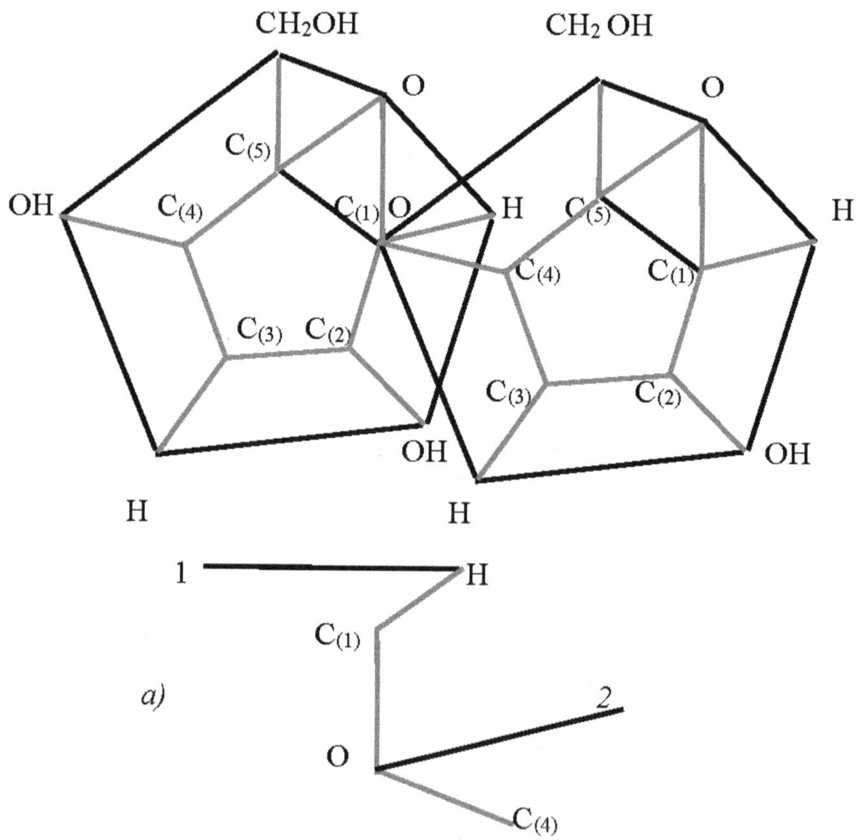

In addition to simple hexoses such as glucose, galactose and mannose, there are a number of derivatives in which either the hydroxyl group of the parent compound is replaced by another, or the carbon atom is oxidized to a carboxyl group. In glucosamine, galactosamine, and mannosamine, the hydroxyl at the C-2 atom in the original sugar molecule is replaced by an amino group. The amino group is almost always condensed with acetic acid, such as in N-acetylglucosamine. This compound is part of many structural polymers, including those that form the cell wall of bacteria. The latter contains another glucosamine derivative - N-acetylmuramic acid. In this compound, the oxygen at the C-3 atom of H-acetylglucosamine is connected by an ethereal bond to lactic acid. The replacement of the hydroxyl group by hydrogen at the C-6 atom in the L-galactose or L-mannose molecule leads to the formation of L-fucose or L-rhamnose, respectively. For all the above connections and similar ones, the construction of simplified three-dimensional models can be applied.

GLYCOPROTEINS

Many proteins contain oligosaccharide sequences. They are attached to the protein part during post-translational modification in the endoplasmic reticulum and Golgi apparatus. Polysaccharides can make up over 70% of the mass of a glycoprotein molecule.

CONCLUSION

A method has been developed for obtaining simplified images of polytopes of higher dimension in a space of lower dimension. This method is used to obtain three - dimensional images of saccharide molecules: pyranoses (the monosaccharide molecules in which the rings of carbon atoms containing one oxygen atom and five carbon atoms) and furanoses (the monosaccharide molecules in which the rings of carbon atoms containing one oxygen atom and four carbon atoms). It is proved that three - dimensional images of pyranose and furanose molecules are fundamentally different from each other. The furanose molecules are a convex three - dimensional body, and the pyranose molecules consist of two convex bodies, in general, forming a non - convex body. This led to a fundamental difference in the structure of polysaccharide chains of pyranose and furanose molecules. If in the furanose spirals each projection of the furanose molecule onto the coordinate planes preserves the general direction of rotation when moving along the spiral, then a complex picture is observed in the chains of pyranose molecules. In the chains of β - anomers, one of the projections is linearly transmitted along the chain, and the projection perpendicular to it along the chain along a curved line. In chains of α - anomers (spirals), it is observed that when moving along a chain of projections in mutually perpendicular planes, the pyranose molecules rotate in opposite directions (Figure 13), (Zhizhin, 2020 c, 2021).

Figure 13. The chain of α - anomers (spirals)

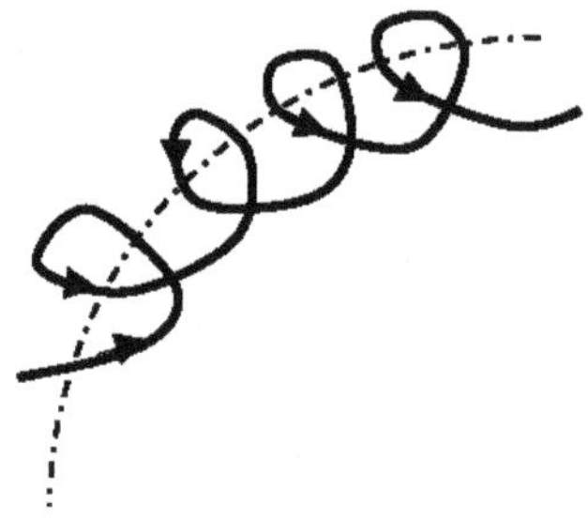

REFERENCES

Koolman. J. & Roehm. K.- H. (2013). Color Atlas of Biochemistry. Georg Thieme Verlag KG.

Lehninger, A. L. (1982). *Principles of Biochemistry* (Vol. 1-3). Worth Publisher, Inc.

Metzler, D. E. (1980). *Biochemistry. The Chemical Reactions of Living Cells* (Vol. 1-3). Academic Press.

Nelson, D. L., & Cox, M. M. (2022). *Lehninger Principles of Biochemistry*. W.H. FREEMAN.

Poincaré, A. (1895). Analysis situs. *J. de é Ecole Polytechnique, 1*, 1 – 121.

Zhizhin, G. V. (2016). The Structure, Topological, and Functional Dimension of Biomolecules. *International Journal of Chemoinformatics and Chemical Engineering*, *5*(2), 44–58. doi:10.4018/IJCCE.2016070104

Zhizhin, G. V. (2018). *Chemical Compound Structures and the Higher Dimension of Molecules: Emerging Research and Opportunities*. IGI Global.

Zhizhin, G. V. (2019a). The structure of polytope of hereditary information. J. *Chemoinformatics and Chemical Engineering*, *8*(2), 7–22.

Zhizhin, G. V. (2019b). The Polytope of Hereditary Information: Structure, Location, Signification. *Biochemistry and Modern Applications.*, *2*, 56–62.

Zhizhin, G. V. (2019 c). *Attractors and higher dimensions in population and molecular biology: emerging research and opportunities*. IGI Global.

Zhizhin, G. V. (2019 d). *The Geometry of Higher – Dimensional Polytopes*. IGI Global.

Zhizhin, G. V. (2020a). Hidden Nucleic Acid Bond Order? *Acta Scientific Biotechnology*, *1*(4), 34–36.

Zhizhin, G. V. (2020b). Interaction of Nucleic Acids: Hidden Order of Interaction. *International Journal of Applied Research in Bioinformatics*, *11*(2), 1–15.

Zhizhin, G. V. (2020c). Spirals of Polysaccharides Simultaneously Rotating in Opposite Directions. *Acta Scientific Biotechnology*, *1*(7), 13–16.

Zhizhin, G. V. (2021). Spatial Models of Sugars and Their Components. *International Journal of Applied Research in Bioinformatics*, *11*(2), 9–23.

Zhizhin, G. V. (2022). *The Classes of Polytopes Higher Dimensions in Physics, Chemistry, and Biology*. IGI Global.

Chapter 5
The Structure of Nucleic Acids in Higher Dimensional Space

ABSTRACT

The types of interaction of nucleic acids are considered, taking into account the higher dimension of the geometric elements that make up nucleic acids. Lists canonical and non-canonical nitrogenous bases that can participate in the creation of a hydrogen bond of nitrogenous bases found in different single-stranded nucleic acids, forming double-stranded nucleic acids. Pairs of non-canonical nitrogenous bases can significantly violate the unique correspondence of nitrogenous bases in pairs (complementarity), which is characteristic of canonical nitrogenous bases. A model of a double-stranded nucleic acid is built considering the spatial model of a five-carbon model of a sugar molecule. When building a model of a double-stranded model of a nucleic acid molecule, a special type of polytope is used: a cross-polytope. The features of the interaction of nucleic acids in ribosomes responsible for the synthesis of protein molecules according to a given program are considered.

INTRODUCTION

In the middle of the last century, there was an idea that nucleic acids (DNA and RNA) ensure the transfer of hereditary information from one generation of living organisms to another (Caspersson, et al., 1941; Brachet, 1942). Moreover, DNA is the main carrier of genetic information, and RNA is an intermediary that receives this information from DNA and implements it in the form of protein biosynthesis. Deoxyribonucleic acid (DNA), as a chemical substance, it was isolated by Johann Friedrich Micher in 1869 from the remains of cells contained in the pus. He singled out a substance that includes nitrogen and phosphorus. When Misher determined

DOI: 10.4018/979-8-3693-0588-1.ch005

that this substance has acid properties, the substance was called nucleic acid (Dahm, 2005). Gradually it was proved that it was DNA, and not proteins, as previously thought, and which is the carrier of genetic information. One of the first decisive proofs was the experiments of Oswald Avery, Colin MacLeod and McLean McCarthy (1944) on the transformation of bacteria.

The structure of the double helix DNA it was proposed by Francis Crick and James Watson in 1953 on the base of the X - ray structural data obtained by Maurice Wilkins and Rosalind Franklin and the "Chargaff rules" according to which in each DNA molecule the strict relationships connecting the quantity of nitrogenous bases of different (Watson, Crick, 1953 a, b). For outstanding contributions to this discovery, Francis Crick, James Watson and Maurice Wilkins were awarded the 1962 Nobel Prize in Physiology or Medicine. Deoxyribonucleic acid (DNA) is a biopolymer, the monomer of which is nucleotide (Albert, et al., 2002; Butler, 2005). Each nucleotide consists of a phosphoric acid residue attached to sugar deoxyribose, to which one of the four nitrogen bases is attached also. The bases that make up the nucleotides are divided into two groups: purines (adenine [A] and guanine [G]) and pyrimidines (cytosine [C] and thymine [T]) are formed by combined five - and six - membered heterocycles.

They managed to show that the DNA isolated from the pneumococci corresponds to the so-called transformation (the acquisition of pathogenic properties by a harmless culture as result of the addition of dead pathogenic bacteria to it). The experiment of American scientists Alfred Hershey and Martha Chase (Hershey - Chase experiment, 1952) with radioactively labeled proteins and bacteriophage DNA showed that only the phage nucleic acid is transmitted to the infected cell, and the new generation of phage contains the same proteins and nucleic acid, as the initial phage (Hershey, Chase, 1952). Deciphering the structure of DNA (1953) has become one of the turning points in the history of biology.

In 1986, Frank-Kamenetskiy in Moscow showed how double - stranded DNA folds into a so-called H - shape, composed not of two but three strands of DNA (Frank – Kamenetskiy, 1986, 1988). Deoxyribonucleic acid (DNA) is a biopolymer, the monomer of which is the nucleotide (Albert et al., 2002; Butler, 2005).

Nucleotides are long polynucleotide chains covalently linked. These chains in the overwhelming majority of cases (except for some viruses possessing single - stranded DNA genomes) are combined pairwise by means of hydrogen bonds into a secondary structure, called the double helix (Watson, Crick, 1953 a, b; Berg, Tymoczko, Stryer, 2002). Each base at one of the chains is connected to one definite base on the second chain. This specific binding is called complementary. Purines are complementary to pyrimidines (that is, they are capable of forming hydrogen bonds with them): adenine forms bonds only with thymine, and cytosine - with guanine. In a double helix, chains are also linked by hydrophobic interactions and stacking, which do not depend on the DNA base sequence (Ponnuswamy, Gromiha, 1994). Complementarity of the double helix means that the information contained in one chain is also contained in another chain. Different base pairs form a different number of hydrogen bonds. In the future, the existence of nucleic acids differing in the length of the period and shape with rotation of the spiral both to the right and to the left was experimentally established (Ha, et al., 2005; Cantor, Schimmel, 1980; Frank – Kamenetskiy, 2010).

Watson and Crick postulated the spiral form of the DNA molecule, but they did not discuss the reasons for the formation of such a DNA molecule. Until now there have been no works explaining the existence of a spiral in the DNA molecule.

In the book (Zhizhin, 2018), the molecules of practically all the elements of the periodic system were studied in detail and it was shown that many of them, including magnesium and calcium, form compounds of higher dimension (see also Zhizhin, 2015 a, b; Zhizhin, Diudea, 2016; Zhizhin, Khalaj, Diudea, 2016). In the works (Zhizhin, 2016 a, b), the structures of biomolecules, which also form compounds of higher dimension, are also studied. When analyzing the metric space of nucleic acid structures, one should use the representations of their constituents in the form of polytopes of higher dimension. This primarily applies to the phosphoric acid residues and the D - ribose molecules since nitrogenous bases have a dimensionality of 2 and do not have a significant effect on the geometric structure of nucleic acids.

The determination of the geometric forms of chemical compounds according to the Euler -Poincaré (Poincare, 1895) relation translates classical stereochemistry in three - dimensional space into sulfur chemistry in a space of higher dimension. Here new discoveries are possible in the laws of chemical compounds. Consideration of the geometry of molecules using spaces of higher dimension makes it possible to explain the facts observed in reality, which previously had no clear explanations. In the book (Zhizhin, 2018), for example, showed that the rotation of the plane of light polarization in tartaric acid is due to the difference in the shapes of the molecules of D - tartaric acid and L - tartaric acid, which is clearly visible in their images in five - dimensional space, since their dimension is five. Pasteur (Pasteur, 1960) also tried to explain the rotation of the polarization plane of light in liquid tartaric acid by the difference in the structures of the crystals of D - tartaric acid and L- tartaric acid. This book received an explanation for the fact that there is no branching in the chain of glucose molecules, if they are bound in β - glycoside linkage. If these molecules are bound in α – glucoside linkage, branching of the chain is possible. It is shown that, because of the higher dimensionality of glucose molecules in the case of a β – glycoside linkage, the centers of possible branching of the chain are blocked by glucose molecules. In this time, with α – glycoside linkage the possible branching points of the chain remain free.

The concept of the geometric correspondence of the structure of substances in the formation of a chemical compound is essential for the form of the compound. However, the complementarity of DNA spirals when connecting spirals cannot be the reason for the formation of these spirals. The geometric shape of the constituent parts of the spiral itself must be important here. They are phosphoric acid residues and *D* - ribose molecules. This paper is devoted to the investigation of this problem. In this case it is necessary to take into account the dimension of these parts, which, as shown in (Zhizhin, 2018), is higher than three. It should be noted that there is no contradiction in the representation of the structure of DNA as a spiral in a three - dimensional space with the highest dimension of its components (Zhizhin, 2019 a, b, g; 2021 c). The point is that according to the geometry of Riemann (Riemann, 1854), a space of higher dimension has a boundary. Therefore, a space of a smaller dimension can be located outside this space.

The higher dimensionality of molecules can be considered as an example of Big Data. The concept of Big Data has become widespread in computer science in recent times (Kosrapohr, 2003). The representation of molecules in the form of polytopes of higher dimension means the handling of Big Data as geometric images of a set of points in a space of higher dimension. A striking example of the use of the concept of Big Data in medicine is a work of J.A. Rodger (Rodger, 2015) in which, using the intellectual analysis of Big Data, a program has been developed to improve the prognosis of survival rates for traumatic brain injury and other injuries on several ships. The totality of injury data for each vessel can be considered as Big Data in view of the many parameters characterizing trauma.

THE DIMENSION OF PHOSPHORIC ACID AND ITS RESIDUE

The phosphoric acid H_3PO_4 has the following building (see, for example, Karapetians, Drakin, 1994) (Figure 1).

On the Figure 2 is presented Fisher`s formula of phosphoric acid.

It is obvious that, used in the DNA molecule, the phosphoric acid residue obtained by detaching one hydrogen atom from it has the building also. In this case, instead of one of the hydroxyl groups attached to the phosphorus atom, there will be a negatively charged oxygen atom. To determine the functional topological dimension of phosphoric acid and its residue, it is necessary to connect the vertices, atoms and functional groups entering into these compounds (Zhizhin, 2016 a, 2018). Then the structure of phosphoric acid and its residue is represented in the form of a polytope in Figure 3.

Figure 1. Special structure of phosphoric acid

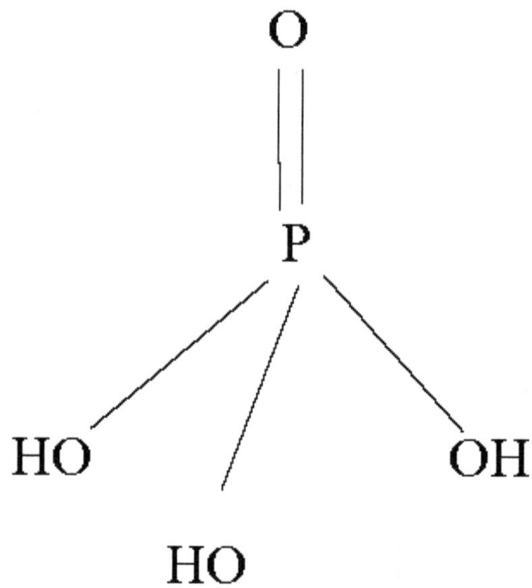

Figure 2. Fisher`s formula of phosphoric acid

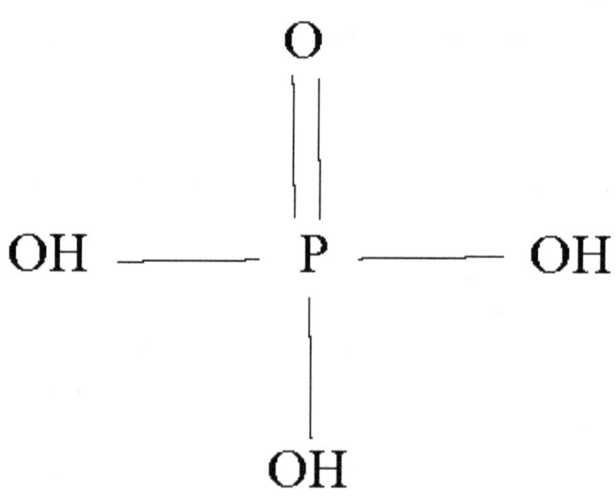

In Figure 5.3 in vertices *a* dispose atom phosphoric; in vertices *c* dispose atom of oxygen with double bond; in vertices *b, d, e* dispose atoms of oxygen with free bond.

Figure 3. The polytope of the phosphoric acid and its residue

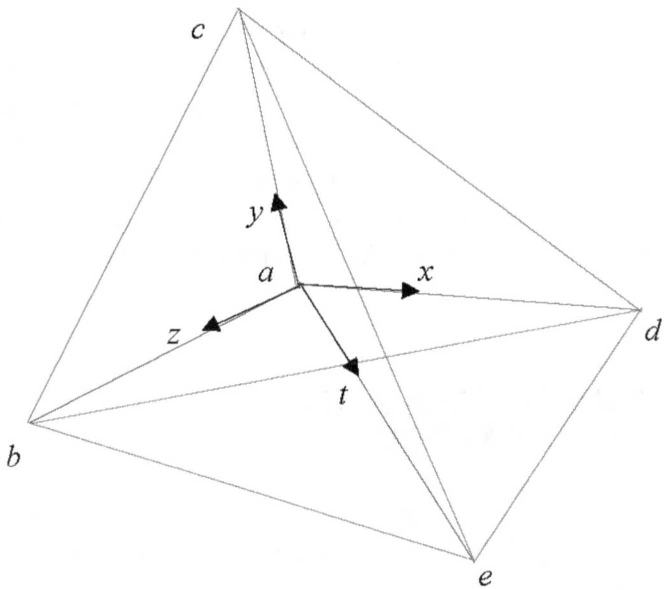

Calculation of the functional topological dimension by the Euler - Poincaré equation (Poincaré, 1895) does not require specifying the exact lengths of polytope edges, and for a given number of the vertices does not depend on the groups of atoms in these vertices (Zhizhin, 2016 a). In Figure 5.3, the edges corresponding to the covalent chemical bonds are marked in red, and the edges that carry only the geometric function, delineating the closed geometric figure, are marked in blue.

It follows from Figure 3 that the polytope of phosphoric acid has five vertices (*a, b, c, d, e*), ten edges(*ac, ab, ad, ae, cd, de, be, ce, cb, bd*), ten plane triangular faces (*cbe, cda, cea, dab, dbe, dce, bae, dae, bce, dcb*), and five three - dimensional faces in the form of tetrahedron (*cbae, caed, cbad, abde, cebd*). These values of elements of different dimensions we must substitute into the Euler - Poincare equation,

$$\sum_{i=0}^{n-1} f_i(-1)^i = 1-(-1)^i, \tag{1}$$

where f_i is the number of elements with dimension *i* in polytope with dimension *n*. In this case there are f_0=5, f_1=10, f_2=10, f_3=5.

Substituting these values into equation (1), can obtain,

$5 - 10 + 10 - 5 = 0$

Thus, equation (1) is satisfied for $n = 4$. This proves that the polytope of phosphoric acid and its residue has a dimension of four. It is worth emphasizing that the polytope of phosphoric acid and its residue is topologically equivalent to a convex polytope of the simplex type with dimension four, since these polytopes completely coincide with the values of the numbers of elements of different dimensions (Zhizhin, 2014 b). The origin of the four - dimensional space can be placed in one of the vertices with sending coordinates along the edges connecting this vertex to other vertices (see Figure 3).

THE DIMENSION OF THE MOLECULES DERIBOSE AND DEOXYRIBOSE

The *D* – ribose is molecule of carbohydrate with five atoms carbon. There are two enantiomer forms of this molecule. Its images in the form of Fisher's formula (Metzler, 1980) are shown in Figure 4 A, B.

Figure 4. A, B. Two enantiomer forms of the molecule D – ribose (formulas of Fisher`s)

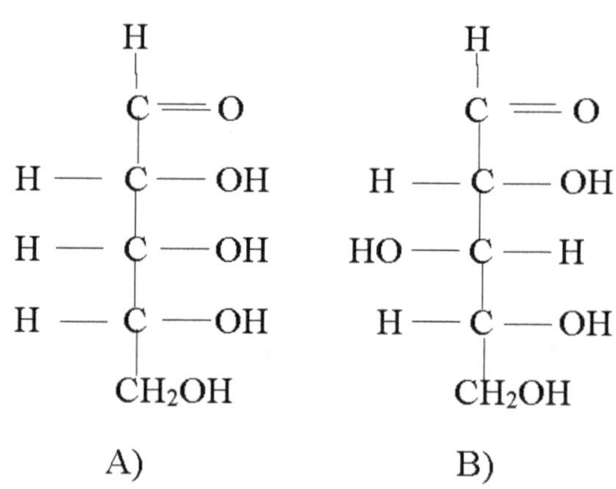

In the DNA and RNA molecules, *D* - ribose as a component enters with a closed carbon chain (in form furanose). Moreover, in DNA molecules, D - ribose takes the form of deoxyribose, in which the hydroxyl group at the second carbon atom in the closed chain is replaced by a hydrogen atom (2 – deoxy – D - ribose). The closed chain of the molecule in Figure 4 is formed due to the rupture of the double bond between oxygen and carbon (considering it the first in the chain), the liberation of the water molecule, and the addition of this oxygen atom to the penultimate (fourth) carbon atom in the carbon chain. The Fischer formula or formula Haworth and their modifications (Metzler, 1980; Lehninger, 1982) may not reflect the spatial structure of the *D* – ribose molecule. For this target the constructs described in the form of convex polytopes with boundary elements which form boundary complex (Grunbaum, 1967). Figure 5 shows the result of the closure of a chain of four carbon atoms through an oxygen atom correspondent Figure 4 A. Chemical bonds are marked as before with red color.

Figure 5. The molecule α - D – ribose with closed a chain of carbon atoms (the form A)

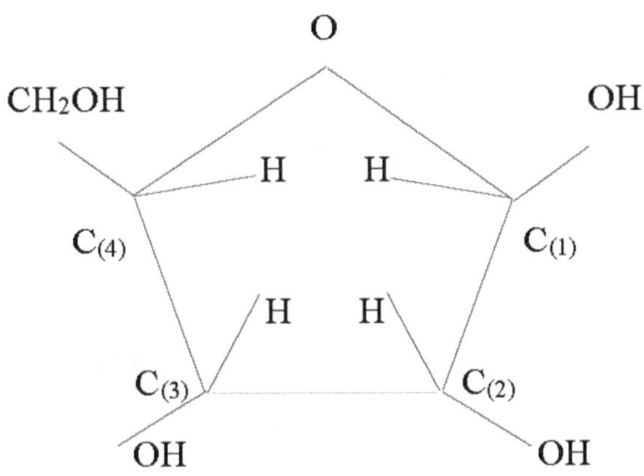

It is easy to see that, in addition to the form of the α - *D* - ribose molecule in Figure 5.5 (a form of *A*), there is another enantiomer form of this molecule. It connected by a symmetry transformation with respect to a straight line passing through an oxygen atom separating the projection of the form of *A* in half (we shall assume that it is a *B* shape). Form *B* of the α - *D* - ribose molecule is shown in Figure 6.

Figure 57 shows the result form of the molecule β – D – ribose with closed a chain of carbon atoms (the form A).

Figure 8 shows the result form of the molecule 2 – deoxy - α – D – ribose with closed a chain of carbon atoms (the form A).

Figure 9 shows the result form of the molecule 2 – deoxy - β – D – ribose with closed a chain of carbon atoms (the form A).

Figure 6. The molecule α - D – ribose with closed a chain of carbon atoms (the form B)

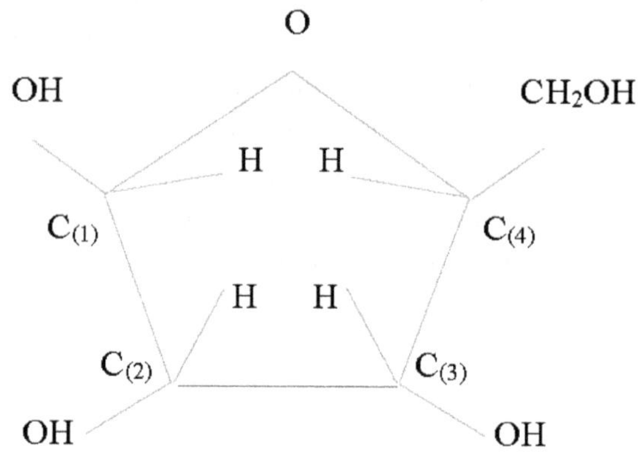

Figure 7. The molecule β – D – ribose (A)

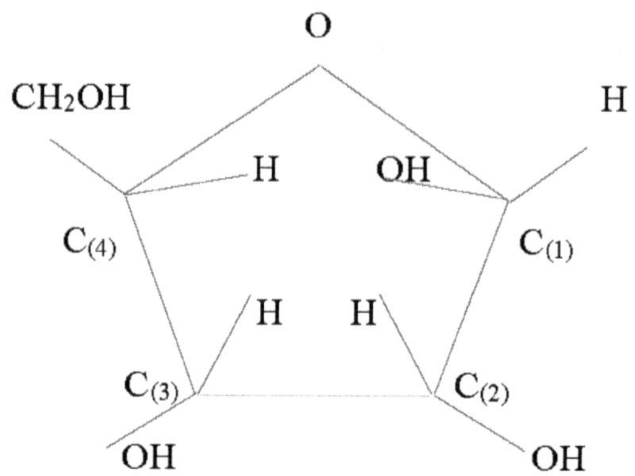

Figure 8. The molecule 2 – deoxy - α – D – ribose (A)

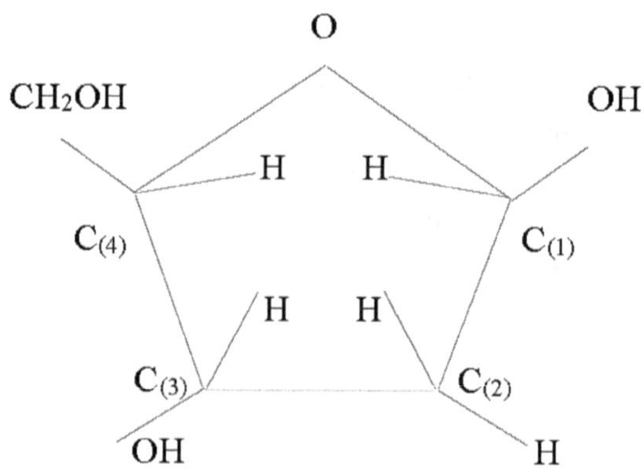

Figure 9. The molecule 2 – deoxy - β – D – ribose (A)

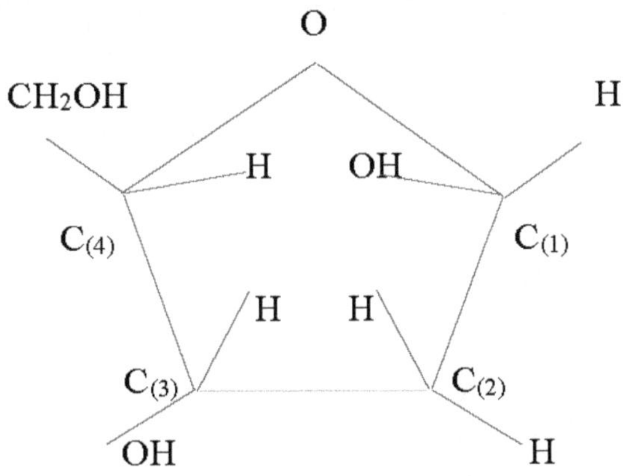

When depicting molecules in Figures 5 - 9, we transfer the hydroxyl groups associated with carbon atoms in the chain to the outer contour of the molecules, given that it is the hydroxyl groups that ensure the implementation of chemical reactions of sugar molecules that really exist, in particular, When depicting molecules in Figures 5 - 9, we transfer the hydroxyl groups associated with carbon atoms in the chain to the outer contour of the molecules, given that it is the hydroxyl groups that ensure the implementation of chemical reactions of sugar molecules that really exist, in particular, in the formation of nucleic acids. Obviously, for each image in Figures 7 - 9 can be obtained by transforming the symmetry of the molecule in the form B, just as it was obtained for Figure 6.

In order to give the molecules on Figures 6 – 9 a kind of polytopes, it is necessary to connect on this Figures each vertex (every atom or functional group, with all other vertices) to the ribs, then can get polytopes of the simplex type, the dimension of which must be determined.

In this chapter, for example, we will represent a molecule 2 – deoxy - α – D – ribose (A) in the form of a polytope (Figure 10)

The polytope in Figure 10 has 13 vertices in which there are one oxygen atom, four hydrogen atoms, three hydroxyl groups, four carbon atoms, compound CH_2OH. It is known that the dimension of a simplex is one less than the number of its vertices (Zhizhin, 2019 a). Therefore, the dimension of the polytope in Figure 10, if the Euler - Poincaré equation (1) holds for it, equals $n=f_0 - 1=12$. The simplex is a simplicial polytope, and all its faces therefore are simplexes. Since all the vertices of a simplex are related to each other, the number of edges is determined by the number of combinations of the number of vertices of two. In this case, the number of edges of the polytope in Figure 10 is $f_1 = C_{n+1}^2 = 78$. The number of two - dimensional faces (triangles) of this polytope is $f_2 = C_{n+1}^3 = 286$.

Figure 10. The molecule 2 – deoxy - α – D – ribose (A) in form of the polytope

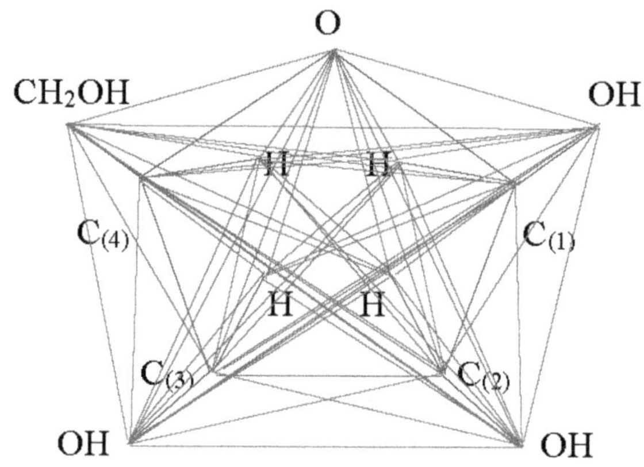

The number of three - dimensional faces (tetrahedrons) of this polytope is $f_3 = C_{n+1}^4 = 715$.

The number of four – dimensional faces (4 – simplexes) of this polytope is $f_4 = C_{n+1}^5 = 1287$.

The number of five – dimensional faces (5 – simplexes) of this polytope is $f_5 = C_{n+1}^6 = 1716$.

The number of six – dimensional faces (6 – simplexes) of this polytope is $f_6 = C_{n+1}^7 = 1716$.

The number of seven – dimensional faces (7 – simplexes) of this polytope is $f_7 = C_{n+1}^8 = 1287$. The number of eight – dimensional faces (8 – simplexes) of this polytope is $f_8 = C_{n+1}^9 = 715$. The number of nine –dimensional faces (9 – simplexes) of this polytope is $f_9 = C_{n+1}^{10} = 286$.

The number of ten – dimensional faces (10 – simplexes) of this polytope is $f_{10} = C_{n+1}^{11} = 78$.

The number of eleven – dimensional faces (11 – simplexes) of this polytope is $f_{11} = C_{n+1}^{12} = 13$. Substituting the values f_i, $(0 \leq i \leq 11)$ in the Euler - Poincaré equation (1), can see that it holds for $n = 12$ for polytope on Figure 10

$$13 - 78 + 286 - 715 + 1287 - 1716 + 1716 - 1287 + 715 - 286 + 78 - 13 = 0.$$

This finally proves that the polytope 2 – deoxy - $\alpha - D$ – ribose (A) has dimension 12 and a simplex type. Obviously, that all polytopes α - *D* - ribose (A), $\beta - D$ – ribose (A), $\alpha - D$ – ribose (B), $\beta - D$ – ribose (B), 2 – deoxy - α – D – ribose (A), 2 – deoxy - $\alpha - D$ – ribose (B), 2 – deoxy - $\beta - D$ – ribose (A), 2 – deoxy - $\beta - D$ – ribose (B) has the same dimension and type, as all have equal the numbers f_i, ($0 \leq i \leq 11$).

THE STRUCTURE OF THE $\alpha - D$ – RIBOSE AND 2 – DEOXY $\alpha - D$ – RIBOSE NUCLEIC ACIDS

The nucleic acid is a chain of *D* – ribose (α - *D* - ribose or 2 – deoxy – *D* – ribose) molecules between which are the phosphoric acid residues that connect these molecules. Let us construct this circuit, taking into account the concept of repulsion of electron pairs (Gillespie, 1972; Gillespie, Hargittai, 1991). According to this concept, tetrahedral coordination of atoms and functional groups in these compounds is carried out around the phosphorus atom in the phosphoric acid residue, and in the carbon atoms in the *D* - ribose molecule also. It is this concept that leads to the higher dimensionality of the remainder of phosphoric acid and the molecule of *D* - ribose. When a phosphoric acid residue is added to the *D* - ribose molecule from the hydroxyl group of the phosphoric acid residue and the hydroxyl group of the *D* - ribose molecule, a water molecule that leaves the compounds forms and an oxygen atom that binds the phosphoric acid residue and the *D* - ribose molecule. The concept of repulsion of electron pairs also operates here, according to which two valence electrons of the oxygen atom maximally repel each other, forming electron pairs with electrons of the phosphorus atom and the carbon atom. Therefore, taking into account that tetrahedral coordination is carried out around the atoms of phosphorus and carbon already, it can assume that the valence bonds of the oxygen atom are located linearly with respect to the oxygen atom. When the second *D* - ribose molecule is added to the phosphoric acid residue, the second hydroxyl group of the phosphoric acid residue with the compound CH_2OH of this *D* - ribose molecule also form molecules of water that leaves the compound and an oxygen atom connecting the phosphorus atom and the functional group CH_2 of the *D* - ribose molecule. By virtue of the concept of repulsion of electron pairs, the linear arrangement of the valence bonds of the oxygen atom can be considered here also. In addition, as the first step within the framework of the notions of the functional dimension of molecules (Zhizhin, 2016 a), can shall assume the linear arrangement of the valence bonds also in the neighborhood of the functional group CH_2 of the *D* - ribose molecule. One will take into account this when constructing a chain of *D* - ribose molecules and phosphoric acid residues. To construct the sequence of *D* - ribose molecules and the phosphoric acid residue in the chain can will use a simplified image of the *D* - ribose molecule of dimension 12. Let us leave only the outer contour and valence bonds necessary for interaction with the environment from the *D* - ribose molecule. At the same time, one will not depict the hydrogen atoms inside the contour of carbon atoms. We approximately to consider that the lengths of chemical bonds carbon - carbon, carbon - oxygen close to each other. The images of the α - *D* – ribose and 2 – deoxy – α – D - ribose molecules of the enantiomer form *A* under these conditions takes the same form (Figure 11 and Figure 12).

Figure 11. The simplified image of the enantiomer A of α - D – ribose molecule

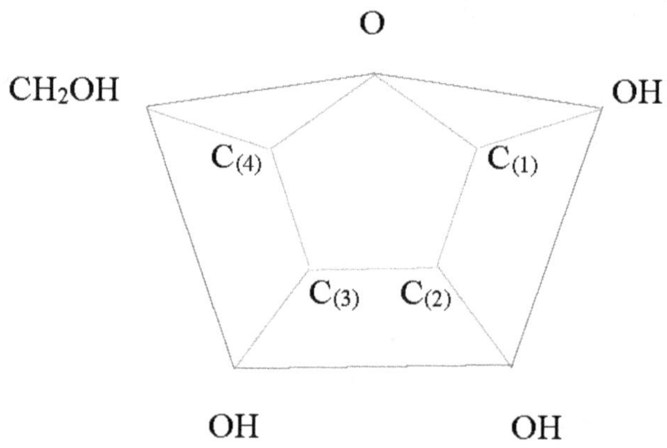

Figure 12. The simplified image of the enantiomer A of 2 – deoxy – α – D - ribose molecules

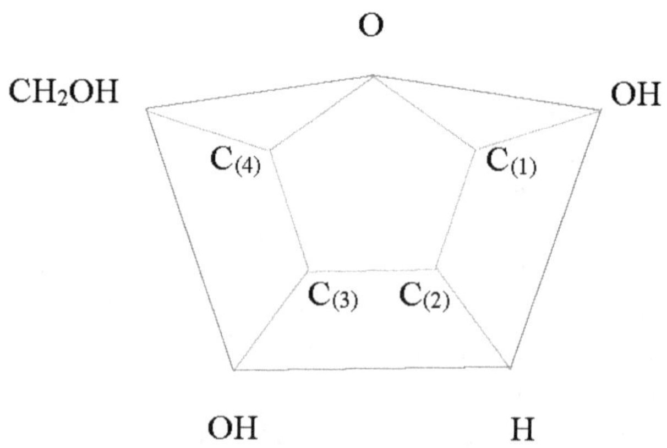

The corresponding images of the α - *D* – ribose and 2 – deoxy – α – D – ribose molecules of the enantiomer form *B* are shown in Figure 13, Figure 14.

Figure 13. The simplified image of the enantiomer B of α - D – ribose molecule

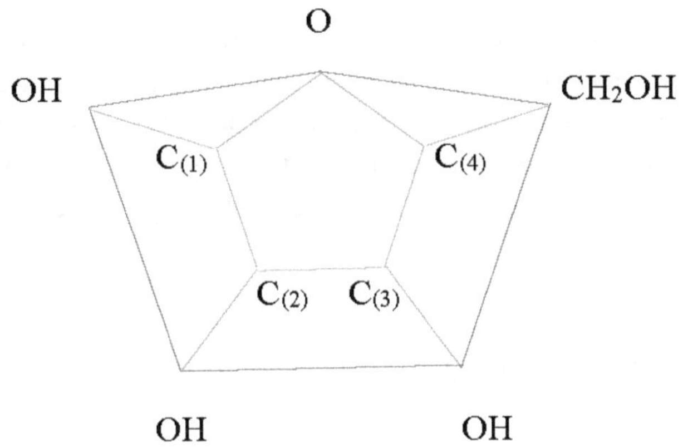

Figure 14. The simplified image of the enantiomer B of 2 – deoxy – α – D - ribose molecules

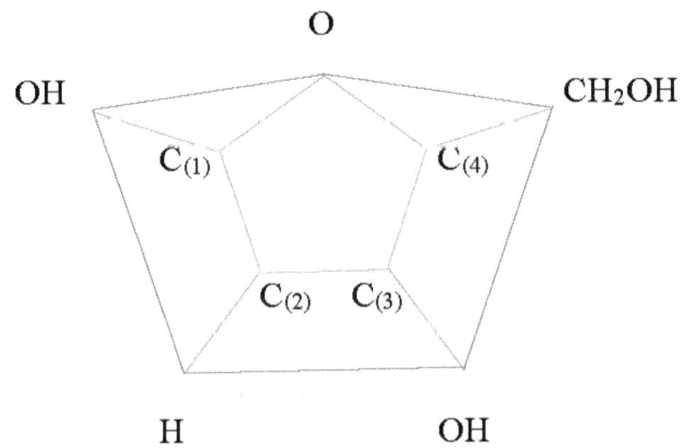

The absence of edges and atoms in Figures 11 – 14 in comparison with Figure 10 does not mean that they really are not. This is only the conventionality of the image.

It should be noted that under the assumed projection conditions, the α - *D* - ribose and 2 – deoxy - α - *D* - ribose of molecules on a two - dimensional plane in which a closed carbon ring is located has a number of properties that are convenient for further constructions. The enclosed carbon ring is a regular pentagon with an internal angle at the vertices $\pi - \frac{2\pi}{5} = 108°$. The angles in the three trapezoids formed by the valence bonds and the outer contour, because of the symmetry and parallelism of the bases, are 54 degrees and 126 degrees. In two obtuse - angled isosceles triangles formed by valence bonds and the outer contour, the angles are 126 degrees and 27 degrees.

Now, in a projection onto a two - dimensional plane, one shall depict a sequence of the simplified images of α - *D* - ribose molecules and phosphoric acid residues, taking into account the separation of water molecules at the junction of phosphoric acid residues and α - *D* - ribose molecules and the linear arrangement of the valence bonds in vicinity of these compounds. In Figure 15 is presented the sequence for α - *D* – ribose molecules in form *A*.

Figure 15. Structure of the nucleic acid molecule with α - D – ribose molecules in form A

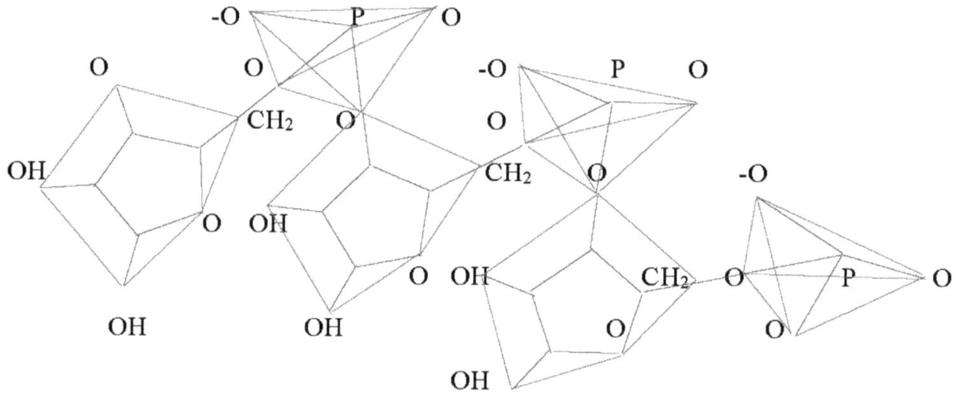

In Figure 16 is presented the sequence structure of the nucleic acid molecule with for *D* – ribose molecules in form *B*.

In Figure 17 is presented the sequence for 2 – deoxy - α - *D* – ribose molecules in form *A*.

In Figure 18 is presented the sequence structure of the nucleic acid molecule with 2 – deoxy - *D* – ribose molecules in form *B*.

Figure 16. Structure of the nucleic acid molecule with D – ribose molecules in form B

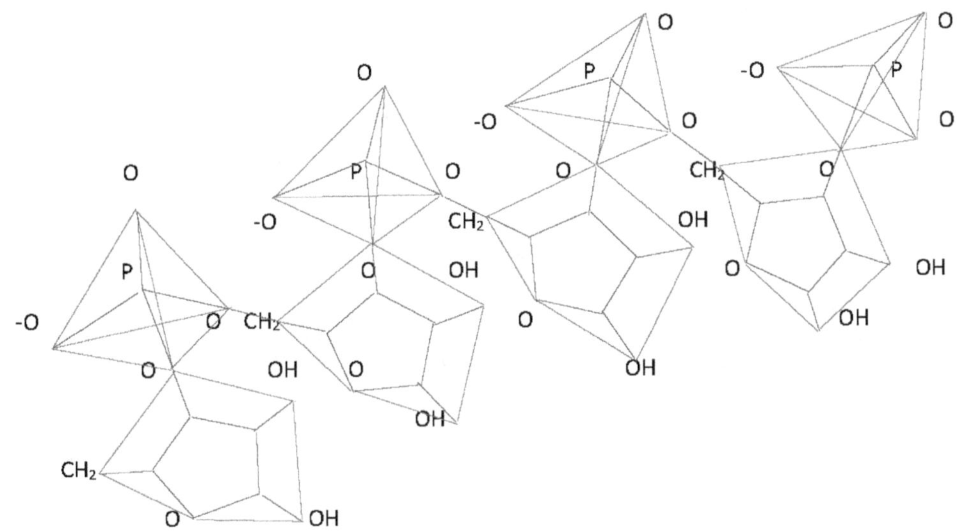

Figure 17. Structure of the nucleic acid molecule with 2 – deoxy - α - D – ribose molecules in form A

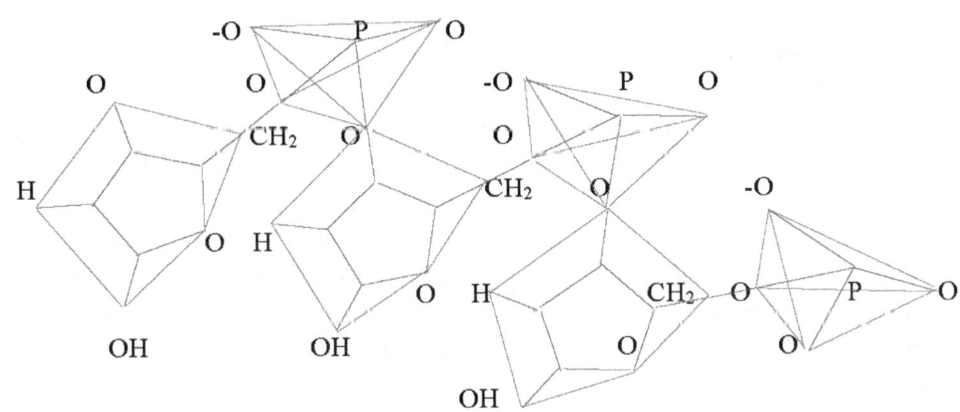

It follows from Figures 15 - 18 that the *D* - ribose molecules in both chains are not linearly distributed. Any sequence of corresponding points of *D* - ribose molecules in chains forms a curved line with rotation to the right side. However, in these sequences there is a fundamental difference. It is a consequence of the use in these two nucleic acid structures of two enantiomeric molecules of α - *D* - ribose: the α - *D* - ribose molecule in form *A* in the chain in Figure 15, the α - *D* - ribose

Figure 18. Structure of the nucleic acid molecule with 2 – deoxy - α - D – ribose molecules in form B

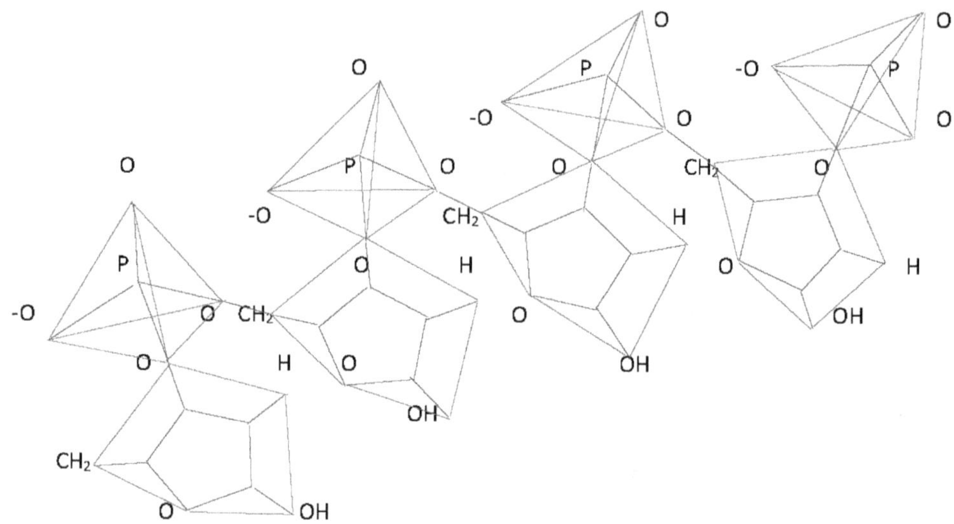

molecule in the *B* form in the chain in Figure 16. This difference consists in the fact that in the chain in Figure 15 the oxygen atom in the carbon ring of the α - *D* - ribose molecule is the extreme right in each α - *D* - ribose molecule. And in the chain in Figure 16 the oxygen atom in the carbon ring of the α - *D* - ribose molecule is the extreme left in the carbon ring of the molecule α - *D* - ribose. It can be argued that the nucleic acid structures in Figures 15 and 16 are enantiomeric nucleic acid structures, which are a consequence of the presence of two enantiomeric structures of α - *D* - ribose molecules in them. The existence of two enantiomeric nucleic acid structures was not noted before since details of the arrangement of atoms in the nucleic acid structures were not considered. Denote the structure of the nucleic acid in Figure 15 of the nucleic acid of the *EA*, and the structure of the nucleic acid in Figure 16 of the *EB*.

The projection of the nucleic acid molecules depicted in Figures 15 and 16 can be described mathematically. This description will also be valid for nucleic acids in which 2 - deoxy – α – D -ribose molecules are used instead of α – D - ribose molecules, since in these cases the only difference between 2 – deoxy – α – D - ribose molecules and α – D - ribose molecules is the replacement at the carbon atom $C_{(2)}$ hydroxyl group per hydrogen atom. This difference does not affect the results of mathematical analysis of the geometry of molecules.

Let us turn to a detailed picture of the connection of two α - *D* – ribose molecules of the form *B* in Figure 16 (Figure 19).

Figure 19. Two α - D – ribose molecules of the form B in the structure of the EB

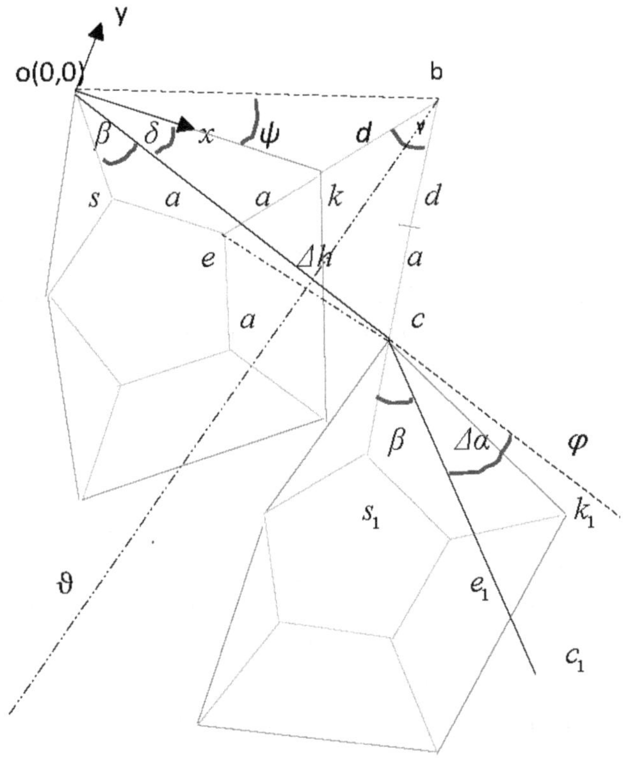

The vertex *b* in this figure corresponds to the phosphorus atom, and the two edges emanating from it correspond to the projections of two valence bonds of the atom phosphorus with two oxygen atoms. They are at some angle γ on the projection. This angle characterizes the location of the residual phosphoric acid in space with respect to the α - D - ribose molecule B at the site of their connection (vertex k*)*. It can be shown that the angle γ *is* linearly related to the angle $\Delta\alpha$ *be*tween two neighboring nuclei in the spiral chain. Indeed, in Figure 5.19, the smallest distance between two corresponding points of two neighboring α - D - ribose molecules B *in* the chain Δh *co*rresponds to the segment OC. *T*he smallest angle $\Delta\alpha$ *cha*racterizing the helical surface corresponds to the angle between the straight line that is the extension of the OC *se*gment and the next segment CC1 *be*tween two molecules of α - D - *r*ibose, originating from the vertex C. T*h*e angle $\Delta\alpha = \angle ocb - \beta$, where $\angle ocb = \angle s_1 c\varphi$, β *i*s the angle between segment OC *a*nd edge OS *i*n the initial molecule α - D – ribose. The same angle is repeated in the next molecule of α - D - ribose. The angle *β=54°-δ*, *whe*re 54° is the angle between edge OS a*nd* edge OK.

The angle $\angle obc = \angle obk + \gamma$, but $\angle obk = \pi - \angle okb - \psi, \angle okb = 126°$ by construction, therefore $\angle obk = 54° - \psi$.

On the other side it is

$$\angle ocb = \pi - \angle cob - \angle obc = \pi - (\delta + \psi) - \angle obk - \gamma = 126° - \delta - \gamma.$$

Therefore $\Delta\alpha = 126° - \delta - \gamma - (54° - \delta) = \dfrac{\pi}{5} - \gamma.$

An amazing result was obtained: the rotation angle between two molecules of α - *D* - ribose in the chain depends linearly only on the angle between the projections of the valence bonds in the phosphoric acid residue, that is, the orientation of the phosphoric acid residue in the space with respect to the α - *D* - ribose molecule,

$$\Delta\alpha = \frac{\pi}{5} - \gamma. \tag{5.2}$$

Concordantly with model of Watson and Crick (Watson, Crick, 1953 a, b) in period of nuclei chain included 10 nuclei. In this chain the angle $\Delta\alpha$=36°. Therefore in this case, according to (5.2), the angle γ *is* also equal to 36 degrees. If in the period of nuclei chain include 12 nuclei (see Ha, et. al., 2005), so $\Delta\alpha$=*30*°, and γ=*42*°.

It is possible to calculate the minimum distance between the corresponding points of neighboring α - D - ribose molecules in the nucleic chain Δh . It follows from Figure 5.19 that $\Delta h = \sqrt{y_c^2 + x_c^2}$.

Values y_c x_c are from the system,

$$4(d+a)^2 \cos^2 \frac{\pi - \gamma}{2} = (y_e - y_c)^2 + (x_e - x_c)^2, (d+a)^2 = (y_b - y_c)^2 + (x_b - x_c)^2, \tag{5.3}$$

where

$$x_e = a(1 + \cos 54°), y_e = -a \sin 54°, x_b = a(1 + 2\cos 54°) + d \sin 36°, y_b = d \cos 36°.$$

The first equation of system (5.3) can obtain from the equality of the square of the length of the segment *ec* to the sum of the squares of the differences of the coordinates of its vertices, on the one hand. On the other hand, the square of the length of the segment *ec* is the square of the doubled product of the cosine of the angle at the base of the isosceles triangle *ecb* by the length of the force side. The second equation of system (5.3) is obtained from the equality of the square of the side length *cb* to square of the sum of the segments *a* and *d*, on the one hand, and to the sum of the squares of the differences of the coordinates of its vertices, on the other hand. Solving system (5.3) for given *a*, *d*, one can find the value of Δh. To find the solution, the squares of the coordinate differences in the right - hand sides of the equations of system (5.3) are opened and the second equation is subtracted from the first. The linear connection of the coordinates of the vertex *c* is obtained. The equation of this connection is substituted into the second equation of system (5.3) and a quadratic equation is obtained for one of the coordinates of vertex *c*. From two roots of the equation, the root corresponding to Figure 19 is chosen. Thus, for each value of the angle γ for given the values of *a*, *d* are the coordinates of the vertex *c*.

It is easy to verify that by mirroring the image on Figure 19 with respect to the bisector $b\vartheta$ of the angle γ passing through the vertex *b*, one can obtain an image of two coupled nucleotides corresponding to another enantiomeric form of nucleic acid (Figure 15). In this way, the calculation of the parameters of the nucleic acid structure according to the system (5.3) corresponds directly to two enantiomeric forms of the nucleic acid structure. In the beginning, in order to find out the properties of the solutions of the system (5.3) corresponding to possible nucleic acid structures, it can numerically study these solutions, arbitrarily varying the value of the angle γ, which is the defining parameter of this system. Note that in this study it confines ourselves to analyzing one chain of a nucleic acid molecule, although, as is known (Watson, Crick, 1953 a, b; Frank – Kamenetskiy, 1988, 2004, 2010), a DNA molecule consists of two or three chains of the DNA molecules connected to each other by hydrogen bonds. Such a connection of chains can be carried out in the future, relying on the obtained regularities of one chain.

Table 1. The characteristic dimensions of the right and left nucleic acid helices

Γ degree	Δα degree	N	Δh Nm	R Nm
36	36	10	0,4677	0,7447
42	30	12	0,501	0,9575
64,5	7,5	48	0,607	4,6397
72	0	∞	0,6394	∞
84	12	30	0,6815	3,2557
90	18	20	0,7	2
102	30	12	0,7337	1,4
108	36	10	0,7639	1,2163

Since, the lengths of chemical bonds between carbon - carbon and carbon - oxygen are ≈ 0.15 nm, can will adopt a=0.15nm. The length of chemical bond phosphor – oxygen can adopt d = 0.18 nm. The results of calculating the characteristics of the nucleic acid structure from the equations of system (5.3) are shown in the graphs (Figure 20) and the numerical values of some parameters are given in Table 1.

Figure 20. Graphs of the change in the distance between nucleotides Δh, the angle of rotation between adjacent nucleotides $\Delta\alpha$, the number of nucleotides in the period N and the radius of the spiral $r = \dfrac{N\Delta h}{2\pi}$, depending on the angle γ in the projection between the valence bonds of the phosphorus atom

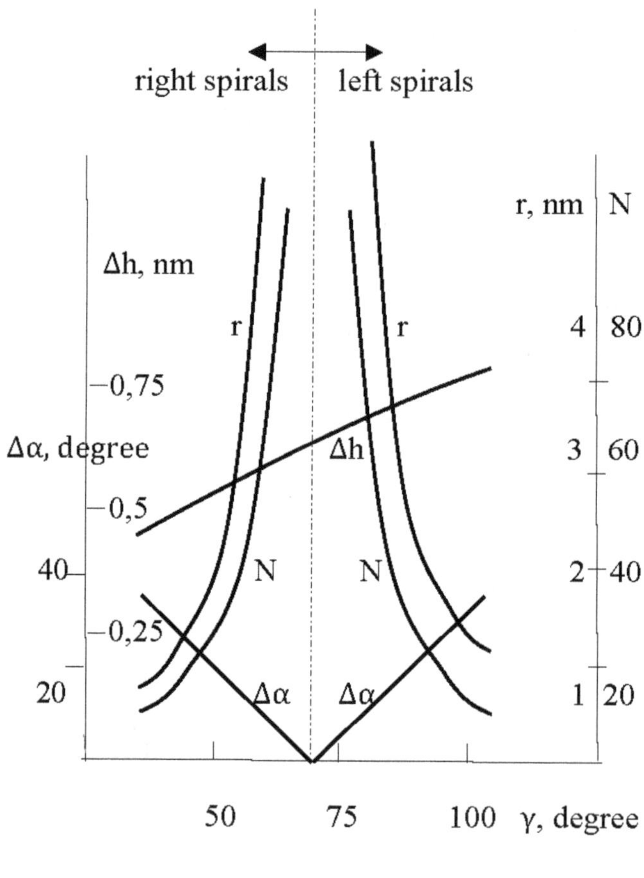

As can be seen from Figure 20 and Table 1 extension of the angle γ between the projections of the valence bonds of the phosphorus atom in the phosphoric acid residue from the angle of 36 degree to the plane of the carbon ring of the α - *D* - ribose molecule increases almost linearly the distance between the nucleotides in the nucleic acid. In this case, the radius of the nucleic acid helix increases nonlinearly asymptotically to infinity in the neighborhood of the value of the angle $\gamma = 72$ degrees. At a value γ of 72 degrees, the nucleic acid molecule is straightened out into a straight line. With further increase in angle γ, the nucleic acid helix begins to rotate to the left. In this range of values, increasing the angle γ also leads to an almost linear increase in the distance between nucleotides, but to a nonlinear decrease in the radius of the nucleic acid helix. This occurs up to an angle $\gamma = 110$ degrees, at which the spiral turns into a closed ring, since there is no rise of nucleotides with the help of phosphoric acid residues. In cases where the angle is greater than 72 degrees $\Delta\alpha = \gamma - 72$. In Figure 21, as an example, an image of two adjacent nucleotides at 90 degrees is presented.

Figure 21. Two nuclei of the nucleic acid structure of the EB for γ = 90 degrees

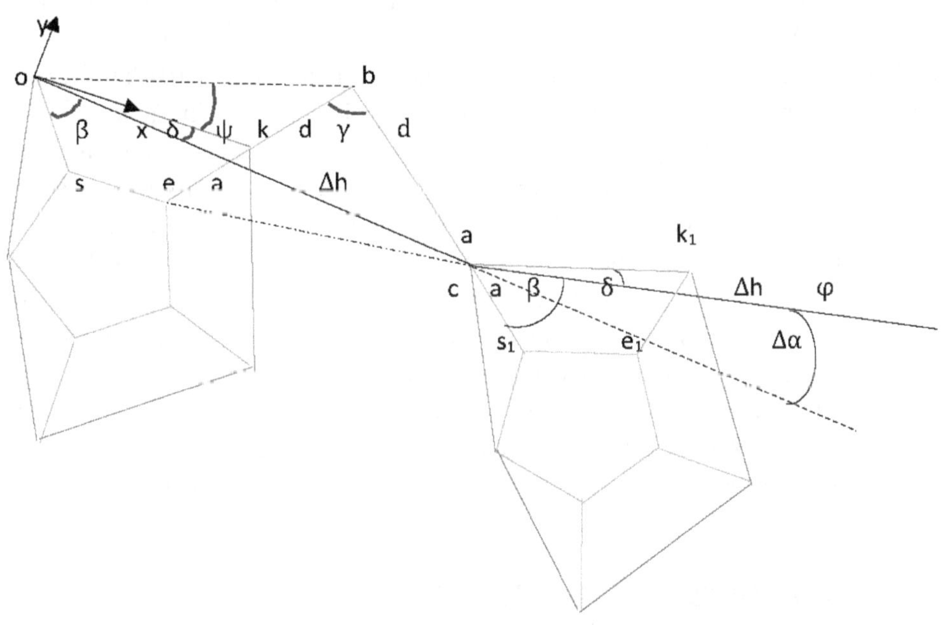

It can be seen from this figure that when γ is greater than 72 degrees, the angle $\Delta\alpha$ between two adjacent segments is laid to the left of the initial segment. This leads to the rotation of the spiral to the left.

Then from Table 1 can see that for $\gamma = 36$ degree the distance between nucleotides in structure nucleic acid $\Delta h = 0,4677$ nm. Where in the number of nucleotides in the period is 10 and the radius of the spiral $r = 0,7447$ nm. To this value of the radius should be added the size of the phosphoric acid residue, which is along the length of the valence bond d and the angle between the valence bonds of 110 degrees $\Delta r = \sin 55°\ d2 \approx 29$nm. Then the radius of the nucleic acid helix is $r + \Delta r \approx 1$ nm. The same nucleic acid radius was observed in experiments (Watson, Crick, 1953 a, b) for 10 nucleotides in the period. This form of nucleic acid structure was designated B. From Table 1 can see that for $\gamma = 42$ degree the distance between nucleotides in structure nucleic acid $\Delta h = 0,5$ nm. Wherein the number of nucleotides in the period is 12 and the radius of the spiral $r = 0,957$ nm. Then the radius of the nucleic acid helix is $r + \Delta r \approx 1,\ 35$ nm. These data roughly correspond to the form of the nucleic acid structure, designated A. It is also known from experiments the existence of a nucleic acid structure with 12 nucleotides in a period with left rotation (Ha, et al., 2005; Frank – Kamenetskiy, 2004). This form of nucleic acid structure was designated Z. From Table 1 it follows that the value of the angle $\gamma = 102$ degrees corresponds to this form. The existence of nucleic acid structures in the form of a closed ring has also been experimentally confirmed (Frank - Kamenetskiy, 2010). This form of nucleic acid structure corresponds to the angle $\gamma = 110$ degrees in Table 1.

Thus, from the variety of solutions of the system (5.3) corresponding to the possible nucleic acid structures, four forms of the nucleic acid structure were experimentally confirmed. This does not mean that other forms of nucleic acid structure that correspond to other solutions of system (5.3) do not exist. It should be assumed that these solutions have not yet been found in experiments, and in the future, when creating the appropriate conditions, they will be found. The received decisions should be considered also at consideration of forms of structures of nucleic acid consisting of three connected spirals (form H) (Frank - Kamenetskiy, 1988). In addition, must not forget that each solution of the system (5.3) corresponds simultaneously to two enantiomeric forms of the nucleic acid structure.

THE THREE–DIMENSIONAL MODEL OF THE NUCLEIC ACID MOLECULE

For clarity, it is convenient to have three - dimensional images of nucleic acids. They will be needed in particular in the next section when analyzing the interaction of nucleic acids. For this purpose, you can use the simplified three - dimensional images of amino acid molecules obtained. However, it should be remembered here that in reality the molecules of amino acids and nucleic acids have a higher dimension. When analyzing a chain of amino acid molecules, taking into account their higher dimension, in the previous section, we considered the projection of the chemical bond connecting the amino acid molecules with the phosphorus atom of the phosphoric acid residue onto the plane of the oxygen - carbon cycle of the amino acid molecule. Obviously, in this case, the chemical bond itself is located in a space of higher dimension. In the simplified 3D model (Chapter 4), the chemical bond between molecules is orthogonal to the plane of the oxygen - carbon cycle, so its projection onto the plane of the oxygen - carbon cycle is a point. With such a simplification, it would be impossible to obtain the analytical conditions for the existence of various forms of nucleic acids, which were found in the previous section, taking into account the highest dimension of space. Recognizing the importance of the obtained conditions, we nevertheless present a simplified three - dimensional image of nucleic acid due to its clarity.

A nucleic acid molecule is a chain of residues of phosphoric acid (a tetrahedron with a center) and sugar molecules (a prism with a pyramid). The chain (for molecules α – D - ribose (A) or and 2 – deoxy – α - *D* – ribose (A) of successively alternating sugar molecules and residues of phosphoric acid in a three - dimensional form is shown in Figure 22.

A nucleic acid molecule with sugar molecules $\alpha - D$ - ribose (B) or and 2 – deoxy – α - D – ribose (B) is shown in Figure 23.

At the junctions of the phosphoric acid residue with sugar, due to the separation of water molecules, an oxygen atom and the functional group CH_2 are formed (Figure 22, Figure 23). In addition, one of the four nitrogenous bases is attached to the $C_{(1)}$ carbon atom instead of the hydroxyl group at the $C_{(1)}$ carbon atom (Figure 22, Figure 23). The yellow color in Figure 22, Figure 23 indicates the covalent bond lines connecting the phosphoric acid residues and the sugar molecules.

Figures 22 and 23 represent two chiral forms of nucleic acids (two chiral forms of DNA and two chiral forms of RNA), since not only five - carbon sugar molecules participate in them in specular reflections, but also phosphoric acid residues are also in specular reflections. Let us consider in more detail one of these forms, presented, for example, in Figure 22. It is significant that the communication lines connecting the tetrahedron with the sugar molecule form, by repulsing the electron pairs, uniform straight segments perpendicular to the bases of the prism. In the center of the tetrahedron, the chemical bond lines form a kink, since in the center of the tetrahedron there are straight lines connecting the center with the vertices of the tetrahedron. Ultimately, this fracture with simultaneous movement along the vertical coordinate leads to the formation of a helix of a nucleic acid molecule. In the projection on the plane of the figure, the covalent bond segments form a broken line close to the circle (polygon). The radius of the circle describing this polygon will be determined by the angle between the projections on the plane of the chemical bonds of the phosphorus atom with the oxygen atoms at its vertices belonging to the polygon. The tetrahedron with the center (the residue of phosphoric acid), connecting with the top of the base of the prism, has a degree of freedom. It can be rotated relative to the connection between the vertex of the prism and the center of the tetrahedron at an arbitrary angle. Therefore, even in the projection on the plane perpendicularity to axes of rotation, the angle between the chemical bonds emanating from the center of the tetrahedron can be arbitrary (Figure 24).

Figure 22. A nucleic acid molecule with sugar molecules α – D - ribose (A) or and 2 – deoxy – α - D – ribose (A)

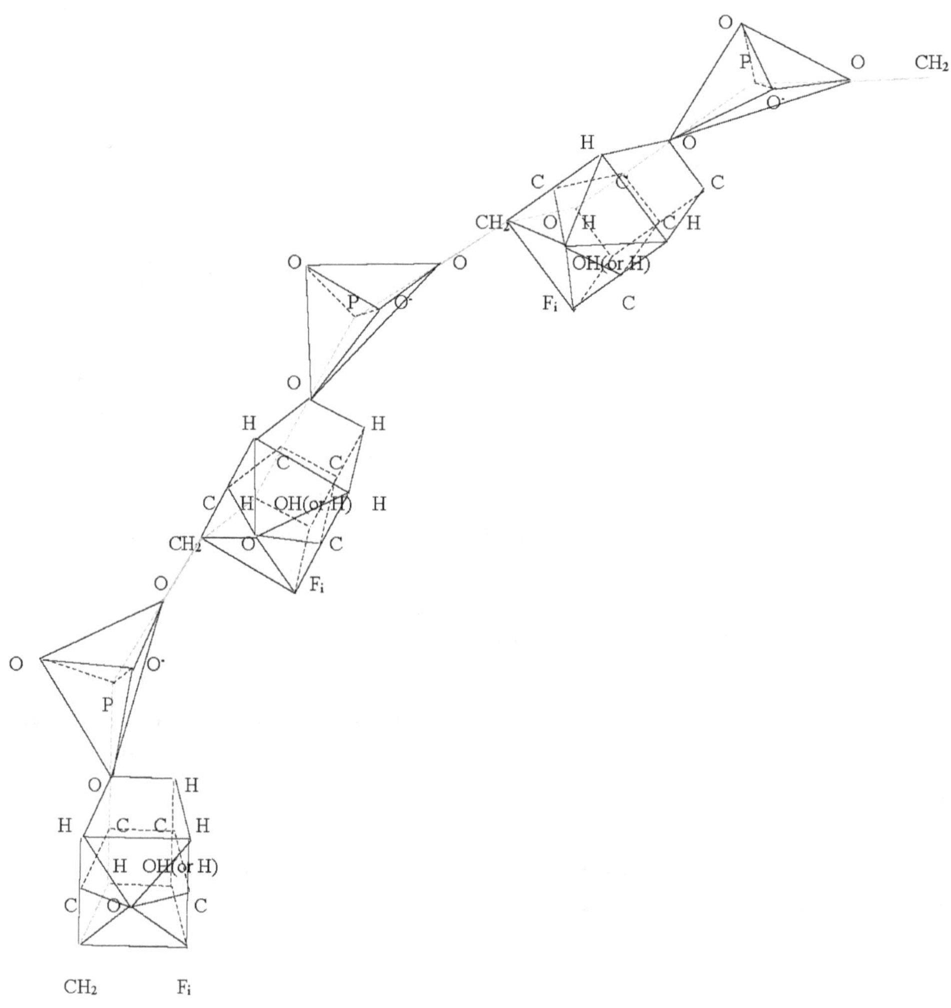

Figure 23. A nucleic acid molecule with sugar molecules α – D - ribose (B) or and 2 – deoxy – α - D – ribose (B)

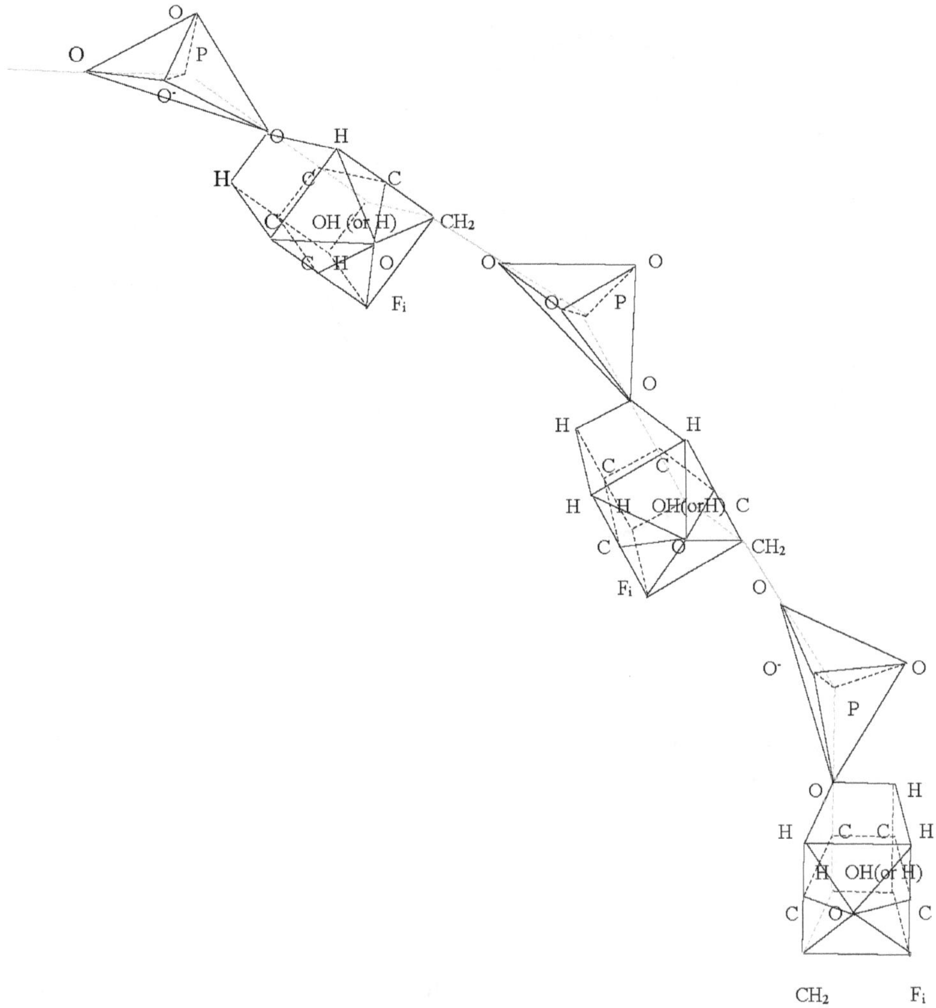

Let this angle be γ. Then the broken line between the centers of the tetrahedrons closes when the product of the number of these rotations n and the angle π - γ is equal to 2π. (On projection in Figure 24 edge of low bases of prism in visible due to the perpendicularity of the generators and bases.) You can find the period of the spiral, given that there is still movement along a line perpendicular to the projection plane. Thus, to determine the radius of a circle, we have the equality, $(\pi - \gamma)n=2\pi$.

Figure 24. The projection of spiral on the plane perpendicularity to axes of rotation

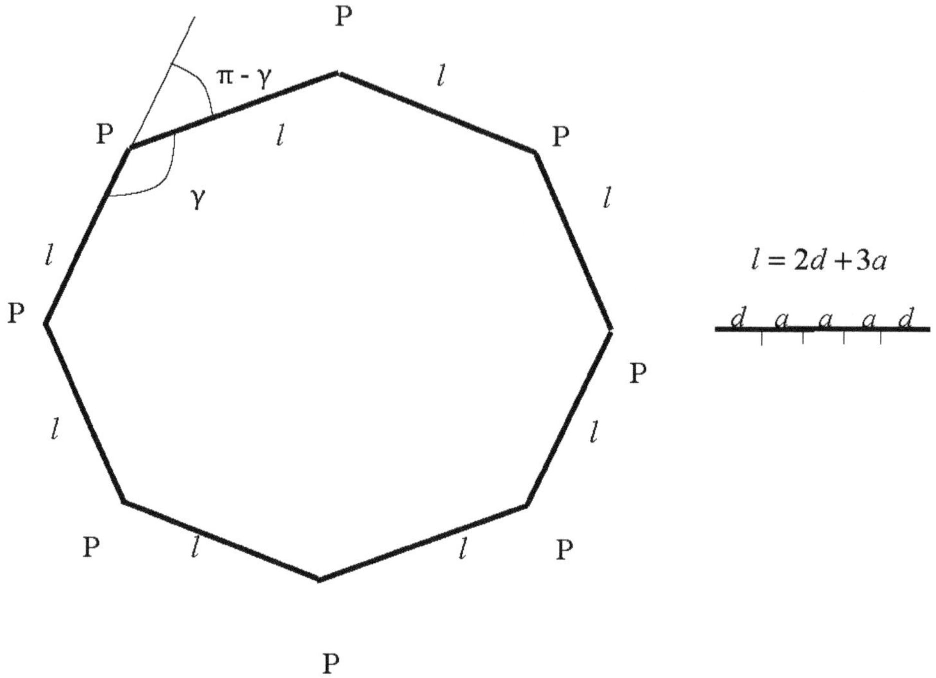

Since the number of turns equals the number of segments that make up a polygon, one has approximately the equality,

$$\frac{2\pi}{\pi - \gamma} = \frac{2\pi R}{l},$$

R is radius of the circle, and l is length of the segment between centers of tetrahedrons.

From the last equality can find the radius of the circle

$$R = \frac{l}{\pi - \gamma}.$$

From Figure 5.22 and Figure 5.24 it follows that l=3a+2d. Subject to the accepted values a and d can get

$R = \dfrac{0.81}{\pi - \gamma}$ nm. For example, in case γ=140° is R=*1*.16 nm.

This radius value is close to the experimental helix radius of nucleic acids measured by Watson and Crick. At this angle between the chemical bonds in the projection of the phosphoric acid residue, the number of nucleotides in the period $\dfrac{2\pi}{\pi - \gamma} = 9$. The number is in satisfactory agreement with the experimental number of nucleotides 10, given that the helix of the helix in the period is somewhat greater than the circumference of the same radius. This confirms the correctness of the three - dimensional model of the nucleic acid molecule.

The helix surface is formed due to the fact that in a tetrahedron with a center the valence bonds connecting the center with the vertices do not lie in the same plane (Zhizhin, 2019 b). Therefore, the valence bonds connecting the two sugar molecules do not lie together in the plane passing through the edge of the prism. The bond leaving the phosphorus atom, approximately 23 degrees, retreats from the plane in which the bond lies, which is part of the phosphorus atom. This causes a chain to shift along a coordinate perpendicular to the circle. The equation of the helix surface in this case can be written as $h=n\Delta h$, $\Delta h=l\tan(\Delta\alpha)$, $\Delta\alpha$ *is* the angle between the previous segment l *and* the subsequent segment l, n *is* the number of turns, h *is* the height of the period. Thus, Δh=*0*.343nm, h=*3*.43nm. This period height value is close to the value found in the Watson and Crick experiments (3.46 nm).

If the valence bond in the tetrahedron connecting its center with the next sugar molecule is rejected in the projection to the left of the valence bond connecting the center of the tetrahedron with the previous sugar molecule, i. e. it is $\gamma>180°$ the nucleic acid is a left - handed helix. If is $\gamma=180°$, then the nucleic acid has a linear appearance.

In ribonucleic acids, a helix return in one molecule is observed (Spirin, 2019). This is possible if the angle γ *is* variable, i. e. in the chain, the angle γ *in*creases to 180 degrees and the achievement of even larger values of this angle. This means that along the chain there is a sequential rotation of the tetrahedron around the valence bond connecting its center with the previous sugar molecule.

Obviously, a similar consideration can be made for the single - stranded nucleic acid molecule shown in Figure 23. The difference between Figure 22 and Figure 23 is that the chain helix in Figure 22 rotates to the right, the chain helix in Figure 23 rotates to the left. In this case, the residues of phosphoric acid at the corresponding positions in the chains are mirrored to each other. Since the phosphoric acid residue has a degree of freedom (rotation around the axis connecting it to the five - carbon sugar molecule), with a different initial state of the phosphoric acid residue in Figure

22, there will be another state of the phosphoric acid residue in Figure 23. This leads to the set of single - stranded nucleic acids (both DNA and RNA) is broken into two sets of chiral forms. Each form in one set contains a chiral form in another set. Moreover, in each set there are possible rotation of the spirals both in the right and in the left direction.

CONCLUSION

The currently known forms of the helical structure of nucleic acid molecules *A, B, Z*, and nucleic acid molecules in the form of a closed ring were experimentally found in the works of Watson, J.D., Crick, F.H.C., Ha, S.C., Frank - Kamenetskiy, M. and other scientists. However, the reasons for the formation of such a different form of nucleic acid molecules have not been clarified. In this chapter is proved that the main reason for the formation of nucleic acid molecules of different shapes is the different geometric realization of the chemical bond between the phosphoric acid residue and the *D* - ribose molecule, as objects of higher dimensionality. In other words, the residues of phosphoric acid can geometrically act as a bridge, connecting the molecules of *D* - ribose with each other. It is shown that the angle γ between the projections of the valence bonds of the phosphorus atom in the phosphoric acid residue on the plane of the carbon ring in the *D* - ribose molecule is a parameter characterizing the various geometric forms of chemical bonding of phosphoric acid residues with *D* - ribose molecules. The angle γ characterizes the orientation of the phosphoric acid residue in the three - dimensional space. This parameter can therefore be regarded as an internal degree of freedom of the structure of the nucleic acid molecule.

A mathematical model describing the geometric features of the chemical bond of these objects is constructed. Calculations on this model showed that the characteristics of known nucleic acid structures, determined experimentally, and calculated from the constructed model, are in good agreement with each other, while the model does not contain any empirical parameters. Calculations show that known nucleic acid structures are particular cases in the model and it can be considered that known examples of nucleic acid structure forms are experimental confirmation of the correctness of the mathematical model. However, according to the mathematical model, its solutions are many other nucleic acid structures that have not been experimentally discovered so far. One can hope that their experimental discovery is ahead.

The existence of two enantiomeric forms of nucleic acid molecules, connected with the existence of two enantiomeric forms of *D* - ribose molecules, is established.

The resulting variety of nucleic acid molecule structures is represented in the form of analytical dependencies, graphs and tables.

They should be taken into account when analyzing the synthesis of proteins, replication and translation of nucleic acid in the work of ribosomes (Stewart, McLachlan, 1975; Spirin, Gavrilova, 1971; Spirin, 1986).

The higher dimensionality of the constituent nucleic acid molecules, which allows to describe mathematically the structure of nucleic acid, requires reconsidering the issues of tight packing of nucleic acid molecules in cells, viruses and bacteria, provided that the nucleic acid chains necessary for the preservation and transfer of genetic information are complementary.

REFERENCES

Albert, B. (2002). *Molecular Biology of the Cell* (4th ed.). Garland Science.

Berg, J., Tymoczko, J., & Stryer, L. (2002). *Biochemistry*. W.H. Freeman & Co Ltd.

Brachet, J. (1941 – 1942). La détection histochimique et la microdosage des acides pentosenucleiques. *Enzymologia, 10*, 87–96.

Butler, J. M. (2005). *Forensic DNA Typing*. Gardners Books.

Cantor, C. R., & Schimmel, P. R. (1980). *Biophysical Chemistry. Part I. The Conformation of Biological Macromolecules*. W.H. Freeman & Co Ltd.

Caspersson, T., Landeström–Hyden, H., Aquilonius, L. (. (1941). Cytoplasmsnukleotide in eiweissproduzierenden Drüsenzellen. *Chromosoma, 2*, 111–131. doi:10.1007/BF00325956

Dahm, R. (2005). Friedrich Miescher and the discovery of DNA. *Developmental Biology, 278*(2), 274–288. doi:10.1016/j.ydbio.2004.11.028 PMID:15680349

Frank - Kamenetskiy, M. (1986). DNA structure: A simple solution to the stability of the double helix? *Nature, 324*.

Frank - Kamenetskiy, M. (1988). H – form DNA and the hairpin – triplex model. *Nature, 333*.

Frank – Kamenetskiy, M.D. (2004). *The DNA Age*. Moscow, Publishing house KDU.

Frank – Kamenetskiy. M.D. (2010). Queen of a living cell. From the structure of DNA to the biotechnological revolution. Moscow: Act – Press.

Gillespie, R. J. (1972). *Molecular Geometry*. Van Nostrand Reinhold Company.

Gillespie, R. J., & Hargittai, I. (1991). *The VSEPR Model of Molecular Geometry*. Allyn and Bacon.

Grunbaum, B. (1967). *Convex Polytopes*. Springer.

Ha, S. C., Lowenhaupt, K., Rich, A., Kim, Y.-G., & Kim, K. K. (2005). Cristal structure of a junction between B – DNA and Z – DNA revels two extruded bases. *Nature*, *437*(7062), 1183–1186. doi:10.1038/nature04088 PMID:16237447

Hershey, A., & Chase, M. (1952). Independent functions of viral protein and nucleic acid in growth of bacteriophage. *The Journal of General Physiology*, *36*(1), 39–56. doi:10.1085/jgp.36.1.39 PMID:12981234

Karapet'yants, M. X., & Drakin, S. I. (1994). *General and Inorganic Chemistry*. Chemistry.

Kosrapohr, (Ed). (2003). *Managing data mining technologies in organizations: techniques and applications*. Hershey, PA: Idea Group Publishing.

Lehninger, A. L. (1982). *Principles of Biochemistry* (Vol. 1-3). Worth Publisher, Inc.

Metzler, D. E. (1980). *Biochemistry. The Chemical Reactions in Living Cells* (Vol. 1-3). Academic Press.

Oswald, A. T., MacLeod, C. M., & McCarty, M. (1944). Studies on the Chemical Nature of the Substance Inducing Transformation of Pneumococcal Types: Induction of Transformation by a Desoxyribonucleic Acid Fraction Isolated from Pneumococcus Type III. *The Journal of Experimental Medicine*, *79*(2), 137–158. doi:10.1084/jem.79.2.137 PMID:19871359

Pasteur, L. (1960). *Selected works*. Publishing House of the Academy of Sciences of the USSR.

Poincare, A. (1895). Analysis situs. *J de e'Ecole Polytechnique.*, *1*, 1–121.

Ponnuswamy, P., & Gromiha, M. (1994). On the conformational stability of oligonucleotide duplexes and tRNA molecules. *Journal of Theoretical Biology*, *169*(4), 419–432. doi:10.1006/jtbi.1994.1163 PMID:7526075

Riemann, G. F. B. (1854). On the Hypotheses Which Lie at the Foundations of Geometry. Gottingen: Gottingen Observatory.

Rodger, J. A. (2015). Discovery of medical Big Data analytics: Improving the prediction of traumatic brain injury survival rates by data mining Patient Informatics Processing Software Hybrid Hadoop Hive. *Informatics in Medicine Unlocked*, *1*, 17–26. doi:10.1016/j.imu.2016.01.002

Spirin, A.C. (2019). *Molecular biology. Ribosome and protein biosynthesis. Moskow: Laboratory of knowledge.*

Spirin, A.S. (1986). *Molecular biology. Structure of the ribosome and protein biosynthesis*. Moscow: High school.

Spirin, A. S., & Gavrilova, L. P. (1971). *The Ribosome*. Science.

Stewart, M., & McLachlan, A. D. (1975). Fourteen actin-binding sites on tropomyosin? *Nature*, *257*(5524), 331–333. doi:10.1038/257331a0 PMID:1161036

Watson, J. D., & Crick, F. H. C. (1953a). Molecular structure of nucleic acids. *Nature*, *171*(4356), 738–740. doi:10.1038/171737a0 PMID:13054693

Watson, J. D., & Crick, F. H. C. (1953b). Genetical implications of the structure of deoxyribose nucleic acid. *Nature*, *171*(4361), 964–967. doi:10.1038/171964b0 PMID:13063483

Zhizhin, G. V. (2014a). On the higher dimensions in nature. *Biosphere*, *6*(4), 313–318.

Zhizhin, G. V. (2014b). *World 4D*. Polytechnic Service.

Zhizhin, G. V. (2015a). The dimensions of supramolecular compounds. *Biosphere*, *7*(2), 149–154.

Zhizhin, G. V. (2015b). Dimensions of compounds in supramolecular chemistry. *Int. J. Chem. Modeling*, *7*(2), 215–238.

Zhizhin, G. V. (2016a). From multidimensional phase spaces to multidimensionality of space in Nature. *Biosphere*, *8*(3), 258–267.

Zhizhin, G. V. (2016b). The structures, topological and functional dimension of biomolecules. *International Journal of Chemoinformatics and Chemical Engineering*, *5*(2), 44–58. doi:10.4018/IJCCE.2016070104

Zhizhin, G. V. (2018). *Chemical Compound Structures and the Higher Dimension of Molecules: emerging Research and Opportunities*. IGI Global. doi:10.4018/978-1-5225-4108-0

Zhizhin, G. V. (2019 a). *The Geometry of Higher - Dimensional Polytopes*. IGI Global. doi:10.4018/978-1-5225-6968-8

Zhizhin, G. V. (2019 b). *Attractors and higher dimensions in population and molecular biology: emerging research and opportunities*. IGI Global. doi:10.4018/978-1-5225-9651-6

Zhizhin, G. V. (2019c). The Metric Structure of Nucleic Acids and Higher Dimension of Their Constituents. *J. Chemoinformatics and Chemical Engineering*, *7*(2), 1–15.

Zhizhin, G. V. (2021). The Structures of DNA Taking into Account the Higher Dimensional of its Components. In Kosrupohr (Ed.) Encyclopedia of Organizational Knowledge, Administration, and Technologies. Hershey, PA: IGI Global.

Zhizhin, G. V. (2022). *The Classes of Polytopes Higher Dimensions in Physics, Chemistry and Biology*. IGI Global.

Zhizhin, G. V., & Diudea, M. V. (2016). Space of nanoworld. In M. V. Putz & C. M. Marius (Eds.), *Sustainable nanosystems, development, properties, and applications* (pp. 221–245). IGI Global.

Zhizhin, G. V., Khalaj, Z., & Diudea, M. V. (2016). Geometrical and topology dimensions of the diamond. In A. R. Ashrafi & M. V. Diudea (Eds.), *Distance, symmetry and topology in carbon nanomaterials* (pp. 167–188). Springer. doi:10.1007/978-3-319-31584-3_12

Chapter 6
Interaction of Nucleic Acids in Higher Dimensional Space and Its Models

ABSTRACT

On the basis of images of higher dimension of nucleic acids, the properties of the interaction of nucleic acids with each other have been studied. It was found that the region of interaction of nucleic acids is a cross - polytope of dimension 13 (the polytope of hereditary information). Its structure indicates that the interaction of nucleic acids has a hidden order, according to which the number of standard amino acids exactly corresponds to the number of possible locations of nitrogenous bases in the polytope of hereditary information. In accordance with the law of conservation of incidence (chapter two), the total flow of information transmitted by the polytope of hereditary information at the local location of the polytope of hereditary information on the chain of double-stranded nucleic acids is calculated. The possibility of creating polytopes of hereditary information in three and four helical nucleic acids is considered.

INTRODUCTION

Continuing research on the dimension of chemical compounds in chemistry, physics and biology, conducted by the author for a number of years (Zhizhin, 2014 a, b; Zhizhin, 2015 a, b; Zhizhin, Diudea, 2016; Zhizhin, Khalaj, Diudea, 2016; Zhizhin, 2018), it was established in the Chapter 5 nucleic acids (DNA, RNA) are a chain of sugar molecules with five carbon atoms in a closed circuit, connected to each other by phosphoric acid residues. In this regard, it is significant that both phosphoric acid residues and sugar molecules have a higher dimension: 4 and 12, respectively (Zhizhin, 2019 c, 2021 a, b, c). This fact was not taken into account neither by the

DOI: 10.4018/979-8-3693-0588-1.ch006

discoverers of nucleic acids, nor by all their subsequent researchers up to the present time (Micher, 1869; MacLeod, McCarthy, 1944; Hershey, Chase, 1952; Watson, Crick, 1953 a, b; Cantor, Schimmel, 1980; Albert, et al, 2002; Butler, 2005; Frank-Kamenetsky, 1986, 1988, 2010; Nelson, Cox, 2022). Apparently, this fact seemed to them unworthy of attention.

However, as shown in the chapter 5, it is enough to take this fact into account and it is natural to take a linear continuation of the valence bonds from the carbon atom in the CH_2 molecule towards the carbon atom in the closed sugar cycle and towards the nearest oxygen atom in the phosphoric acid residue (Figures 15 - 18), as soon as the sugar molecules line up in a spiral. The basis for this is purely geometric regularities in the sugar molecule and in the phosphoric acid residue. At the same time, it was found that the type of nucleic acid depends on the angle between the valence bonds coming from the carbon atoms of the closed cycle, the sugar molecules closest to each other (Zhizhin, 2019 c, 2021 a, b, c). Therefore, there is no need to postulate the helical form of the nucleic acid chain, since it is a consequence of geometric patterns between valence bonds sugar molecules, taking into account their higher dimension.

It is important that the characteristic sizes of different types of nucleic acids obtained using this approach practically coincide with the experimental sizes of the corresponding nucleic acids (Zhizhin, 2019 c, 2021 a, b, c). It is important for the future that one of the hydroxyl groups of the sugar molecule is replaced by some functional group F_i (A - adenine, G -guanine, C - cytosine, T - thymine, U - uracil) which is a nitrogenous base (Figures 22, 23). It is considered with sufficient accuracy that in nitrogenous bases the atoms are located in a two-dimensional plane. However, the geometric characteristics of these planes, which determine their orientation in multidimensional space, are still unknown. In addition, non-canonical nitrogenous base pairs are known that can link nucleic acids. Their possible use will violate the one-to-one correspondence (complementarity) of nitrogenous bases in pairs and may lead to deviation of the planes of nitrogenous bases in a pair from each other.

It is also important that the three-dimensional view of nucleic acids used in this work for greater clarity is based on the preliminary construction of nucleic acids of the highest dimension and, in a simplified, mathematically justified manner, takes into account all the features of the geometry of the highest dimension. Models of double-stranded nucleic acids take into account that the single-stranded nucleic acids forming them are directed towards each other. Therefore, they are described using polytopes of a special type: cross-polytopes (Zhizhin, 2019 a). Three-dimensional images of nucleic acids, used earlier, did not take into account the features of their images in a space of higher dimension. Therefore, it should be considered that they do not reflect the real structure of nucleic acids.

TYPES OF INTERACTION OF NUCLEIC ACIDS

Nitrogenous bases attached to sugar molecules, due to the hydrogen interaction of the bases, provide the possibility of interaction of nucleic acids with each other, forming double-stranded nucleic acids (double-stranded DNA and double-stranded RNA). This reveals the possibility of the formation in DNA of the following pairs of nitrogenous bases linked by a hydrogen bond A: T, T: A, G: C, C: G (the colon indicates a non-covalent bond).

In double-stranded RNA pairs, thymine is replaced by uracil. These pairs of nitrogenous bases are called canonical (Watson, Crick, 1953 a, b). The existence of these pairs of nitrogenous bases is ensured by the active role of nitrogen atoms in them. In this case, two parallel bonds appear between thymine and adenine (Figure 1).

Figure 1. Two hydrogen bonds between thymine and adenine

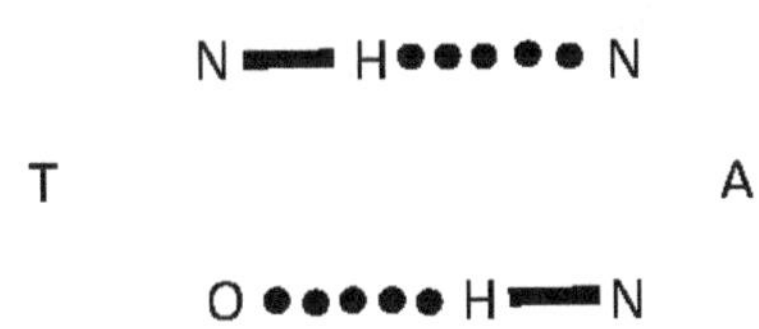

And there are three parallel bonds between cytosine and guanine (Figure 6.2)

Figure 2. Three hydrogen bonds between cytosine and guanine

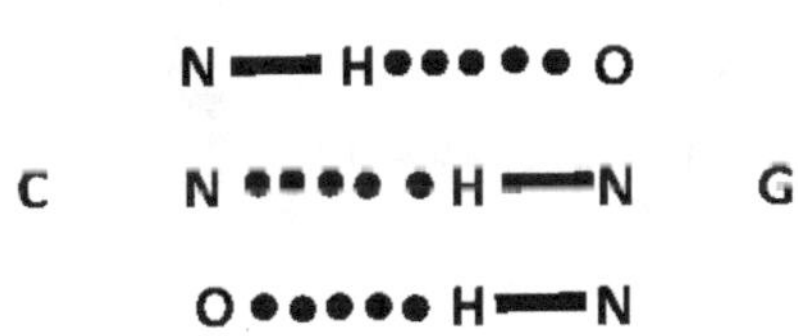

In addition, non-canonical bonds can also arise between nitrogenous bases (Spirin, 2019).

Each of these bonds has two parallel hydrogen bonds. Hydrogen bonds between nitrogenous bases can be single.

Obviously, two or three parallel hydrogen bonds define a plane coinciding with the plane of nitrogenous bases. However, in the case of a single hydrogen bond, the planes of the bound nitrogenous bases may not coincide with each other.

Non-canonical pairs of nitrogenous bases of a different type, the so-called Hoogsteen purine-purine interactions with non-parallel hydrogen bonds, can occur at the edges of the double-stranded regions (Figure 3 a, b and Figure 4 a, b)

A feature of double-stranded RNA sections is the formation of sections of single-stranded RNA that do not fit into the double helix and form loops on the side of the helix.

The interaction of nucleic acids ensures the transfer of information about the types and arrangement of nitrogenous bases in the original single-stranded DNA of another DNA helix in double-stranded DNA with the accuracy of matching (complementarity) of nitrogenous bases in Watson-Crick pairs (replication). Double-stranded DNA unwinds and stores this information on the second strand, which is then used to obtain RNA containing (taking into account the reverse transition) information about nitrogenous bases that was contained in the original single-stranded DNA.

Figure 3. Base pairing in triplex DNA
*a) T = A * T, b) C = G * C⁺*

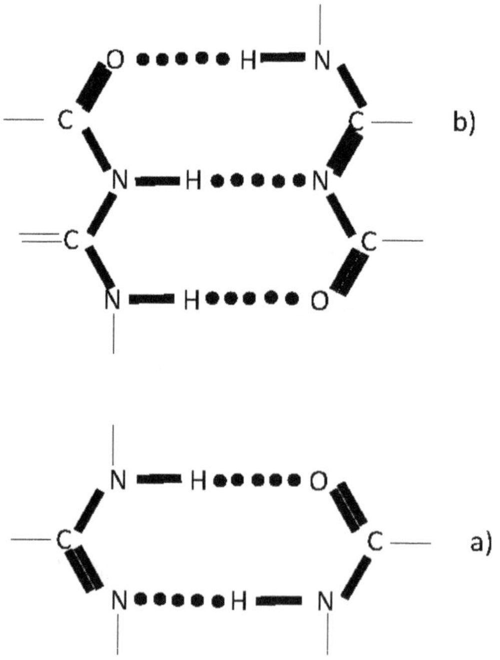

Figure 4. Non-canonical pairs of nitrogenous bases involving the nitrogen atom of the purine ring
(Spirin, 2019)
a) U: A, b) G: G

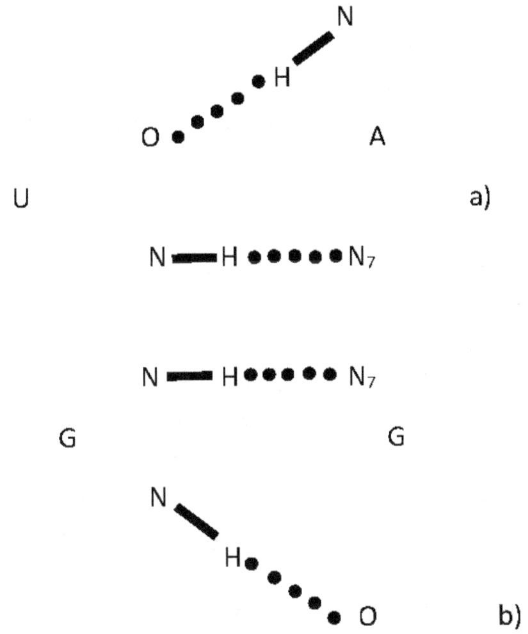

When genes are expressed, RNA molecules act as intermediaries, carrying information encoded in DNA that is determined to the ribosome. Process to ribosome the amino acid sequence of a functional protein. Given that eukaryotic DNA is mainly located in the nucleus, and protein synthesis occurs in the cytoplasm on ribosomes, genetic information from the nucleus to the cytoplasm should be transferred not by DNA, but by other molecules. These molecules are RNA molecules. They are located in the cell nucleus and in the cytoplasm, and an increase in protein synthesis is accompanied by an increase in the content of RNA in the cytoplasm and an increase in the rate of its turnover. In 1961, François Jacob and Jacques Monod (1961) provided a correct picture of this process and coined the term “messenger RNA” (mRNA) for those cellular RNAs that carry genetic information from DNA to ribosomes. The process of mRNA formation on a DNA template is called transcription. Once mRNA reaches the ribosome, it itself becomes a template that defines specific amino acid sequences in polypeptide chains. One mRNA molecule can encode one or more polypeptide chains. If mRNA encodes only one polypeptide, then it is called monocistronic, if two or more polypeptides, then it is called polycistronic. A cistron can be identified with a gene. However, mRNAs transcribed from DNA are always slightly longer than necessary simply to encode the polypeptide sequence (or sequences). An additional non-coding RNA sequence contains regions that regulate protein synthesis.

Messenger RNA is only one of the classes of cellular RNA. Another type of RNA is transfer RNA. These are adapter molecules that take part in protein synthesis, covalently attaching an amino acid at one end. Each tRNA binds complementary to mRNA so that the amino acids are inserted into the growing polypeptide chain in the correct order. Ribosomal RNAs are components of ribosomes. There are also many different types of RNA with specialized functions, including those with enzymatic activity (ribozymes).

Paired chains in all types of RNA and DNA are antiparallel.

By the second half of the 1950s, it was established that protein synthesis in living cells is carried out by ribonucleoprotein particles - ribosomes. The main component of ribosomes is ribosomal RNA, it consists of transfer RNA that transports selected amino acids to messenger RNA, which is synthesized and edited as a result of RNA processing. This is the third type of nucleic acid interaction. It was found that due to the interaction of side groups (nitrogenous bases) with each other, a single - stranded RNA polymer folds onto itself, forming secondary and tertiary structures (Spirin, 2009). The formation of these structures also occurs due to complementary interactions result of the interaction of nucleic acids, the central dogma of molecular biology is fulfilled: the flow of hereditary information is transmitted from DNA to RNA and further to protein. At present, in all types of interaction of nucleic acids, it is believed that nitrogenous bases are linearly sequential along the polymeric chain of acids. But since there are four traditional types of nitrogenous bases in each chain, and there are 20 types of amino acids in proteins, it has come to be recognized that each amino acid corresponds to three nitrogenous bases. This assumption is based on the concept of the genetic code, the deciphering of which establishes these correspondences (Gamov, 1954; Crick, 1959). However, in recent works (Zhizhin, 2019 a, b, c, 2020 a, b; 2021 a, b, c) it was found that the region of interaction of nucleic acids is not a linear section of the chain, but a polytope of higher dimension (a polytope of hereditary information), at the tops of which nitrogenous bases are located. In this regard, the question of the interaction of nucleic acids requires additional consideration.

POLYTOPES WITH ANTIPARALLEL EDGES

In single - stranded and double - stranded nucleic acids (RNA, DNA), the constituents of acids (residues of phosphoric acid and sugar molecules) interact with each other

(Watson, Crick, 1953a, b; Spirin, Gavrilova, 1971; Frank – Kamenetskiy, 1986, 1988). Phosphoric acid residues are connected by divalent metal ions, mainly magnesium ions, due to the interaction of negative charges of phosphoric acid residues with positive charges ions (Spirin, Gavrilova, 1971). This interaction is essential for the stability of nucleic acid structures, especially in the ribosomes. Sugar molecules interact with each other due to the hydrogen bond between the nitrogenous bases attached to the sugar molecules. The constituents of nucleic acids being geometric forms interact with each other to form new geometric forms - new polytopes. However, it is not known how flat nitrogenous bases are oriented in space, whether their orientation depends on the type of nitrogenous base. Currently there is no information on this. There is also no information on how exactly the metal ions are located, connecting the phosphoric acid residues. It should be remembered here that the adopted three - dimensional model of the components and the nucleic acid molecule itself is only a model for visual perception. As it was shown earlier, the phosphoric acid residue is a polytope of dimension 4, and the sugar molecule has a dimension of 12.

The movement of triangles along a helix leads to the formation of polytopes with antiparallel edges. Consider an arbitrary triangle *ABC* on the plane. Choose some point O' on the plane outside the triangle to his left. Let this point be the base of the axis of the helix passing through the triangle. Rotate the *ABC* triangle 180 degrees to the right, moving it up the helix, parallel to the initial plane. In the projection on the plane, both triangles *ABC* and the displaced triangle $A'B'C'$ will be located as shown in Figure 5.

Figure 5. Polytopes of dimension three with anti – parallel edges

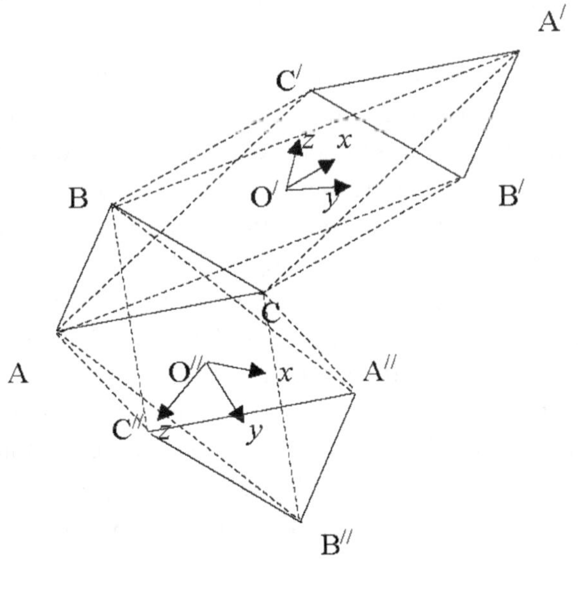

It is easy to see that the edges of the triangle *ABC* and $A'B'C'$ are antiparallel. It can now connect in space the vertices of the triangle *ABC* with the vertices of the triangle $A'B'C'$ so that there are no connections of the vertices with the same letters. In a projection on the plane the connections are represented by dotted segments. It can be seen that the connecting segments also break up into pairs of antiparallel segments. Let us now verify that the image $ABCA'B'C'$ in Figure 5, along with the dotted segments, is a projection of a three - dimensional convex polytope. We use the Euler – Poincaré equation (5.1) (Poincaré, 1895) for this aim.

The shape $ABCA'B'C'$ in Figure 65 has 6 vertices, 12 edges, 8 triangular faces (rectangles are not faces by construction, since connecting, for example, vertex *A* with vertices B', C' it turns out to be exactly the triangle $AB'C'$). Substituting these values of elements of different dimensions into equation (5.1), can find 6 – 12 +8 = 2, i. e. the Euler – Poincaré equation holds in this case for $n = 3$. This proves that the resulting figure is a convex polytope of dimension 3 (if the figure were not convex, the Euler – Poincaré equation would be violated). The point O' in Figure 5 coincides with the center of the three - dimensional figure $ABCA'B'C'$ as diagonal figures pass through it and point O' located on the axis of the helix. Point O' can be considered as the origin of three - dimensional space. Coordinate axes *x, y, z* in this direction emanates from directions AA', BB', CC' respectively. Three pairs of these axes define the coordinate planes of the space of this shape. Choose some point O'' on the plane outside the triangle below it. Let this point be the base of the axis of the helix passing through the triangle. Rotate the *ABC* triangle 180 degrees to the left, moving it up the helix, parallel to the initial plane. In the projection on the plane, both triangles *ABC* and the displaced triangle $A''B''C''$ will be located as shown in Figure 6.5.

It is easy to see that the edges of the triangle *ABC* and $A''B''C''$ are antiparallel. It can now connect in space the vertices of the triangle *ABC* with the vertices of the triangle $A''B''C''$ so that there are no connections of the vertices with the same letters. In a projection on the plane the connection are represented by dotted segments. It can be seen that the connecting segments also break up into pairs of antiparallel segments. Let us now verify that the image $ABCA'B'C'$ in Figure 6.5, along with the dotted segments, is a projection of a three - dimensional convex polytope. Obviously, a shape $ABCA''B''C''$ has as many vertices, edges, and flat faces as a shape $ABCA'B'C'$. Therefore, it satisfies the Euler – Poincaré equation (5.1) with dimensionality $n = 3$, i. e. it is a convex three - dimensional polyhedron. The point O'' in Figure 5 coincides with the center of the three - dimensional figure $ABCA''B''C''$ as diagonal figures pass through it and point O'' located on the axis of the helix. Point O' can be considered as the origin of three - dimensional space. Coordinate axes *x, y, z* in this direction emanates from directions AA'', BB'', CC'' respectively. The order of the coordinate axes *x, y, z* in the figures $ABCA'B'C'$ and $ABCA''B''C''$, as can be seen from

Figure 5, coincide. This suggests that both figures $ABCA'B'C'$ and $ABCA''B''C''$are topologically one and same figure - the wrong octahedron.

Interestingly, to transform an arbitrary tetrahedron $ABCD$ into a tetrahedron $A'B'C'D'$with anti-parallel edges, it is not enough to rotate it along a helix by 180 degrees. To do this, you must turn the helix together with the tetrahedron and move the tetrahedron along the helix in the opposite direction, also rotating it 180 degrees. In the initial state, the tetrahedron on the initial helix and the tetrahedron on the reversed helix, after its rotation by 180 degrees, will have anti -parallel edges. Both tetrahedrons can $ABCD$ and $A'B'C'D'$ be present by the Figure 6 for rotation to the right. On Figure 66 the point O' is the point of rotation.

Figure 6. Polytope of dimension four with anti-parallel edges

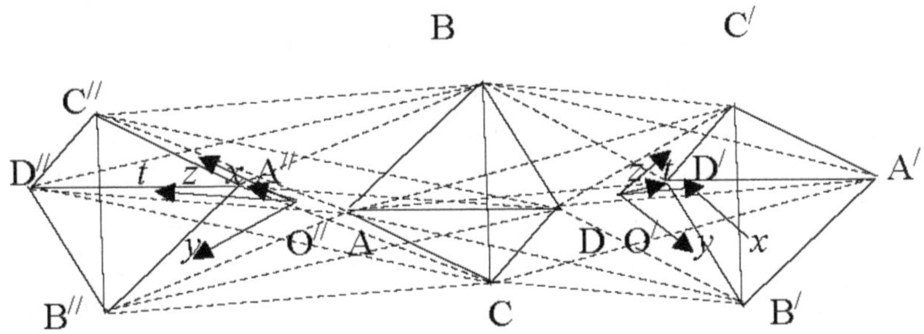

Now connect the vertices of the tetrahedrons so that the connecting edges (dotted segments) do not have the same letters. The resulting figure (along with dotted edges) has 8 vertices ($f_0 = 8$), 24 edges ($f_1 = 24$), 24 triangular faces ($f_2 = 24$), and 8 tetrahedrons ($f_3 = 8$). Substituting these values into equation (5.1), can find 8 - 24 + 24 - 8 = 0. Consequently, the Euler – Poincaré equation is satisfied in this case for $n = 4$. Thus, the polytope $ABCD\ A'B'C'D'$ in Figure 6.6 has dimension 4. It is easy to see (Zhizhin, 2019 a) that this is 4 - cross – polytope (Grunbaum, 1967; Zhizhin, 2014 b). The point O' in Figure 6 coincides with the center of the fourth - dimensional figure $ABCD\ A'B'C'D'$ as diagonal figures pass through it and point O' located on the axis of the helix. Point O' can be considered as the origin of fourth - dimensional space. Coordinate axes x, y, z, t in this direction emanates from directions AA', BB', CC', DD' respectively. Six pairs of these axes define the coordinate planes of the space of this shape.

Choose some point O'' on the plane outside the tetrahedron $ABCD$ below it. Let this point be the base of the axis of the helix passing through the tetrahedron. Rotate the $ABCD$ tetrahedron 180 degrees to the left, moving it up the helix, parallel to the initial plane. In the projection on the plane, both tetrahedrons *ABCD* and the displaced tetrahedron $A^{//}B^{//}C^{//}D^{//}$ will be located as shown in Figure 6. It is easy to see that the edges of the tetrahedrons *ABCD* and $A^{//}B^{//}C^{//}D^{//}$ are antiparallel. It can now connect in space the vertices of the tetrahedron *ABCD* with the vertices of the tetrahedron $A^{//}B^{//}C^{//}D^{//}$ so that there are no connections of the vertices with the same letters. In a projection on the plane the connection is represented by dotted segments. It can be seen that the connecting segments also break up into pairs of antiparallel segments. Let us now verify that the image *ABCD* $A^{//}B^{//}C^{//}D^{//}$ in Figure 6, along with the dotted segments, is a projection of a fourth - dimensional convex polytope. Obviously, a shape *ABCD* $A^{//}B^{//}C^{//}D^{//}$ has as many vertices, edges, and flat faces as a shape *ABCD* $A^{/}B^{/}C^{/}D^{/}$. Therefore, it satisfies the Euler – Poincaré equation (5.1) with dimensionality $n = 4$. Thus, the polytope *ABCD* $A^{//}B^{//}C^{//}D^{///}$ in Figure 6 has dimension 4. It is easy to see (Zizhin, 2019 a) that this is 4 - cross - polytope.

The point $O^{//}$ in Figure 6.6 coincides with the center of the 4 – cross - polytope as diagonal figures pass through it and point $O^{//}$ located on the axis of the helix. Point $O^{//}$ can be considered as the origin of fourth - dimensional space. Coordinate axes *x, y, z, t* in this direction emanates from directions $AA^{//}, BB^{//}, CC^{//}, DD^{//}$ respectively. From Figure 6.6 it follows that the sequence of alternation of coordinate axes *x, y, z, t* in a 4 - cross – polytope *ABCD* $A^{/}B^{/}C^{/}D^{/}$ differs from the sequence of alternation of coordinate axes *x, y, z, t* in a 4 - cross – polytope *ABCD* $A^{//}B^{//}C^{//}D^{//}$. Thus, a surprising fact emerged: the figures *ABCD* $A^{/}B^{/}C^{/}D^{/}$ and *ABCD* $A^{//}B^{//}C^{//}D^{//}$, being 4 - cross - polytopes, are topologically different from each other. It is impossible to move from one of them to another by continuous transformation since they have a different order of alternation of vertices.

SIMPLIFIED 3D IMAGE OF A DOUBLE-STRANDED NUCLEIC ACID

When analyzing the interaction of nucleic acids, simplified ideas were established about the interaction of nitrogenous bases connected to sugar molecules located in different helices of a double-stranded nucleic acid molecule. Assuming that each nitrogenous base is located in a two-dimensional plane, the interacting nitrogenous bases are also considered to be located in the same two-dimensional plane. Moreover, how this plane is located relative to the other parts of the interacting molecules remains unknown. The situation is sharply complicated after the establishment of the highest dimension of the components of nucleic acid. Given that geometric representations play an extremely important role in the analysis of the interaction of spatial objects and obtaining scientifically based conclusions about these interactions, it is necessary to take into account the different dimensions of the objects participating in the interaction. This is primarily necessary for a correct assessment of the patterns of transmission of hereditary information, which will be considered in the next chapter.

Figure 7. Double–stranded nucleic acid molecule with simplified images of sugar molecules

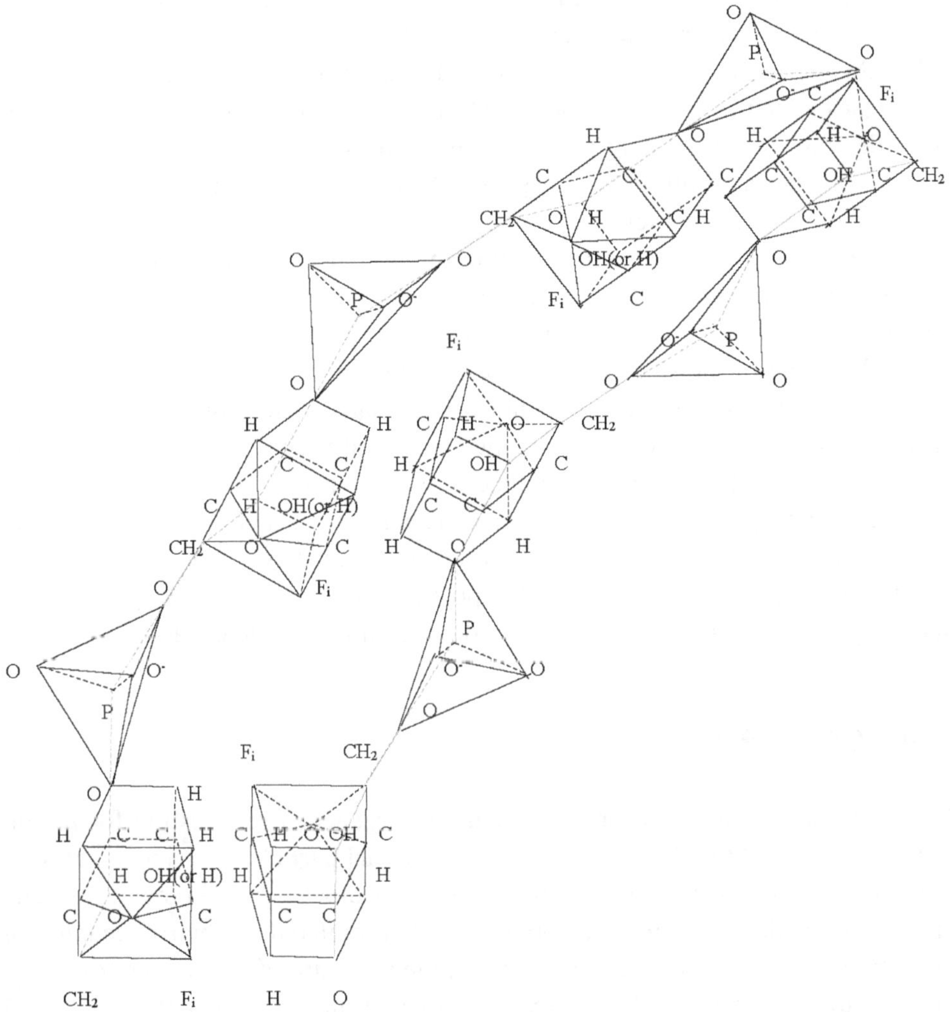

In chapter 3, it was proved that the dimension of the sugar molecule, which includes five carbon atoms, is 12 and a simplified three-dimensional image of this molecule is given (Figure 10 of chapter 4). It was obtained in this chapter from an image of higher dimension, provided that the lengths of the chemical bonds carbon

- carbon and carbon - oxygen are close (which is quite justified) and the assumption of a symmetrical arrangement of chemical bonds coming from the carbon atoms of the carbon cycle, relative to the plane of the carbon cycle (Zhizhin, 2019 a). The corresponding patterns of nucleic acids in right- and left-handed helices are presented in Chapter 5 (Figure 22, Figure 23). The vertices of the sugar molecule, indicated in these figures by the symbol F_i, represent the points of attachment of nitrogenous bases to sugar molecules. How the planes of nitrogenous bases can be located relative to sugar molecules is still unclear.

For this definition, it is necessary to add to each helix a second helix with its own sugar molecules. Figure 7 shows 3D images of double-stranded nucleic acid molecule with simplified images of sugar molecules.

How the planes of nitrogenous bases can be located between the helices will be discussed in the next chapter.

From Figure 7 it follows that if we assume the existence of a nucleic acid consisting of three single-stranded nucleic acids, then it is obvious that the three helices in this molecule should be split into two pairs. In each pair, the geometric elements will be antiparallel. One of the spirals will be common to these pairs. Two of these three helixes will be in a parallel position and will not satisfy the possibility of locating nitrogenous bases between them. Also, single-stranded nucleic acids must be arranged so as not to interfere with the placement of nitrogenous bases between nucleic acid helices in an anti-parallel position. If a nucleic acid consists of four single-stranded nucleic acids, then in this common nucleic acid there must exist four pairs of double-stranded nucleic acids in an antiparallel position. Moreover, each single-stranded nucleic acid belongs to only two pairs of double-stranded nucleic acids.

RIBOSOMAL RNA

Ribosomes are particles in the cells of living organisms responsible for protein synthesis. From 50 to 70% of the mass of ribosomal particles are ribosomal RNA (rRNA). They perform the main function of ribosomes - protein synthesis. Like all single-stranded polyribonucleotide`s, rRNA molecules control changes in ionic strength and temperature by changing the structure of ribosomes from fully unfolded to very compact. Ribosomes are divided into two parts (two subunits). The small subunit of the ribosome (30S or 40S) contains one molecule of high molecular weight rRNA, which is designated as 16S-rRNA in the case of bacterial ribosomes, and 16S-like rRNA in other cases. The large ribosomal subunit (50S or 60S) contains a high molecular weight rRNA called 23S-rRNA in the case of bacteria and 23S-like rRNA otherwise (Hall, Slayter, 1959; Huxley, Zubay, 1960; Vasiliev, 1971; Nanomura, Blobel, Sabatini, 1971; Lake, 1976). Ribosomal RNAs form complex structures of the primary, secondary, tertiary, and quaternary levels in ribosomes, describing various types of RNA interactions: perfect hairpins (double helixes with end loops), double helix hairpins with protrusions, with branches, end loops, pseudoknots, single-stranded helices, helix-helix interactions, spiral-loop interactions, etc.(Doty, Boedtker, Fresco, Haselkorn, Litt, 1959; Endelman, Moore, Schoenborn, 1975; Korostelev, Trakhanov, Laurberg, Noller, 2006; Spirin, Kiselev, Shakulov, Bogdanov, 1963; Gongadze, 2001). Details of the description of these interactions can be found in the monograph (Spirin, 2019). In this context, we note that all these structures are quite complex and certainly have a higher dimension. The processes in ribosomes are very complex and cannot be accurately described.

However, it should be noted the exceptional role of magnesium ions in maintaining the integrity of the ribosome, in the unification of all nucleic acids in the ribosome in protein production. It has been experimentally observed that as soon as the concentration of magnesium ions in the ribosome becomes less than a certain critical value, the ribosome breaks up into parts. It is not very convincing to fully explain this by the standard divalence of magnesium ions. Apparently, the influence of the vacant electron orbitals of magnesium in the space of higher dimension takes place here, as was discussed in the first chapter.

THE DIMENSION OF SPACE INTO A RIBOSOME

Nucleic acids are the basic substances serving to store and transmit hereditary information. The main stage of this transfer of hereditary information in the creation of protein occurs in ribosomes. The components of both bacterial and eukaryotic ribosomes, which make up their bulk, are RNA and proteins (Spirin, 1986; Spirin, Gavrilova, 1971). From the geometry point of view, as it was shown in (Zhizhin, 2018 a), proteins are chains of centered tetrahedrons, in the vertices of which carbon, hydrogen and function groups (R, CO, NH, NH_2, COOH) are located. The dimension of these monomeric units is at least 4. As is known (Spirin, 1986; Spirin, Gavrilova, 1971), the presence of magnesium ions Mg^{2+} is necessary for existence of ribosomes. If the concentration of these ions is less than a certain critical ribosome break up into subunits. With a further decrease in the concentration of ions Mg^{2+} and subunit become unstable and decay. Magnesium ion Mg^{2+} plays the role of bridges, connecting the negative charged sections of various compounds (Metzler, 1980; Stewart, McLachlan, 1975). This leads to an increase in dimensionality of compounds. Let us consider, for example, two residues of phosphoric acid in an ATP molecule that are connected to one another at the oxygen atom. In the ATP molecule, three phosphoric acid residues form a chain of three centered tetrahedrons, each of which has a dimensionality of 4 (Zhizhin, 2018 b). Suppose that two neighboring phosphoric acid residues in the chain, in addition to the oxygen atom compound, have one more compound at the magnesium atom. This can occur with a sufficient concentration of hydrogen ions and the separation of hydrogen ions and separation of hydroxyl groups from phosphorus atoms to from water molecules (Figure 8).

Figure 8. Fischer`s formula of two phosphoric acid residues connected via oxygen and magnesium atoms

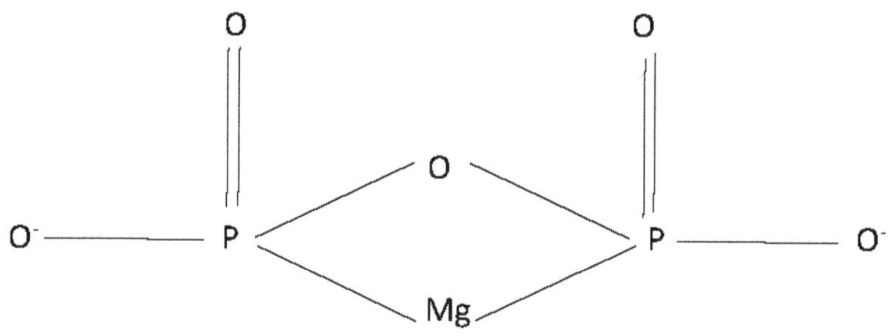

Then the two residues of phosphoric acid form a closed polytope. Indeed, let`s imagine using the already mentioned procedures atoms the compound in Figure 5.8 as a polytope Figure 9.

Figure 9. The polytope corresponded of two phosphoric acid residues connected via oxygen and magnesium atoms

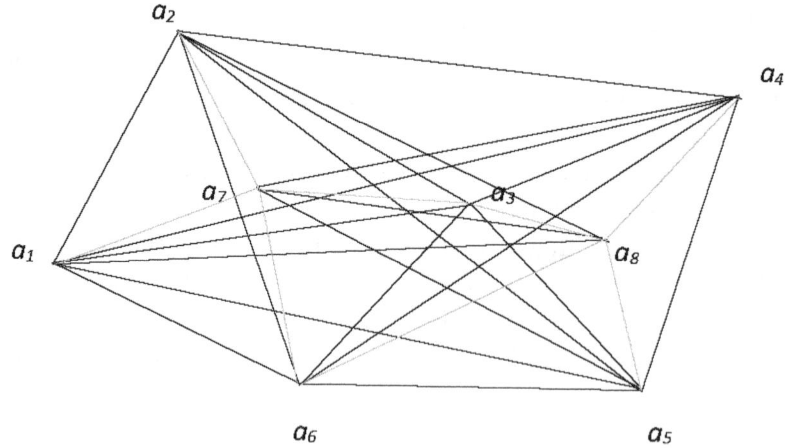

In Figure 9, phosphorus atoms are located at the centers a_7,a_8 of tetrahedrons $a_1a_2a_3a_6$, $a_3a_4a_5a_6$. These tetrahedrons have a common edge a_3a_6, at the vertices of which there are a magnesium atom (a_6) and oxygen atom (a_3). The remaining vertices contain oxygen atoms belonging to individual phosphoric acid residues. The polytope on Figure 9 has eight vertices (f_0=8). Each vertex connected with all another vertices. Therefore (Zhizhin, 2019), the polytope on Figure 6.9 is simplex and this dimension equal 7.

Thus, the presence of a significant amount of magnesium ions in ribosome leads to the cross-linking of both protein monomers and RNA nuclei. In this case, the dimensionality of the entire space is significantly increased. Consequently, even while not knowing the pattern of specific atomic distributions in the ribosome space it can be confidently asserted that the process of transferring hereditary information in the synthesis of proteins, so important for all living organisms, occurs in a space of very high dimension.

CONCLUSION

The chapter discusses various interactions of nucleic acids in the process of transmission of hereditary information. They all result from hydrogen bonding between nitrogenous bases covalently attached to five-carbon sugar molecules in nucleic acids. In double-stranded DNA, the connection between the helices is carried out mainly with the help of the canonical pairs of nitrogenous bases adenine - thymine, guanine - cytosine. The elements in these pairs can be interchanged. In double-stranded RNA, thymine is replaced by uracil in canonical pairs of nitrogenous bases. This interaction of nitrogenous bases (and, consequently, the interaction of nucleic acids) is provided by the chemical structure of nitrogenous bases. In all interactions, a nitrogen atom is involved, which is included in all nitrogenous bases. With the help of correspondence (complementarity) of nitrogenous bases in interaction pairs, the information embedded in the original DNA is transmitted during the transition from the original DNA to subsequent nucleic acids mRNA, rRNA, tRNA, mRNA up to the formation of a protein in ribosomes according to the given information in the original DNA. However, in addition to canonical pairs of nitrogenous bases, there are non-canonical interactions of nitrogenous bases and, consequently, non-canonical pairs of nitrogenous bases, the transmission of information through which differs from the information transmitted by canonical pairs of nitrogenous bases.

This chapter lists both canonical and non-canonical base pairs. It is significant that in one part of the pairs of nitrogenous bases, their bond is carried out using two parallel hydrogen chemical bonds, in the other part of the pairs of nitrogenous bases, their bond is carried out using three parallel hydrogen chemical bonds. However, in non-canonical pairs of nitrogen bases there are pairs in which the bond is carried out by two non-parallel hydrogen bonds, and there are even pairs of nitrogen bases with single hydrogen bonds. Thus, under the action of non-canonical pairs of nitrogenous bases, there is no unambiguous coding of the information contained in the original DNA. In addition, non-canonical pairs of nitrogenous bases make it possible to assert the possibility of the arrangement of nitrogenous bases in a pair in different planes. Indeed, if nitrogenous bases are considered flat, then in canonical pairs of nitrogenous bases, in which the bond is carried out by two and three parallel hydrogen bonds, the planes of nitrogenous bases must coincide, forming a certain single plane of the pair. In non-canonical pairs, there is no reason to assert this, especially for pairs in which the bond is carried out by a single hydrogen bond.

It is important that the interaction of single-stranded nucleic acids in double-stranded nucleic acids occurs between helices directed towards each other. This complicates the process of reading information from one helix to another. It is shown that the features of such a reading are described by the geometry of cross-polytopes (a special kind of higher-dimensional polytopes). The characteristics of such polytopes are given. In particular, for the first time, simplified images of double-stranded nucleic acids were constructed, in which five-carbon sugar molecules are presented as three-dimensional polyhedron that reflect all the main features of these molecules in full, i.e. in the form of polytopes of dimension 12. Certain vertices of these three-dimensional polyhedron are the points of attachment of nitrogenous bases to them. It is essential that, as follows from the constructed images, the points of connection of nitrogenous bases to sugar molecules are located in a space of higher dimension. The location of nitrogenous bases, taking into account their geometry, must be studied separately. This will be done in the next chapter.

The interactions of nucleic acids in ribosomes, which carry out the creation of protein molecules according to the program specified in the original DNA, are quite complex. Descriptions of the processes of these interactions, despite the results already available, require further development, taking into account the higher dimension of the chemical compounds of nucleic acids. The exceptional role of magnesium ions in these processes, which ensure the stability of ribosomes at a sufficient concentration of these ions in ribosomes, can be noted. It can be assumed that this stability is affected by the possibility that magnesium ions have vacant electron orbitals in higher-dimensional space.

It can be noted that the penetration of large dimensions into genetics does not contradict the trend of penetration of large numbers into medicine, physics, and biology (Kosrapohr, 2003; Rodger, 2015; Zhizhin, 2022 a, b).

REFERENCES

Albert, B. (2002). *Molecular Biology of the Cell* (4th ed.). Garland Science.

Berg, J., Tymoczko, J., & Stryer, L. (2002). *Biochemistry*. W.H. Freeman & Co Ltd.

Brachet, J. (1941 – 1942). La détection histochimique et la microdosage des acides pentosenucleiques. *Enzymologia*, *10*, 87–96.

Butler, J. M. (2005). *Forensic DNA Typing*. Gardners Books.

Cantor, C. R., & Schimmel, P. R. (1980). *Biophysical Chemistry. Part I. The Conformation of Biological Macromolecules*. W.H. Freeman & Co Ltd.

Caspersson, T., Landeström–Hyden, H., & Aquilonius, L. (1941). Cytoplasmsnukleotide in eiweissproduzierenden Drüsenzellen. *Chromosoma*, *2*, 111–131. doi:10.1007/BF00325956

Doty, P., Boedtker, H., Fresco, J. R., Haselkorn, R., & Litt, M. (1959). Secondary structure in ribosomes acids. *Proceedings of the National Academy of Sciences of the United States of America*, *45*(4), 482–499. doi:10.1073/pnas.45.4.482 PMID:16590404

Engelman, D. M., Moore, F. B., & Schoenborn, B. P. (1975). Neutron scattering measurements of separation and shape of proteins in 30S ribosomal subunits of *Escherichia coli S2 – S5, S5 – S8, S3 – S7. Proceedings of the National Academy of Sciences of the United States of America*, *72*(10), 3888–3892. doi:10.1073/pnas.72.10.3888 PMID:1105567

Frank - Kamenetskiy, M. (1986). DNA structure: A simple solution to the stability of the double helix? *Nature, 324*, 325.

Frank - Kamenetskiy, M. (1988). H – form DNA and the hairpin – triplex model. *Nature, 333*, 214.

Frank – Kamenetskiy, M.D. (2004). *The DNA Age*. Moscow, Publishing house KDU.

Frank – Kamenetskiy. M.D. (2010). Queen of a living cell. From the structure of DNA to the biotechnological revolution. Moscow: Act – Press.

Gamow, G. (1954). Possible relation between deoxyribonucleic acid and protein structures. *Nature*, *173*(4398), 318. doi:10.1038/173318a0

Gongadze, G.M. (2001). $5S_p$ RNA and ribosome. *The successes of modern biochemistry, 51*, 162 - 165.

Grunbaum, B. (1967). *Convex Polytopes*. Springer.

Hall, C. E., & Slayter, H. C. (1959). Electron microscopy of ribonucleoprotein particles from *E., coli. Journal of Molecular Biology, 1*(4-5), 329–332. doi:10.1016/S0022-2836(59)80014-0

Hershey, A., & Chase, M. (1952). Independent functions of viral protein and nucleic acid in growth of bacteriophage. *The Journal of General Physiology, 36*(1), 39–56. doi:10.1085/jgp.36.1.39 PMID:12981234

Huxley, H. E., & Zubay, G. (1960). Electron microscope observations on the structure of microsomal particles from *Escherichia coll. Journal of Molecular Biology, 2*(1), 10–18. doi:10.1016/S0022-2836(60)80003-4

Jacob, F., & Monod, J. (1961). Genetic regulatory mechanisms in the synthesis of proteins. *Molecular Biology, 3*(3), 318–356. doi:10.1016/S0022-2836(61)80072-7 PMID:13718526

Korostelev, A., Trakhanov, S., Laurberg, M., & Noller, H. F. (2006). Crystal structure of a 70S ribosome- tRNA complex reveals functional interaction and rearrangements. *Cell, 126*(6), 1065–1077. doi:10.1016/j.cell.2006.08.032 PMID:16962654

Kosrapohr, (Ed). (2003). *Managing data mining technologies in organizations: techniques and applications.* Hershey, PA: Idea Group Publishing.

Lake, J. A. (1976). Ribosome structure is determined by electroscope of *Escherichia coli* small subunits, large subunits, and monomeric ribosomes. *Journal of Molecular Biology, 105*(1), 131–159. doi:10.1016/0022-2836(76)90200-X PMID:792456

Lehninger, A. L. (1982). *Principles of Biochemistry* (Vol. 1-3). Worth Publisher, Inc.

Metzler, D. E. (1980). *Biochemistry. The Chemical Reactions in Living Cells* (Vol. 1-3). Academic Press.

Nelson, D. L., & Cox, M. M. (2022). *Lehninger Principles of Biochemistry*. W.H. Freeman.

Nonomura, Y., Blobel, G., & Sabatini, D. (1971). Structure of liver ribosomes studied by negative staining. *Journal of Molecular Biology, 60*(2), 303–324. doi:10.1016/0022-2836(71)90296-8 PMID:4106575

Oswald, A. T., MacLeod, C. M., & McCarty, M. (1944). Studies on the Chemical Nature of the Substance Inducing Transformation of Pneumococcal Types: Induction of Transformation by a Desoxyribonucleic Acid Fraction Isolated from Pneumococcus Type III. *The Journal of Experimental Medicine, 79*(2), 137–158. doi:10.1084/jem.79.2.137 PMID:19871359

Pasteur, L. (1960). *Selected works*. Publishing House of the Academy of Sciences of the USSR.

Poincare, A. (1895). Analysis situs. *J de e'Ecole Polytechnique.*, *1*, 1–121.

Ponnuswamy, P., & Gromiha, M. (1994). On the conformational stability of oligonucleotide duplexes and tRNA molecules. *Journal of Theoretical Biology*, *169*(4), 419–432. doi:10.1006/jtbi.1994.1163 PMID:7526075

Riemann, G. F. B. (1854). On the Hypotheses Which Lie at the Foundations of Geometry. Gottingen: Gottingen Observatory.

Rodger, J. A. (2015). Discovery of medical Big Data analytics: Improving the prediction of traumatic brain injury survival rates by data mining Patient Informatics Processing Software Hybrid Hadoop Hive. *Informatics in Medicine Unlocked*, *1*, 17–26. doi:10.1016/j.imu.2016.01.002

Spirin, A.C. (2019). *Molecular biology. Ribosome and protein biosynthesis. Moskow: Laboratory of knowledge.*

Spirin, A.S. (1986). *Molecular biology. Structure of the ribosome and protein biosynthesis*. Moscow: High school.

Spirin, A. S., & Gavrilova, L. P. (1971). *The Ribosome*. Science.

Spirin, A. S., Kiselev, N. A., Shakulov, R. S., & Bogdanov, A. A. (1963). Study of the structure of ribosomes. Reversible unfolding of ribosomal particles into ribonucleoprotein strands and a folding pattern. *Biochemistry*, *28*, 920–930. PMID:14092472

Stewart, M., & McLachlan, A. D. (1975). Fourteen actin-binding sites on tropomyosin? *Nature*, *257*(5524), 331–333. doi:10.1038/257331a0 PMID:1161036

Vasiliev, V. D. (1971). Electron microscopy study of 70 S ribosomes of *Escherichia coli*. *FEBS Letters*, *14*(4), 203–205. doi:10.1016/0014-5793(71)80617-8 PMID:11945758

Watson, J. D., & Crick, F. H. C. (1953a). Molecular structure of nucleic acids. *Nature*, *171*(4356), 738–740. doi:10.1038/171737a0 PMID:13054693

Watson, J. D., & Crick, F. H. C. (1953b). Genetical implications of the structure of deoxyribose nucleic acid. *Nature*, *171*(4361), 964–967. doi:10.1038/171964b0 PMID:13063483

Zhizhin, G. V. (2014a). On the higher dimensions in nature. *Biosphere*, *6*(4), 313–318.

Zhizhin, G. V. (2014b). *World 4D*. Polytechnic Service.

Zhizhin, G. V. (2015a). The dimensions of supramolecular compounds. *Biosphere*, *7*(2), 149–154.

Zhizhin, G. V. (2015b). Dimensions of compounds in supramolecular chemistry. *Int. J. Chem. Modeling*, *7*(2), 215–238.

Zhizhin, G. V. (2016a). From multidimensional phase spaces to multidimensionality of space in Nature. *Biosphere*, *8*(3), 258–267.

Zhizhin, G. V. (2016b). The structures, topological and functional dimension of biomolecules. *International Journal of Chemoinformatics and Chemical Engineering*, *5*(2), 44–58. doi:10.4018/IJCCE.2016070104

Zhizhin, G. V. (2018). *Chemical Compound Structures and the Higher Dimension of Molecules: emerging Research and Opportunities*. IGI Global. doi:10.4018/978-1-5225-4108-0

Zhizhin, G. V. (2019 a). *The Geometry of Higher - Dimensional Polytopes*. IGI Global. doi:10.4018/978-1-5225-6968-8

Zhizhin, G. V. (2019 b). *Attractors and higher dimensions in population and molecular biology: emerging research and opportunities*. IGI Global. doi:10.4018/978-1-5225-9651-6

Zhizhin, G. V. (2019c). The Metric Structure of Nucleic Acids and Higher Dimension of Their Constituents. *J. Chemoinformatics and Chemical Engineering*, *7*(2), 1–15.

Zhizhin, G. V. (2021 a). The Structures of DNA Taking into Account the Higher Dimensional of its Components. In Kosrupohr (Ed.) Encyclopedia of Organizational Knowledge, Administration, and Technologies. Hershey, PA: IGI Global.

Zhizhin, G. V. (2021b). Hidden Order of Interacting Nucleic Acids. *Journal of Nature Science and Sustainable Technology*, *15*(4), 1–5.

Zhizhin, G. V. (2021c). Interaction of Nucleic Acids: Hidden Order of Interaction. *International Journal of Applied Research in Bioinformatics*, *11*(2), 1–8. doi:10.4018/IJARB.2021070101

Zhizhin, G. V. (2022 a). *The Classes of Polytopes Higher Dimensions in Physics, Chemistry and Biology*. IGI Global.

Zhizhin, G. V. (2022b). The Properties of Biomaterials Hidden in the Spaces of Higher Dimension. *Annals of Advanced Biomedical Sciences*, *5*(2), 1–9. doi:10.23880/aabsc-16000189

Zhizhin, G. V., & Diudea, M. V. (2016). Space of nanoworld. In M. V. Putz & C. M. Marius (Eds.), *Sustainable nanosystems, development, properties, and applications* (pp. 221–245). IGI Global.

Zhizhin, G. V., Khalaj, Z., & Diudea, M. V. (2016). Geometrical and topological dimensions of the diamond. In A. R. Ashrafi & M. V. Diudea (Eds.), *Distance, symmetry and topology in carbon nanomaterials* (pp. 167–188). Springer. doi:10.1007/978-3-319-31584-3_12

Chapter 7
Polytope of Hereditary Information

ABSTRACT

Variants of damage to DNA and the polytope of hereditary information under the influence of radiation, chemical reactions, and the possibility of restoring DNA functions, including with the help of methyl groups, are considered. The geometric features of nucleic acids during methylation of their nitrogenous bases, which is typical for viral nucleic acids in infectious diseases, were studied. It was found that methylation of viral nucleic acids leads to the transition of their nitrogenous bases into the space of higher dimension and sharply increases (more than 2000 times) the intensity of information processes in the field of interaction of nucleic acids (in the polytopic of hereditary information). The activity of methylated viral nucleic acids can be countered by activation, i.e. a fairly high degree of methylation, nucleic acids of the organism.

INTRODUCTION

There are two known nucleic acids: deoxyribonucleic acid (DNA) and ribonucleic acid (RNA). Direct experiments by Avery, McLeod, and McCarthy (1944) proved that DNA can be the carrier of the hereditary traits of an organism, i.e. carrier of genetic information. At the same time, RNA is an intermediary that receives this information from DNA and implements it in the process of protein biosynthesis. Watson and Crick (1953 a, b) proposed a model of the macromolecular structure of DNA (Dahm, 2005). This is a double helix, where two DNA polymer strands are twisted relative to each other around a common axis and are held together due to pair interactions of nitrogenous bases covalently bonded to five - carbon sugar molecules. DNA is a biopolymer, monomer of which is the nucleotide (Albert, et

DOI: 10.4018/979-8-3693-0588-1.ch007

al., 2002; Berg, Tymoczko, Stryer, 2002; Butler, 2005). In 1986, Frank-Kamenetsky in Moskow showed how a double-stranded DNA fold into a so-called H-shape, composed not of two but three strands of DNA (Frank-Kamenetskiy, 1986, 1988). The existence of nucleic acids differing in the length of the period and shape with rotation of the spiral both to the right and to the left was experimentally established (Ha, et al., 2005; Cantor, Schimmel, 1980; Frank-Kamenetskiy, 2010). In turn, sugar molecules are linked to each other by phosphoric acid residues, forming a chain. Bases are large circular molecules with nitrogen atoms. DNA nucleotides (a sugar molecule, a phosphoric acid residue and a base) have one of the four bases adenine (A), guanine (G), cytosine (C) and thymine (T). In RNA, thymine is replaced by uracil (U). Cytosine, thymine and uracil have one ring of atoms and are called pyrimidine bases. Adenine and guanine each have two rings and are called purine bases. It is significant that nitrogenous bases are practically flat. The reproduction (reduplication) of the DNA structure is based on the principle of complementarity: in a double helix, two DNA polymer chains are linked to each other by hydrogen bonds in the formation of pairs G: C, C: G, A: T, T: A (the colon denotes a hydrogen bond). If two chains of the double helix diverge, then a new complementary chain can be built on each of them. At the same time, opposite the G of the original chain, C of the new chain is established, opposite C of the old chain - G of the new chain, opposite A - T, and opposite T - A. As a result, two child double helixes are obtained, completely identical to the original parent. The same principle of complementarity provides a mechanism for RNA replication on a DNA template. This is another type of interaction between nucleic acids. The only difference is that RNA polymerizes only on one of the two separated strands of the DNA double helix. During RNA synthesis, opposite A of the DNA chain becomes U (uridylic ribonucleotide), instead of T during DNA synthesis. The replicating strand of RNA is thus an exact copy of the opposite strand of DNA, with T replaced by Y. As a result of this replication, RNA is formed as a flexible single - stranded polymer.

In the previous chapter, it was shown that approximately five-carbon sugar molecules, which are in the corresponding state in a double-stranded nucleic acid molecule, can be represented as two three-dimensional special polyhedrons with antiparallel edges. At the same time, it is not forgotten that, in fact, sugar molecules have a higher dimension. The dimension of molecule is defined in relation of Poincare (Poincare, 1895). Obviously, together two sugar molecules make up a single body, especially since they are connected by nitrogenous bases. In this chapter, this single body will be considered as a polytope of higher dimension, because we do not forget that sugar molecules have a higher dimension. For what follows (in particular, for the theory of heredity), the properties of this polytope are of exceptional importance. It should be noted here that experimental works have recently appeared in biology, according to which genetic memory can be associated not only with memory embedded in the sequence of genes, i.e. in the sequence of nucleotides, or even more simply in the sequence of nitrogenous bases in nucleic acid. We are talking about the so-called epigenetic memory associated with the presence of certain atoms in the nucleic acid chain that are not included in the genes (De Candolle, Lamarck, 2016; Lindquist, et al.,2016; Mancuso, et al.,2014; Sanbonmatsu, et al.,2016; Mancuso, 2017). This is consistent with the ideas about the influence on heredity of the presence (incident) of various atoms in the nucleic acid. Therefore, when analyzing heredity, it is necessary to use the law of conservation of incidence, established in the second chapter of this work.

Figure 1. The two sugar molecules in a simplified form with antiparallel edges

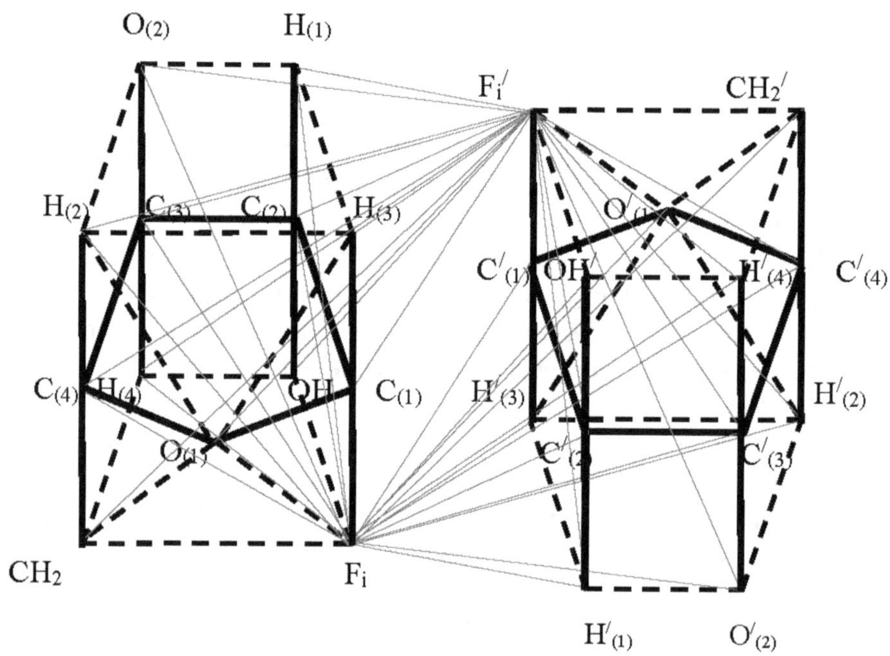

THE POLYTOPE OF HEREDITARY INFORMATION DOUBLED STRANDED NUCLEIC ACID

Let us consider in detail the formation of a polytope of two sugar molecules with antiparallel edges. Here, as in the case of the tetrahedron, to form a polytope with antiparallel edges from two sugar molecules, you must have one sugar molecule on one helix to turn this helix together with the sugar molecule and move the sugar molecule along this reversed helix in opposite direction to the original helix direction. When the sugar molecule rotates 180 degrees to the right while moving, then the original sugar molecule and the sugar molecule on the reversed helix are two polytopes with antiparallel edges. Both of these sugar molecules in a simplified form are represented in Figure 1.

When the five - carbon sugar molecule is rotated to the left by 180 degrees, the nitrogenous bases are on opposite sides of both molecules, so that for their connection it is necessary to cross the entire set of atoms of two molecules. This is un existent, therefore this option is not considered. When the sugar molecule is rotated 180 degrees to the left, it is possible to connect the sugar molecules through nitrogenous bases only between two different chains of nucleic acids or in the case of turning the chain itself in the opposite direction.

In full, the sugar molecules have a dimension of 12, in the corresponding polytope each vertex must have an edge connection with all the other vertices. Knowledge of this now one need. For the formation of a polytope of dimension 13, it is necessary to connect each vertex of one polytope with the vertices of another polytope so that there are no vertex connections with the same letters. All connecting edges break into pairs of antiparallel edges. At the same time, a set of coordinate a two - dimensional planes emanates from the center of the formed polytope as from the origin of coordinates. Their number is equal to the number of combinations from 13 to 2, i. e. 48 coordinate planes. In order to clarify the possible geometrical circumstances of the connection of helices in double - helix nucleic acid molecules with nitrogenous bases, we are primarily interested in the coordinate planes containing these nitrogenous bases $F_i, F_i^/$. There are exactly 12 such coordinate planes in the obtained polytope of dimension 13. They are depicted in Figure 1 by blue solid lines and are indicated below by the vertices of the polytopes contained in them:

$$\begin{array}{cccccccccccc} F_i^/ H_{(1)} & F_i^/ O_{(2)} & F_i^/ H_{(2)} & F_i^/ C_{(3)} & F_i^/ C_{(2)} & F_i^/ H_{(3)} & F_i^/ C_{(4)} & F_i^/ H_{(4)} & F_i^/ OH & F_i^/ C_{(1)} & F_i^/ O_{(1)} & F_i^/ CH_2 \\ H_{(1)}^/ F_i\text{'} & O_{(2)}^/ F_i\text{'} & H_{(2)}^/ F_i\text{'} & C_{(3)}^/ F_i\text{'} & C_{(2)}^/ F_i\text{'} & H_{(3)}^/ F_i\text{'} & C_{(4)}^/ F_i\text{'} & H_{(4)}^/ F_i\text{'} & (OH)^/ F_i\text{'} & C_{(1)}^/ F_i\text{'} & O_{(1)}^/ F_i\text{'} & (CH_2)^/ F_i \end{array}.$$

Other edges of the polytope of dimension 13 are not shown in Figure 1, so as not to ignite the figure. In the center of each parallelogram, indicated by its four vertices, is the origin of coordinates and the corresponding pair of coordinate axes (they are not shown). To identify the different hydrogen and oxygen atoms at vertices of polytope, they indicated by numbers in brackets at the lower indices.

It is surprising that the number of coordinate planes containing vertices is exactly equal to the number of possible compounds of nitrogenous bases 12 (Spirin, 2019)

A:U, U:A, G:C, C:G, G:U, G:A, U:U, U:C, A·A, A·C, U:G, A:G (1)

Since each coordinate plane designated by the vertices of the parallelograms has a specific atomic environment, it can be assumed that each of the 12 possible compounds of nitrogenous bases is located on one particular coordinate plane out of 12 possible. This solves the question of the possible orientation of the bond of nitrogenous flat bases in nucleic acids using ideas about the high dimensionality of the constituent nucleic acids. It is also surprising that in order to create 13 – cross - polytopes, providing the connection with nitrogenous bases $F_i, F_i^{/}$, nature specially created double - stranded nucleic acids with oppositely directed spirals. This is realized in DNA and RNA when creating regions with inverted helices.

In a variety of nucleic acid molecules, the issue of chain interaction is important. In ribosomes, RNA interacts with each other due to bivalent metal ions (mainly magnesium ions). Positively charged magnesium ions attract negative charges of phosphoric acid residues, ensuring the stability of the ribosomes. In double - stranded nucleic acids, the helices are connected to each other by means of nitrogenous bases complementarily interacting with each other by a hydrogen bond. However, the magnesium ions and nitrogen bases in nucleic acids could not be specifically located. It has been established that magnesium ions and flat nitrogenous bases can be located inside special polytopes of higher dimension. Here knowledge is needed of the higher dimension of phosphoric acid residues and sugar molecules. Such polytopes are polytopes with antiparallel edges, i. e. cross - polytopes of higher dimension. Binding agents are located on the free coordinate planes of these polytopes in the vicinity the center of the polytope.

In this case, the two - dimensional coordinate plane on the boundary of the polytope should contain the objects to be joined. In the case of magnesium ions, there are four specific coordinate planes inside the 5 – cross - polytope, in which an ion can accommodate, combining negative charges. In the case of nitrogenous bases, the existence of 12 coordinate planes inside a cross - polytope of dimension 13, in which flat nitrogenous bases can be located, connecting the helix of nucleic acids, was discovered. Exactly as much as there are options for combining nitrogenous bases. It is given, that each coordinate plane of these 12 planes has a specific environment of atoms. It should be assumed that only one of the 12 possible compounds of nitrogenous bases is placed in each of these planes. It is surprising that the existence of higher - dimensional polytopes with anti - parallel edges is possible only in the case of the opposite direction of interacting helices, and this is exactly what nature provides in the double - helix DNA and in the RNA segments with self - inversion of the helix in the opposite direction.

To build the polytopic of hereditary information, we need (Zhizhin, 2019 a) four – dimensional polytope with anti – parallel edges to supplement with edges that create a closed and convex figure. In this case, it is not necessary to connect the vertices symmetrically relative to the center of figure with the edges. Each vertex will be connected by an edge to the other vertices. In this case, the polytope will be a cross -polytope and its dimension is equal to half the number of vertices. Indeed, according to (Zhizhin, 2019 b), there is a relation between f_i the number of dimension elements i in the cross - polytope and the dimension d of the cross - polytope itself:

$$f_i(d) = 2^{1+i} C_d^{d-1-i}. \tag{2}$$

At the vertex $i = 0$, therefore, according to (6.2) $f_0 = 2C_d^{d-1} = 2d, d = f_0 / 2 = 13$.

Thus, the polytope of hereditary information has dimension 13. When portraying this polytope, we will use the technique that is used in portraying cross-polytopes. We distribute all 26 vertices on the circle so that the vertices opposite in Figure 1 remain opposite and there is no edge between them (Figure 2).

Figure 2. The polytope hereditary information

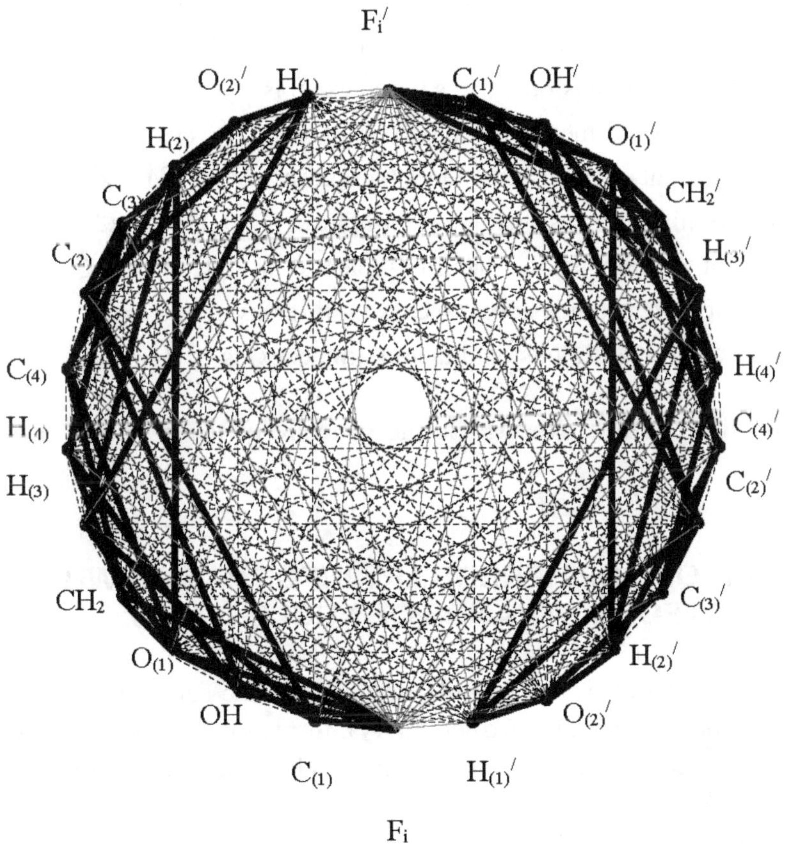

The edges corresponding to the chemical bonds are indicated in this figure by thick black solid lines. Antiparallel edges highlighting coordinate planes with vertices corresponding to nitrogenous bases are indicated by bluey lines. The remaining edges are indicated by thin dash-dotted black lines. We emphasize that there is a chemical bond between the nitrogenous bases, but it is not covalent. Therefore, this connection is not indicated by an edge.

When the polytope of hereditary information moves along the common axis of the double helix of nucleic acids in accordance with the sequence of bases in DNA, the connection between the vertices $F_i^{/}, F_i^{//}$ is carried out by one of the twelve base pairs, which occupies one of the 12 coordinate planes of the polytope.

By relation (equation 2), one can determine the number of elements of different dimensions in the polytope of hereditary information.

The number of edges is $f_1(13) = 2^2 C_{13}^{11} = 312$,
The number of triangles is $f_2(13) = 2^3 C_{13}^{10} = 2288$,
The number of tetrahedrons is $f_3(13) = 2^4 C_{13}^{9} = 10560$,
The number of fourth-dimensional faces (simplexes) is $f_4(13) = 2^5 C_{13}^{8} = 41184$,
The number of fives-dimensional faces (simplexes) is $f_5(13) = 2^6 C_{13}^{7} = 109824$,
The number of six-dimensional faces (simplexes) is $f_6(13) = 2^7 C_{13}^{6} = 219648$,
The number of seven-dimensional faces (simplexes) is $f_7(13) = 2^8 C_{13}^{5} = 164736$,
The number of eight – dimensional faces (simplexes) is $f_8(13) = 2^9 C_{13}^{4} = 183040$,
The number of nine-dimensional faces (simplexes) is $f_9(13) = 2^{10} C_{13}^{3} = 292864$,
The number of ten-dimensional faces (simplexes) is $f_{10}(13) = 2^{11} C_{13}^{2} = 159744$,
The number of eleven-dimensional faces (simplexes) is $f_{11}(13) = 2^{12} C_{13}^{1} = 26624$,
The number of twelve-dimensional faces (simplexes) is $f_{12}(13) = 2^{13} = 4096$.

The obtained numbers determine the structure of the polytope of hereditary information.

THE POLYTOPE OF HEREDITARY INFORMATION WITH THREE AND FOUR STRANDS OF NUCLEIC ACID

If a nucleic acid includes three single-stranded nucleic acids, then the hereditary information polytope includes two hereditary information polytopes located between two pairs of helices, since one of these helices is located between two other helices (Nelson and Cox, 2022).

If a nucleic acid consists of four single-stranded nucleic acids, then the polytope of hereditary information includes four polytopes of double-stranded nucleic acids.

Consequently, the number of geometric elements of each dimension in the polytope of hereditary information in a nucleic acid with three single-stranded nucleic acids is at least doubled. Respectively. the number of geometric elements of each dimension in the polytope of hereditary information with four single-stranded nucleic acids is increased by at least four times.

The combination of hereditary information polytopes belonging to each pair of interacting single-stranded nucleic acids into one hereditary information polytope seems to be incorrect, since non-antiparallel edges appear in this case. They violate the geometry of polytopes of hereditary information, which is characteristic of interacting nucleic acids.

THE LAW OF CONSERVATION OF INCIDENTS IN POLYTOPES OF HEREDITARY INFORMATION WITH A FEW SINGLE STRANDED NUCLIEC ACIDS

The law of conservation of incidence in polytopes of hereditary information operates independently for each pair of nucleic acids bound by nitrogenous bases. Therefore, the law of conservation of incidence considered in this section is valid for any pair of linked nucleic acids.

The monograph (Zhizhin, 2019 c) introduced the concept of the incidence coefficients of elements of lower dimension with respect to elements of the higher dimension and elements of higher dimension with respect to elements of the lower dimension. The first characterizes the number of elements of a certain higher dimension to which the given element of a lower dimension belongs. The second characterizes the number of elements of a given lower dimension that are included in a particular element of a higher dimension. Here we must remember that the vertices of geometric elements of various dimensions are atoms, molecules or functional groups. Therefore, the incidence of geometric elements to a friend means contacted between particles of the matter, including living matter. The contact between particles of matter can be interpreted as the transfer of information on material structures, including biological structures. We introduce the notation: $k_{d_j^i d_u}$ is the number of elements of dimension u, which include an element of dimension $j\,(u > j)$ with number i. Thus, $k_{d_j^i d_u}$ is the incidence factor of element i with dimension j relative to elements with dimension u. We introduce the notation also: $k_{d_j d_u^i}$ is the number of elements of dimension j, which included in element i with a dimension

$u\,(u > j)$. Thus, $k_{d_j d_u^i}$ is the incidence factor of element i with dimension u relatively to elements with dimension j.

The smallest dimension of the cross - polytope is 4. From (2) it follows that in this polytope there are 8 vertices, 24 edges, 32 two - dimensional faces, 16 three - dimensional faces. The factors of incidents (from smaller dimension to larger) are:

$$k_{d_0^i d_1} = 6, i = 1 \div 8; k_{d_0^i d_2} = 12, i = 1 \div 8; k_{d_0^i d_3} = 8, i = 1 \div 8; k_{d_0^i d_4} = 1, i = 1 \div 8; k_{d_1^i d_2} = 4, i = 1 \div 24;$$
$$k_{d_1^i d_3} = 4, i = 1 \div 24; k_{d_1^i d_4} = 1, i = 1 \div 24; k_{d_2^i d_3} = 2, i = 1 \div 32; k_{d_2^i d_4} = 1, i = 1 \div 32; k_{d_3^i d_4} = 1, i = 1 \div 16.$$

Sum up the incidence coefficients for all vertices, edges, two - dimension faces and three - dimension faces of the 4 - cross – polytope:

$$\sum_{i=1}^{8} k_{d_0^i d_1} + \sum_{i=1}^{8} k_{d_0^i d_2} + \sum_{i=1}^{8} k_{d_0^i d_3} + \sum_{i=1}^{8} k_{d_0^i d_4} + \sum_{i=1}^{24} k_{d_1^i d_2}$$
$$+ \sum_{i=1}^{24} k_{d_1^i d_3} + \sum_{i=1}^{24} k_{d_1^i d_4} + \sum_{i=1}^{32} k_{d_2^i d_3} + \sum_{i=1}^{16} k_{d_3^i d_4} + \sum_{i=1}^{32} k_{d_2^i d_4} = 544. \tag{3}$$

The factors of incidents (from larger dimension to smaller) are

$$k_{d_0 d_1^i} = 2, i = 1 \div 24; k_{d_0 d_2^i} = 3, i = 1 \div 32; k_{d_0 d_3^i} = 4, i = 1 \div 16; k_{d_0 d_4} = 8; k_{d_1 d_2^i} = 3, i = 1 \div 32;$$
$$k_{d_1 d_3^i} = 6, i = 1 \div 16; k_{d_1 d_4} = 24; k_{d_2 d_3^i} = 4, \mathrm{i} = 1 \div 8; k_{d_2 d_4} = 32; k_{d_3 d_4} = 16.$$

Sum up the incidence coefficients for all elements of the 4 – cross - polytope with dimension larger of zero:

$$\sum_{i=1}^{24} k_{d_0 d_1^i} + \sum_{i=1}^{32} k_{d_0 d_2^i} + \sum_{i=1}^{32} k_{d_1 d_2^i} + \sum_{i=1}^{16} k_{d_0 d_3^i} + \sum_{i=1}^{16} k_{d_1 d_3^i} + \sum_{i=1}^{16} k_{d_2 d_3^i} + k_{d_0 d_4} + k_{d_1 d_4} + k_{d_2 d_4} + k_{d_3 d_4} = 544. \tag{4}$$

Comparing (4) and (3) you can see that the sum of incidents in a 4 - cross - polytope from elements with a lower dimension to elements with a higher dimension is equal to the sum of incidents from elements with a higher dimension to elements with a lower dimension. Thus, the sum of incidents retains its value when changes the direction of the relationship between the elements (the law of conservation of incidents, (Zhizhin, 2019 d)).

Theorem 7.1. In any cross - polytope of dimension n, the sum of all incidents of elements of a lower dimension with respect to all elements of a higher dimension is equal to the sum of all incidents of elements of a higher dimension with respect to all elements of a lower dimension and equals the sum of the series:

$$f_0(n)3^{n-1} + f_1(n)3^{n-2} + f_2(n)3^{n-3} + ... + f_{n-1}(n)3^0, f_d(n) = C_n^{n-1-d} 2^{1+d}, d = 0 \div (n-1).$$

P r o o f. According to equation (2), each cross - polytope of dimension n has $2n = f_0(n)$ vertices. The peculiarity of a cross - polytope is that each of its vertices has an opposite vertex, with which it is not connected by an edge. Moreover, there is one edge between this vertex and all other vertices. We subtract from the total number of vertices two vertices (the selected vertex and its opposite) 2n - 2. This is the possible number of edges emanating from the selected vertex. Thus, the incidencecoefficientoftheedgesofanyvertexis $k_{d_0^i d_1} = 2(n-1) = 2\,C_{n-1}^1, i = 1 \div f_0(n)$. A cross - polytope of any dimension can be depicted as a projection on a two - dimensional plane (Zhizhin, 2019 c). In this image, all its vertices are located on a circle, with the selected vertex and its opposite vertex located symmetrically relative to the center of the circle. A mentally drawn line through these two vertices halves the circle (Figure 2). Therefore, the number of variants the location of the edges from the selected vertex to the other vertices in one of the halves of the circle is n - 1, i. e. the number of combinations of n - 1 vertices one by one. Since there are two halves of a circle, then for the total number of vertex selection options for edge formation, this number of combinations should be multiplied by 2. This is the meaning of the expression for the incidence rate of any vertex in the n - cross - polytope with respect to the edge. To further prove the theorem 1 and clarify the nature of the formation of the coefficients of incidence in the n - cross - polytope, we will use this technique. We arrange the vertices of the 4 - cross - polytope in two lines, so that the vertices unlinked by an edge form vertical pairs: each vertex in the top line is not connected by an edge to the vertex in the bottom line located strictly under this vertex in the top line:

$$\begin{matrix} d_0^1 & d_0^2 & d_0^3 & d_0^4 \\ d_0^5 & d_0^6 & d_0^7 & d_0^8 \end{matrix} . \tag{5}$$

We choose some vertex in the top line (5), for example d_0^1. It cannot be connected by an edge with a vertex d_0^5 in the bottom line, but it can be connected by an edge with other vertices of the bottom line. The number of such options is 3, i. e. with n

$= 4$ is $3 = C_{n-1}^{1} = C_3^1$. However, a vertex d_0^1 may be connected by an edge with any of the remaining three vertices in the top line (7.5). Therefore, the incidence coefficient of a vertex with respect to an edge in 4 - cross - polytope $k_{d_0^i d_1} = 2\,C_{n-1}^{1} = 6, i = 1 \div 8$. When considering the belonging of a vertex d_0^1 to two - dimensional faces, it is necessary to determine the number of options for the participation of two vertices in the top line (5) (without a vertex d_0^1) and in the bottom line (5) (without a vertex d_0^5). This will be the number $2C_3^1$. In addition, as the vertices of the triangle, there may be vertices located in the upper and lower lines (4) in a cross - section way. There are 6 such variants in (5), i. e. more $2C_3^1$. Therefore, the incidence coefficient of a vertex with respect to a two - dimensional element in 4 - cross - polytope is $k_{d_0^i d_2} = 2^2\,C_{n-1}^{2} = 12, i = 1 \div 8$. Multiplying this number by the number of vertices in 4 - cross - polytope you can get the total number of incidents of vertices to two - dimensional elements in 4 - cross - polytope equal to 96. This number coincides with the corresponding number, defined earlier in Figure 2. Combination d_0^1 with three the different vertices from (5) can get the incidence coefficient of vertex to the three - dimension body (tetrahedron) for condition absent in combination opposite vertices. Can show that this combination is 8, $k_{d_0^1 d_3} = 2^3 C_3^3 = 8, i = 1 \div 8$. Multiplying this number by the number of vertices in 4 - cross - polytope you can get the total number of incidents of vertices to three - dimensional elements in 4 - cross - polytope equal to 64. This number coincides with the corresponding number, defined earlier in Figure 2. Combination two vertices, for example d_0^1 and d_0^2 (the edge $d_0^1\, d_0^2$) with two the different vertices from (5) you can get the incidence coefficient of edge to three - dimension body (tetrahedron) for condition absent in combination opposite vertices. Can show that this combination is 4, $k_{d_1^1 d_3} = 2^2 C_2^2 = 4, i = 1 \div 8$. Multiplying this number by the number of edges in 4 - cross - polytope you can get the total number of incidents of edges to three - dimensional elements in 4 - cross - polytope equal to 96. This number coincides with the corresponding number, defined earlier in Figure 2. Combination two vertices (the edge $d_0^1\, d_0^2$) with one the vertex from (5) can get the incidence coefficient of edge to triangle for condition absent in combination opposite vertices. Can show that this combination is 4, $k_{d_1^1 d_2} = 2C_2^1 = 4, i = 1 \div 8$. Multiplying this number by the number of edges in 4 - cross - polytope you can get the total number of incidents of edges to triangle in 4 - cross - polytope equal to 24. This number coincides with the corresponding number, defined earlier in Figure 2. In general case can write of $2n$ vertices of the n - cross - polytope in two lines

$$\begin{matrix} d_0^1 & d_0^2 & d_0^3 & \dots & d_0^n \\ d_0^5 & d_0^6 & d_0^7 & \dots & d_0^{2n} \end{matrix}, \tag{6}$$

where vertex in the top line is not connected by an edge to the vertex in the bottom line located strictly under this vertex in the top line. The incidence coefficient of a vertex with respect to an edge in n - cross - polytope is $k_{d_0^i d_1} = 2\,C_{n-1}^1, i = 1 \div f_0(n)$. The incidence coefficient of a vertex with respect to a two - dimensional element in n - cross - polytope is $k_{d_0^i d_2} = 2^2\,C_{n-1}^2, i = 1 \div f_0(n)$. The incidence coefficient of a vertex with respect to a three - dimensional edge in n - cross - polytope is $k_{d_0^i d_3} = 2^3\,C_{n-1}^3, i = 1 \div f_0(n)$. Go on can to say the incidence coefficient of a vertex with respect to a (n – 1) - dimensional element in n - cross - polytope $k_{d_0^i d_{n-1}} = 2^{n-1}\,C_{n-1}^{n-1}, i = 1 \div f_0(n)$. Obviously, that the incidence coefficient of a vertex with respect to n - cross - polytope is $k_{d_0^i d_n} = 1, i = 1 \div f_0(n)$.

Multiplying of the incidence coefficients of a vertex with respect to different dimension elements in n - cross - polytope and sum the product you can get the common express for the number of incident vertices to elements of different dimension in a n - cross – polytope:

$$f_0(2^0 + 2^1 C_{n-1}^1 + 2^2 C_{n-1}^2 + \dots + 2^{n-1} C_{n-1}^{n-1}) = f_0(n) \sum_{i=0}^{n-1} 2^i C_{n-1}^i = f_0(n) 3^{n-1}.$$

In this way you can get the common express for the number of incident edges to elements of different dimension more one in n - cross - polytope

$f_1(2^0 + 2^1 C_{n-2}^1 + 2^2 C_{n-2}^2 + \dots + 2^{n-2} C_{n-2}^{n-2}) = f_1(n) \sum_{i=0}^{n-2} 2^i C_{n-2}^i = f_1(n) 3^{n-2}$, and go on.

In result, one can get the common express for the sum of all incidents of elements of a lower dimension with respect to all elements of a higher dimension in a n - cross – polytope:

$$f_0(n) 3^{n-1} + f_1(n) 3^{n-2} + \dots + f_{n-1} 3^0 = \sum_{i=0}^{n-1} f_i(n) 3^{n-1-i}, f_i(n) = C_n^{n-1-i} 2^{1+i}. \tag{7.7}$$

Let $k_{d_h d_j^i}$ be the number of elements of dimension h in a n - cross - polytope belonging to some one element of dimension j ($h<j$), that is d_j^i . Obviously, this number is equal to $f_h(j)_s$ for simplex, and $i=1\div f_j(n)_{cr}$ for n - cross - polytope. So, elements of a cross - polytope are simplexes the product $f_h(j)_s f_j(n)_{cr}$ corresponds to the number of elements of dimension h belonging to all elements of dimension j for a simplex. This product is equal to $f_d(n) = C_{n+1}^{d+1}$

$$C_{j+1}^{h+1} 2^{1+j} C_n^{n-1-j} = 2^{1+j} \frac{(j+1)!}{(h+1)!(j-h)!} \frac{n!}{(n-1-j)!(1+j)!}.$$

Let us compare this number with the number of elements of dimension j, which have elements of dimension h in the n - cross – polytope:

$$k_{d_h^i d_j} f_h(n)_{cr} = C_n^{n-1-h} 2^{1+h} C_{n-h-1}^{j-h} 2^{j-h} = 2^{1+j} \frac{n!}{(n-1-h)!(1+h)!} \frac{(n-h-1)!}{(n-j-1)!(j-h)!}.$$

Obviously, these numbers are equal to each other. This proves that the number of elements of dimension h in a n - cross - polytope belonging to all dimensions of j ($j>h$) in a n - cross - polytope is equal to the number of elements of dimension j, which have elements of dimension h as elements. Thus, the total number of incidents of elements of a smaller dimension with respect to elements of a higher dimension is equal to the total number of incidents of elements of a higher dimension with respect to elements of a smaller dimension. The total number defined expression (7). Q.E.D.

The polytope of hereditary information is a cross-polytope of dimension 13. Substituting the value n = 13 and the values of $f_i(n)$ calculated in the previous section in expression (7), we find the total value of the incident flow in the polytope of hereditary information

$$\sum_{i=0}^{12} f_i(13) 3^{12-i} = 1.78 \cdot 10^9 .$$

A significant value of the total incidence stream in the polytope of hereditary information indicates an intensive exchange of information between elements of the polytope of hereditary information. For example, this value is 400 times larger than the incident flux in a simplex of dimension 13 (Zhizhin, 2019 c). This may explain the recently discovered epigenetic principle of the transmission of hereditary information without changing the sequence of genes in DNA and RNA molecules.

HIDDENED NUCLEIC ACID BOND ORDER

It was established that a polytope of hereditary information exists between two nucleic acids in each region of their bond with nitrogen bases (Figure 2). The dimension of this polytope is 13, its type is a cross-polytope. This polytope has 12 flat coordinate planes, i. e. exactly as many different pairs of connected nitrogenous bases (1). The polytope of hereditary information is the result of combining two sugar molecules located in two nucleic acids in the antiparallel position. In this segment, examining the geometry of the polytopic of hereditary information, it was found that out of 12 coordinate planes of 13 - cross - polytope pairs of flat nitrogenous bases can be located in only five coordinate planes (Zhizhin, 2020, 2021 b, d; 2022 a).

Investigation of the Possible Location of Canonical Pairs of Nitrogenous Bases in the Polytope of Hereditary Information

A simplified image of two sugar molecules with antiparallel ribs is shown in Figure 1. The blue color in Figure 1 indicates 12 coordinate planes passing through the vertices of the polytope, including through the vertices between which flat nitrogenous bases could be located in the coordinate planes. From Figure 1 it follows, however, that not all 12 coordinate planes can have flat nitrogenous bases. They could be located only between the vertices of the polytope belonging to different sugar molecules closest to each other, i.e. between the planes $OH\text{-}H_{(3)}\text{-}H_{(1)}\text{-}F_i$ and $OH'\text{-}H_{(3)}{}' - H_{(1)}{}' - F_i'$. Between the distant peaks located on the opposite planes of sugar molecules, the location of flat nitrogenous bases is difficult, due to the obstacles encountered in the form of sugar molecule atoms and covalent chemical bonds. Therefore, out of the 12 possible coordinate planes, only five coordinate planes designated by the vertices of the polytope:

$$\begin{matrix} F_i'H_{(1)} & F_i'C_{(2)} & F_i'H_{(3)} & F_i'C_{(1)} & F_i'OH \\ H_{(1)}'F_i, & C_{(2)}'F_i, & H_{(3)}'F_i, & C_{(1)}'F_i, & (OH)'F_i \end{matrix}. \tag{8}$$

If among 12 pairs (1) canonical (Watson, Crick, 1953 a, b) base pairs for linked DNA molecules are distinguished, then there are only 4 of these pairs:

$$A{:}U,\ G{:}C,\ C{:}G,\ U{:}A. \tag{9}$$

If RNA molecules are linked, then in (9), base U must be replaced with base T.

Considering that there are 4 canonical base pairs, and there are 5 planes in which each pair can be located, the total number of base pairs location options is 20, i. e.

exactly as many different amino acids exist. Thus, each variant of the arrangement of base pairs corresponds to a certain amino acid connected in the ribosome to transport RNA. As is known (Spirin, 2019), the attachment of an amino acid to transport RNA carries out a specific enzyme that recognizes the amino acid and its corresponding transport RNA when they are combined. In this case, transport RNA is associated with messenger RNA, and between these RNAs at the junction point there again exists a polytopic of hereditary information. Thus, information on the state of messenger RNA is transmitted by transport RNA, on which the given amino acid joins. As we see in this process, there is no need to use the concept of a genetic code to describe it. The latter was introduced as an attempt to explain the difference in the number of canonical base pairs from the number of amino acids. However, this concept together with the concept of the codon does not explain anything. Taking into account the dimension of the polytope of genetic information allows us to determine the internal (hidden) order in the location of the bases. The situation is similar to the situation with the discovery of quasicrystals. X- ray diffraction patterns of some intermetallic compounds revealed a lack of translational symmetry in them (Shechtman, et al., 1984). This caused a stormy reaction of the scientific community. Such intermetallic compounds were called quasicrystals. However, in works of author (Zhizhin, 2014, 2015 a, b; 2016 a, b; 2018) it was shown that translational symmetry immediately appears if we consider diffractograms as projections onto a two-dimensional plane of a structure of higher dimension (a hidden order of diffractograms is revealed).

Currently, there are no results of special experiments to determine a unique correspondence between the location of this pair of bases in the polytopic of hereditary information and a specific amino acid. However, one can qualitatively confirm the existence of five different arrangements of base pairs using experimental data on the correspondence of codons (nucleotide triplets) to amino acids (Spirin, 2019). If we assume that the average nucleotide in the codon plays the main influence in this correspondence, then five amino acids correspond to triplets with an average nucleotide *U: Phe*, *Leu*, *Ile*, *Met*, *Val*. Since each amino acid corresponds to one of the positions of nucleotide *U*, it turns out that these directions exist 5. A similar situation is repeated with triplets with an average nucleotide *C*. They correspond to 5 amino acids: *Ser*, *Pro*, *Thr*, *Ala*, *Cus*. For triplets with an average nucleotide *G* there correspond the amino acids: *Ser*, *Cys, Trp*, *Arg*, *Gly*. Triplets with an average nucleotide A correspond to 7 amino acids: *Tyr*, *His*, *Glu*, *Asn*, *Lys*, *Gln*, *Asp*. Some violations of the strict unambiguous equality of 5 amino acid numbers to each nucleotide can be explained by the influence of lateral nucleotides in triplets and noncanonical parasitic bases.

It should be noted that taking into account the highest dimension of the polytopic of hereditary information is also necessary in the analysis of pandemics in which viral RNA penetrates the cell and forces the ribosomes to produce proteins necessary for viruses to reproduce.

CONCLUSION

The currently known forms of the helical structure of DNA molecules *A, B*, *Z*, and DNA molecules in the form of a closed ring were experimentally found in the works of Watson, J.D., Crick, F.H.C., Ha, S.C., and other scientists.

The representations of the sugar molecule and the residue of phosphoric acid in the form of polytopes of higher dimension are used. Based on these ideas and their simplified three - dimensional images, a three-dimensional image of nucleic acids is constructed. The geometry of the neighborhood of the compound of two nucleic acid helices with nitrogen bases has been investigated in detail. It is proved that this neighborhood is a polytope with anti-parallel edges of dimension 13 (13 – cross - polytope). This polytope is called the polytope of hereditary information. The geometry of the polytope of hereditary information is investigated. It is shown that in the flat coordinate planes of the polytope of hereditary information there are located flat complementary hydrogen compounds of the nitrogenous bases of two nucleic acid helices. It turned out that the number of these coordinate planes (12) is exactly as many as there are various options for hydrogen compounds of nitrogenous bases. An image of the polytope of hereditary information with a specific indication of its coordinate planes is constructed. It is shown that not all 12 coordinate planes can have flat nitrogenous bases. Only five coordinate planes can have nitrogenous bases. Considering that there are 4 canonical base pairs, and there are 5 planes in which each pair can be located, the total number of base pairs location options is 20.

It is essential that if a nucleic acid molecule contains several helices of nucleic acids, then the polytope of the hereditary information of the nucleic acid consists of several polytopes of the hereditary information of pairs of chemically linked single-stranded nucleic acids. Combining polytopes of hereditary information of pairs of nucleic acids linked by a hydrogen bond into one polytope contradicts the geometry of interacting nucleic acids. The law of conservation of incidence is only valid within a pair of chemically linked nucleic acids.

REFERENCES

Albert, B. (2002). *Molecular Biology of the Cell* (4th ed.). Garland Science.

Berg, J., Tymoczko, J., & Stryer, L. (2002). *Biochemistry*. W.H. Freeman & Co Ltd.

Brachet, J. (1941-1942). La detection histochimique et la microdosage des acides pentosenucleiques. *Enzymologia*, *10*, 87–96.

Butler, J. M. (2005). *Forensic DNA Typing*. Gardners Books.

Cantor, C. R., & Schimmel, P. R. (1980). *Biophysical Chemistry. Part I. The Conformation of Biological Macromolecules*. W.H. Freeman & Co Ltd.

Dahm, R. (2005). Friedrich Miescher and the discovery of DNA. *Developmental Biology*, *278*(2), 274–288. doi:10.1016/j.ydbio.2004.11.028 PMID:15680349

Frank - Kamenetskiy, M. (1986). DNA structure: A simple solution to the stability of the double helix? *Nature, 324*.

Frank - Kamenetskiy, M. (1988). H – form DNA and the hairpin – triplex model. *Nature, 333*.

Frank – Kamenetskiy. M.D. (2010). Queen of a living cell. From the structure of DNA to the biotechnological revolution. Moscow: Act – Press.

Ha, S. C., Lowenhaupt, K., Rich, A., Kim, Y.-G., & Kim, K. K. (2005). Cristal structure of a junction between B – DNA and Z – DNA revels two extruded bases. *Nature*, *437*(7062), 1183–1186. doi:10.1038/nature04088 PMID:16237447

Hawkes, E. J., Hennelly, S. P., Novikova, I. V., Irwin, J. A., Dean, C., & Sanbonmatsu, K. Y. (2016). COOLAIR antisense RNAs from evolutionarily conserved elaborate secondary structures. *Cell Reports*, *16*(12), 3087–3096. doi:10.1016/j.celrep.2016.08.045 PMID:27653675

Lindquist. (2016). Luminidependens (LD) is an Arabidopsis protein with prion behavior. (2016). *Proceedings of the National Academy of Sciences of the United States of America*, *113*(21), 6065–6070. doi:10.1073/pnas.1604478113 PMID:27114519

Mancuso, S. (2017). *PLANT REVOLUTION. Le piante hanno già inventato il nostro futuro*. GUIUTI EDITORE.

Nelson, D. L., & Cox, M. M. (2022). *Lehninger Principles of Biochemistry*. W.H. Freeman.

Poincare, A. (1895). Analysis situs. *J de e'Ecole Polytechnique.*, *1*, 1–121.

Ponnuswamy, P., & Gromiha, M. (1994). On the conformational stability of oligonucleotide duplexes and tRNA molecules. *Journal of Theoretical Biology*, *169*(4), 419–432. doi:10.1006/jtbi.1994.1163 PMID:7526075

Rasonable, R. R. (2011). Antiviral Drugs for Viruses Otber Than Humanimmunodeficiency virus. *Mayo Clinic Proceedings*, *86*(10), 1009–1026. doi:10.4065/mcp.2011.0309 PMID:21964179

Riemann, G. F. B. (1854). On the Hypotheses Which Lie at the Foundations of Geometry. Gottingen: Gottingen Observatory.

Shechtman, D., Blech, I., Gratias, D., & Cahn, J. W. (1984). Metallic phase with longer range orientational order and no translational symmetry. *Physical Review Letters*, *53*(20), 1951–1953. doi:10.1103/PhysRevLett.53.1951

Spirin, A. S. (Ed.). (1990). *Molecular biology. The structure and biosynthesis of nucleic acids*. High School.

Spirin, A. S., & Gavrilova, L. P. (1971). *The Ribosome*. Science.

Watson, J. D., & Crick, F. H. C. (1953a). Molecular structure of nucleic acids. *Nature*, *171*(4356), 738–740. doi:10.1038/171737a0 PMID:13054693

Watson, J. D., & Crick, F. H. C. (1953b). Genetical implications of the structure of deoxyribose nucleic acid. *Nature*, *171*(4361), 964–967. doi:10.1038/171964b0 PMID:13063483

Zhizhin, G. V. (2014). *World - 4D*. Polytechnic Service.

Zhizhin, G. V. (2015a). The dimensions of supramolecular compounds. *Biosphere*, *7*(2), 149–154.

Zhizhin, G. V. (2015b). Dimensions of compounds in supramolecular chemistry. *Int. J. Chem. Modeling*, *7*(2), 215–238.

Zhizhin, G. V. (2016a). From multidimensional phase spaces to multidimensionality of space in Nature. *Biosphere*, *8*(3), 258–267.

Zhizhin, G. V. (2016b). The structures, topological and functional dimension of biomolecules. *International Journal of Chemoinformatics and Chemical Engineering*, *5*(2), 44–58. doi:10.4018/IJCCE.2016070104

Zhizhin, G. V. (2018 a). *Chemical Compound Structures and the Higher Dimension of Molecules: emerging Research and Opportunities*. IGI Global. doi:10.4018/978-1-5225-4108-0

Zhizhin, G. V. (2018b). The Metric Structure of Nucleic Acids and the Higher Dimension of Their Constituents. *International Journal of Chemoinformatics and Chemical Engineering*, *7*(2), 1–15. doi:10.4018/IJCCE.2018070101

Zhizhin, G. V. (2019a). The Polytope of Hereditary Information the Structure, Location, Signification. *Biochemistry and Modern Applications*, *2*(1), 56–62. doi:10.33805/2638-7735.124

Zhizhin, G. V. (2019b). The Structure of the Polytope of Hereditary Information. *International Journal of Chemoinformatics and Chemical Engineering*, *8*(2), 7–22. doi:10.4018/IJCCE.2019070102

Zhizhin, G. V. (2019 c). *The Geometry of Higher – Dimensional Polytopes*. IGI Global. doi:10.4018/978-1-5225-6968-8

Zhizhin, G. V. (2019d). The Law of Conservation of Incidents of the Space of Nanoworld. *International Journal of Chemoinformatics and Chemical Engineering*, *8*(1), 25–46. doi:10.4018/IJCCE.2019010103

Zhizhin, G. V. (2019 e). *Attractors and Higher Dimensions in Population and Molecular Biology: Emerging Research and Opportunities*. IGI Global. doi:10.4018/978-1-5225-9651-6

Zhizhin, G. V. (2020). Hidden Nucleic Acid Bond Order? *Acta Scientific Biotechnology*, *1*(4), 34–36.

Zhizhin, G. V. (2021 a). The Structure of DNA Taking into Account the Higher Dimension of Its Components. In *Encyclopedia of Information Science and Technology* (5th ed., pp. 730–747). Ed. Mehdi Khosrow- Pour. doi:10.4018/978-1-7998-3479-3.ch050

Zhizhin, G. V. (2021b). Interaction of Nucleic Acids: Hidden Order of Interaction. *International Journal of Applied Research in Bioinformatics*, *11*(2), 1–8. doi:10.4018/IJARB.2021070101

Zhizhin, G. V. (2021c). Spatial Models of Sugars and Their Components. *International Journal of Applied Research in Bioinformatics*, *11*(2), 9–22. doi:10.4018/IJARB.2021070102

Zhizhin, G. V. (2021d). Hidden Order of Interacting Nucleic Acids. *Journal of Nature Science and Sustainable Technology*, *15*(4), 1–9.

Zhizhin, G. V. (2022a). The Properties of Biomaterials Hidden in the Spaces of Higher Dimension. *Annals of Advanced Biomedical Sciences.*, *5*(2), 1–9. doi:10.23880/aabsc-16000189

Zhizhin, G. V. (2022 b). *Non - Euclidean Geometry in Materials of Living and Non – Living Matter in the Space of the Highest Dimension*. Nova Science Publisher.

Zhizhin, G. V. (2022 c). *The Classes of Polytopes Higher Dimensions in Physics, Chemistry, and Biology. Emerging Research and Opportunities*. IGI Global.

Zhizhin, G. V., & Diudea, M. V. (2016). Space of nanoworld. In M. V. Putz & C. M. Marius (Eds.), *Sustainable nanosystems, development, properties, and applications* (pp. 221–245). IGI Global.

Zhizhin, G. V., Khalaj, Z., & Diudea, M. V. (2016). Geometrical and topology dimensions of the diamond. In A. R. Ashrafi & M. V. Diudea (Eds.), *Distance, symmetry and topology in carbon nanomaterials* (pp. 167–188). Springer. doi:10.1007/978-3-319-31584-3_12

Chapter 8
Methylated Polytope of Hereditary Information

ABSTRACT

The established ideas about the genetic code are considered, taking into account the achievements of scientists of the previous period. The shortcomings and contradictions of these ideas are noted, which are mainly related to the assumption of the three-dimensionality of biomolecules and the linearity of nucleic acids. Since in recent years it has been clearly shown that these assumptions are far from reality, a new concept of the genetic code has been put forward, considering the higher dimension of sugar molecules, phosphoric acid residues and nucleic acids. It is shown that nitrogenous bases are located in two-dimensional coordinate planes of the polytope of hereditary information, and the number of these coordinate planes is exactly equal to the number of possible types of amino acids, i.e., each coordinate plane corresponds to some amino acid.

INTRODUCTION

The covalent structure of purines and pyrimidines, as well as nucleotides included in the polytope of hereditary information, undergo spontaneous changes. Changes in the structure of DNA that leads to permanent changes in the genetic information encoded in it are called mutations. There are numerous data that testify in favor of the existence of a close relationship between the accumulation of mutations in the body and the processes of aging and carcinogenesis (Golubev, 2015). Virtually all life forms are exposed to high-energy radiation that can cause chemical changes in DNA (Nelson, Cox, 2022, V.2). It is known that radiation with wavelengths from 200 to 400 nm (UV), which makes up a significant part of the solar spectrum, can

DOI: 10.4018/979-8-3693-0588-1.ch008

cause the formation of pyrimidine dimers and other chemical changes in DNA. We are under the constant influence of ionizing radiation in the form of cosmic rays, which can penetrate deep into the earth, as well as radiation coming from radioactive elements such as radium, pluton, uranium, radon. X-rays, used in medical and dental examinations and in radiation therapy for cancer and other diseases, are another form of ionizing radiation (Melekhova, 2010; Polesskay, 2007).

Free radical reactions that ensure the respiration of organisms (Nelson, Cox, 2022) are accompanied by the generation of reactive oxygen species (ROS), which can initiate chain free radical processes that have a permanent destructive effect on cells. This action is enhanced by unfavorable environmental factors that cause oxidative stress in cells, which is accompanied by membrane destruction due to lipid peroxidation (Melekhova, 2010; Schaich, 2005).

DNA can also be damaged by chemically active substances released into the environment from industrial waste. These substances are able to change certain DNA bases. The most important source of mutations is oxidative damage to DNA. Reactive oxygen species - hydrogen peroxide, hydroxyl radicals and superoxide radicals - are formed under the action of radiation or as by-products of aerobic metabolism (Polesskay, 2007). These compounds can damage DNA in any of the many reactions that make up a large and complex group: from the oxidation of deoxyribose residues and nitrogen bases to chain scission. Cells have a fairly developed defense system against reactive oxygen species; however, some oxidizing agents bypass this defense system. Repair systems significantly reduce the extent of DNA damage. Some nucleotide bases in the DNA molecule are enzymatically methylated. Methylation is usually limited to certain sequences or regions of DNA. In some cases, the function of methylation is well understood, while in others its functional significance remains unclear. There are two main systems that provide DNA methylation. One of them serves as part of a defense mechanism that allows the cell to distinguish its own DNA from foreign by its own DNA with methyl groups and the destruction of foreign DNA that does not contain methyl groups. Another system methylates adenosine residues. The addition of methyl groups is carried out by the methylase component of the system, which corrects mismatched pairs of nitrogenous bases (Nelson, Cox, 2022, V.3).

Despite the key role of free radical chain reactions in maintaining the normal life of organisms and in the development of a number of pathological processes, the theoretical foundations for the study of free radical chain reactions, taking into account their characteristics in biological systems, are still insufficiently developed. A distinctive feature of free radical chain reactions in living systems is a wide variety of types of free radicals with very different molecular weights and activities: reactive oxygen and nitrogen species; radicals of high molecular weight organic compounds, such as lipids and proteins, etc. (Nelson, Cox, 2022). Created by N.N. Semenov, the theory of chain reactions (Semenov, 1934) referred primarily to reactions in the gas phase. In this theory, on the basis of simple physical assumptions, expressions were obtained for the length of the chain and the rate of the chain reaction. Moreover, when deriving these formulas, free radicals of different types were identified, attention was paid only to the total number of radicals. If in the analysis of gas chain reactions this approximation can be considered acceptable as a first step, then in the analysis of free radical chain reactions in biological systems, due to the large difference in the properties of free radicals, it should be considered incorrect.

This work is devoted to the mathematically substantiated derivation of general expressions for the chain length and the rate of a free radical chain reaction, taking into account the difference in the types of free radicals involved in the reaction.

A chain reaction can be considered as a sequence of random events (formation of radicals), the probabilities of outcomes of which depend on the states of the system only at the previous step of the reaction. Such sequences of random events are called Markov chains. The mathematical theory of Markov chains is currently quite well developed (see, for example, (Karlin, 1971). Elements of the theory of Markov chains were used in the description of radical liquid-phase polymerization (Berlin, et al., 1978) . However, the application of the theory of Markov chains to the description of free radical chain reactions in biological systems encounters certain difficulties. First of all, they consist in the fact that a free radical in a cell, entering into a reaction, can immediately create (in accordance with the reaction scheme) with the same probability several free radicals of various types. Such formation of radicals does not occur in radical polymerization, and similar processes are not described in the general and special theories of Markov chains (see, for example, (Harris, 1966)).

An alternative approach to the analysis of sequences of discrete intermolecular interactions that make up biochemical processes is the Gillespie algorithm (Gillespie, 1977), which has gained popularity in biochemistry in recent years (see, for example, (Belov, 2011)). In this approach, the system of differential equations operating with concentrations and rate constants is replaced by probabilistic algebraic equations. However, as shown in (Hayor, Jayaprakash, 2006) by a numerical comparison of various research methods, for the analysis of fast reactions with a large number of particles participating in the reaction, this method becomes inefficient, since it requires a very large computation time. Some attempts to speed up the calculation procedure in these cases do not fundamentally change the difficulties that arise (Gillespie, 1977, 2003) . Therefore, for the free radical chain reactions considered in this work, which are just characterized by a large number of active particles and a high reaction rate, the Gillespie algorithm is ineffective.

In this paper, we propose a new method for the mathematical description of free radical chain reactions in the biosphere with the formation (at each step of the reaction) from one radical at once several different free radicals. Using this method, a mathematical model of free radical chain reactions in the cells of living organisms under oxidative stress has been built.

GRAPHS OF FREE RADICAL CHAIN REACTIONS

Graphs of Chain Reactions Without Chain Termination

At the stage of continuation (spreading) of the chain, each active particle (radical) generates an active particle of a different kind. Therefore, not taking into account the possible chain break for the time being, an unbranched chain can be represented as follows,

$$A_1 \to A_2 \to A_1 \to A_2 \to \ldots\ldots\ldots$$

If the chain is branched, then at least three types of active particles take part in it. Moreover, in this reaction there are both stages of chain continuation and chain branching stages. Let a branched chain reaction have one stage of chain continuation and two stages of chain branching (as, for example, in the oxidation of hydrogen in the simplest description)

chain branching stages: $A_1 \to A_2 + A_3, A_3 \to A_2 + A_1$ (1)

chain continuation stage: $A_2 \to A_1$. (2)

We will consider A_1 it the initial active particle. Then this branched chain reaction can be represented as a graph of the tree type with a root corresponding to a particle A_1 (Zhizhin, 2012), where each branching level of the tree corresponds to a certain step of the chain reaction. Taking into account (1), (2), it is possible to determine the number of active particles of various kinds at any level of branching. Let $\alpha_n, \beta_n, \gamma_n$ b_e the corresponding numbers of particles A_1, A_2, A_3 a_t the level n. Then, taking into account (1), (2) we have the equalities,

$$\alpha_{n+1} = \beta_n + \gamma_n, \beta_{n+1} = \alpha_n + \gamma_n, \gamma_{n+1} = \alpha_n. \quad (3)$$

Since under the accepted conditions α_0=1, β_0=0, γ_0=0, then from equalities (3) we obtain

$$\begin{aligned}&\alpha_1 = 0, \beta_1 = 1, \gamma_1 = 1; \alpha_2 = 2, \beta_2 = 1, \gamma_2 = 0;\\&\alpha_3 = 1, \beta_3 = 2, \gamma_3 = 2; \alpha_4 = 4, \beta_4 = 3, \gamma_4 = 1;\\&\alpha_5 = 4, \beta_5 = 5, \gamma_5 = 4; \alpha_6 = 9, \beta_6 = 8, \gamma_6 = 4;\\&\alpha_7 = 12, \beta_7 = 13, \gamma_7 = 9;\end{aligned} \tag{4}$$

Summing up the active particles at each branching level of the graph, we find that the number of active particles with an increase in the branching level (reaction course) forms a Fibonacci series,

1, 2, 3, 5, 8, 13, 21, 34, 55, 89, 144, (5)

where each number is the sum of the previous two numbers.

It is easy to see from (8.4) that the numbers of particles, as functions, also form a Fibonacci series shifted relative to series (8.5) by two steps

0, 1, 1, 2, 3, 5, 8, 13, 21, 34, 55, 89, 144, (6)

and the numbers of particles and depending on form, respectively, the series,

1, 0, 2, 1, 4, 4, 9, 12, (7)

0, 1, 0, 2, 1, 4, 4, 9, 12, (8)

It can be seen that the series (7) is obtained from the Fibonacci series (8.6) by alternating the operations of adding and subtracting one from the corresponding numbers of the Fibonacci series, starting from the first number. And row (8) is row (7) with an added zero at the beginning. If the particle A_2 is considered as the initial particle, then it can be shown (by constructing the corresponding graph) that the sum of active particles forms a Fibonacci series (6) with a missing zero at the beginning of the series. If the particle A_3 is considered as the initial particle, then the sum of active particles forms series (5). The sums of a separate type of particles in these cases also represent series (5) - (7), differing only in the initial numbers.

In the general case, the number of active particles at each branching level of the chain reaction can be obtained by introducing the transition matrix from particles to particles in one step. If we consider reaction (8.1), then this matrix has the form

$$P = \begin{matrix} & A_1 & A_2 & A_3 \\ A_1 & 0 & 1 & 1 \\ A_2 & 1 & 0 & 0 \\ A_3 & 1 & 1 & 0 \end{matrix} \quad . \tag{9}$$

Units in the corresponding places of the matrix P indicate the reliability of the transformation of some particles into others for an elementary step of the reaction. It should be noted that the matrix P does not characterize the change in the state of the system as a whole, as happens in Markov chains, but characterizes the transition from one particle to another. Therefore, the sum of the numbers horizontally in the matrix does not have to be equal to one, as is required in the accepted description of Markov chains. The state of the system in this case is characterized by a set of numbers equal to the number of particles A_1, A_2, A_3 at a given level of reaction branching n. To find the results of transitions for n steps, it is necessary to calculate the n-th degree of the matrix P and multiply P^n by the row vector of the initial state of the system. For example, if in the initial state we have one particle A_1, then the row vector of the initial state has the form,

$$\begin{matrix} A_1 & A_2 & A_3 \\ (1, & 0, & 0) \end{matrix} .$$

Then, after two steps, the state of the system will be described by the row vector,

$$(1, 0, 0)P^2 = (1, 0, 0) \begin{vmatrix} 2 & 1 & 0 \\ 0 & 1 & 1 \\ 1 & 1 & 1 \end{vmatrix}, \ \alpha_2=2, \ \beta_2=1, \ \gamma_2=0.$$

This coincides with the parameter values $\alpha_2, \beta_2, \gamma_2$ in (4).

If, in the initial state, n_{A_1} particles were formed in a unit volume of gas, then after two steps of the reaction in this volume there will be $2n_{A_{10}}$ particles A*1*, $n_{A_{10}}$ particles A_2, and zero particles A_3. After n steps, respectively, the number of particles A_1,A_2,A_3 is found from the row vector $(n_{A_{10}}, 0, 0)P^n$.

Graph of a Branched Chain Reaction With Chain Termination

Let now with a certain probability the destruction of active particles be possible either as a result of the interaction of radicals with other molecules, or as a result of recombination. In this case, elementary reactions can also occur with probabilities other than unity. When constructing a chain reaction graph (1), you should add edges corresponding to chain breaks and indicate the corresponding probability for each edge (Zhizhin, 2012). There are relations between the probabilities of the elementary stages of the reaction $\delta + \varphi = 1$, $\xi + \psi = 1$, $\nu + \mu = 1$, where φ is the probability of the death of the radical A_1, ψ is the probability of the death of the radical A_2, μ is the probability of the death of the radical A_3, δ is the probability of the reaction $A_1 \rightarrow A_2 + A_3$, ξ is the probability of the reaction $A_2 \rightarrow A_1$, ν is the probability of the reaction $A_3 \rightarrow A_2 + A_1$, a particle that is not a radical ("dead" particle) is denoted by the sign $\varnothing$. The probabilities of chain continuation and its branching correspond to the activities of free radicals.

The reliable number of particles of a certain sort at the n-th branching level of the tree in this case is equal to the integer part of the sum of the probabilities of paths leading to a particle of this sort at the n -th branching level. It coincides with the corresponding number in the graph (Zhizhin, 2012) only at $\delta = \xi = \nu = 1$, and it is the smaller this number (the number of the Fibonacci series), the more the probabilities of chain termination at each elementary stage of the reaction differ from zero. The distribution of active particles at the n-th branching level of the reaction tree can be determined using the transition matrix from one active particle to another in one step. This matrix has the form,

$$
P = \begin{array}{c|cccc|} & A_1 & A_2 & A_3 & \phi \\ A_1 & 0 & \delta & \delta & \varphi \\ A_2 & \xi & 0 & 0 & \psi \\ A_3 & \nu & \nu & 0 & \mu \\ \phi & 0 & 0 & 0 & 0 \end{array}
$$

In order to determine the distribution of active particles at the n-th branching level of the reaction tree, if in the initial state there were $n_{A_{10}}$ particles A_1, it is necessary to multiply the row vector of the initial state by the n-th power of the matrix *P*.

Recently, the question of counteracting viral infections, their spread and development has become acute. Various stages of the virus's development from capsids to return to capsids are studied (Agol, et al., 1990; Novikova, et al., 2002; Novikova, 2007; Spirin, 1990; Wurm, et al., 2001; Wulan, et al., 2015; Lundberg, et al., 2013; Landberg, et al., 2013; Crotty, et al., 2002). Various drugs are being developed to inhibit various stages of viral population development (Gonzalez, et al., 2008; Dong, et al., 2020; De Clerey, 2004; Crotty et al., 2002; Caly et al., 2012; Razonable, 2011). However, we are forced to state that viruses have amazing properties of attacking a living organism, penetrating it, suppressing the protective properties of the organism, using the life-supporting functions of the organism to our advantage, reproducing inside the organism, often leading ultimately to the death of the organism. The drugs developed against viruses can often only temporarily extinguish the infection, and viruses are modified and, in a new form, attack living organisms again. What allows viruses to often be the winner in the fight against living organisms? It should be remembered here that viruses are equal ancient players in the development of life. It is assumed that they preceded the emergence of universal LUCA (Last Universal Common Ancestor) protocells, since structural and functional similarities were found between archaeal viruses and some DNA-containing eukaryotic viruses (Novikova, 2007). This leads to the idea that it was viruses that contributed to the creation of life on Earth.

During their existence (about 5 billion years), viruses have developed complex systems of control, information transfer, recognition of dangers and ways to overcome them. Viruses participated in the creation of DNA and are themselves carriers of genetic information. Communication of information is always an important issue in a complex self-regulating system. Following the model of Watson and Crick (Watson and Crick, 1953 a, b), the transfer of genetic information in a double-stranded DNA molecule occurs as a result of the complementary nitrogenous bases of the two helices of the molecule. At the same time, the genetic information is incorporated not only in the sequence of arrangement along the spirals of various nitrogenous bases (nucleotide), but also in the geometric characteristics of complementary compounds: distances between atoms in compounds (centers of glycosidic bonds, locations of glycosidic centers, angles between glycosidic bonds) (Spirin, 2019).

When constructing the DNA molecule, Watson and Crick did not consider the question about the arrangement of flat nitrogenous bases in the space of the DNA molecule. It was recently discovered (Zhizhin, 2019 a, b) that the vicinity of nitrogenous bases at the place of their connection in DNA is a polytope of dimension 13 (polytope of hereditary information), on the coordinate two-dimensional planes of which nitrogenous bases can be located. In this case, the geometric characteristic of the complementary compound of nitrogenous bases is the geometric characteristic of the structure of the hereditary information polytope, i.e. polytope of higher dimension. The main parameter here is incidence (Zhizhin, 2019 c), i.e. belonging of one element of some dimension of the polytope to other elements of another dimension. The structure of the polytope will include a listing of all such incidents. The law of conservation of incidents was established (Zhizhin, 2019 d), according to which the integral sum of incidents from elements of low dimension to elements of high dimension is equal to the integral sum of incidents from elements of high dimension to elements of low dimension. An analytical expression for calculating the integral sum of incidents over the polytope is obtained. The integral sum of incidents over the polytope is interpreted as the flow of information between the elements of the polytope. It is shown that an increase in the dimension of the polytope leads to a sharp increase in the value of the information flow between the elements of the polytope.

When analyzing the geometry of the hereditary information polytope, it was found (Zhizhin, 2020) that out of 12 coordinate planes of this polytope, nitrogenous bases can be located only in five coordinate planes. Considering that in each nucleic acid there are only four different classical nitrogenous bases, the total number of possible positions of nitrogenous bases in the space of the highest dimension of the nucleic acid molecule is twenty. This number coincides with the number of different amino acids required for protein synthesis. It follows from this that when considering the complementary compound of nitrogenous bases in the space of higher dimensions, it is not necessary to enter a genetic code, since the contradiction between the soap number of nitrogenous bases (4) and a large number of amino acids (20) has disappeared.

A distinctive feature of the DNA of a number of viruses is the presence of methylated nitrogenous bases in them, i.e. a methyl group is attached to nitrogenous bases. For example, in bacterial DNA methylated bases are 5′ - methylcytosine, 6′ - methylaminopurine (Novikova, 2007; Agol, et al., 1990). The origin of such bases is the enzymatic methylation of the already synthesized DNA strand. This process is carried out by virus-specific methylases, which use the methyl groups of S-adenosylmethionine of the host cell as a donor (Novikova, 2007).

In this work, the effect of methylation of nitrogenous bases on an increase in the intensity of transmission of hereditary information is theoretically investigated. This is due to an increase in the dimension of the hereditary information polytope during its methylation. It was shown that the flow of information during methylation of the hereditary information polytope increases 2000 times as compared with the flow of information in the absence of methylation. This may explain the high degree of organization of viruses when they attack living organisms.

It should be noted that the experimentally intensifying effect of methylation on the transmission of hereditary information is currently observed in plants (Hawkes, et al, 2016; Lindquist, et al., 2016; Mancuso, 2017).

NUCLIEC ACIDS METHYLATION

The concept of nucleic acids as chains containing only four types of nucleotides is somewhat simplified. In DNA, as follows from various sources, there is a certain number of methylated nitrogen bases, i.e. bases containing a methyl group – CH_3. For example, the Fisher formula of cytosine $C_4H_5N_3O$ has the form (Figure 1).

Figure 1. Fisher's formula for cytosine

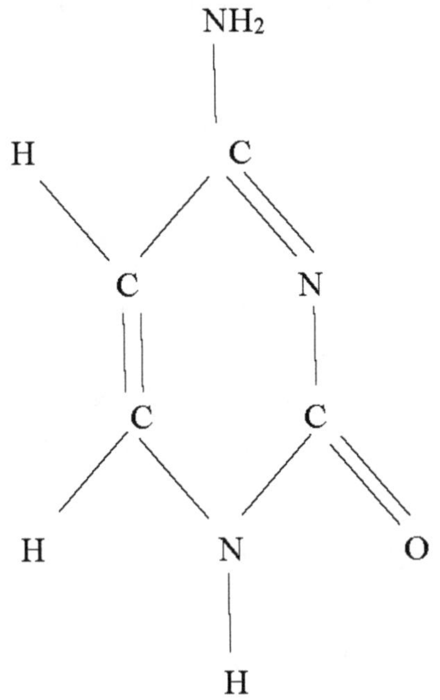

Figure 2. Fisher formula for Methylated Cytosine

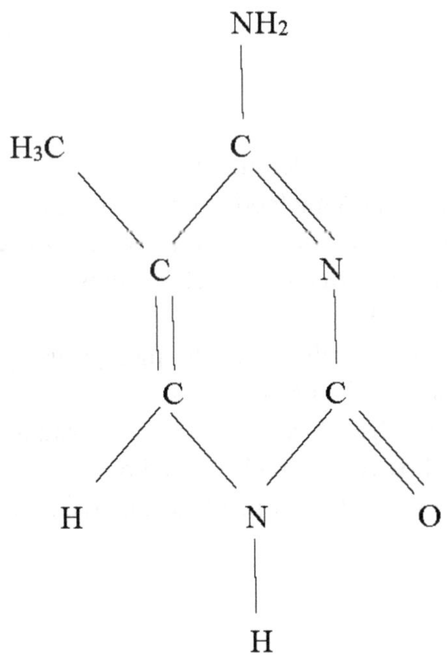

Fisher's formula for methylated cytosine (5 - methylcytosine) is shown in Figure 2.

The nitrogenous base adenine (aminopurin) $C_5H_5N_5$ has the Fisher formula (Figure 3)

Figure 3. Fisher Adenine formula

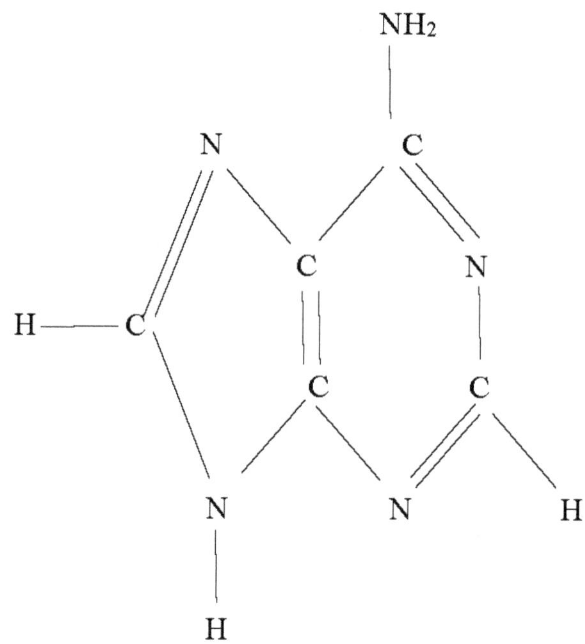

Methylated adenine (6-methyladenine) has the Fisher formula (*Figure 4*)

Methylation occurs after polynucleotide synthesis. It is used by organisms in its fight against viruses. However, viruses also use methylation to counteract the defenses of organisms. For example, viruses in nucleic acids contain oxymethylcytosine instead of cytosine (Figure 5)

Here, a hydroxyl group is included in the methyl groups. Emerging hydroxyl groups may carry one or two glucose residues.

One of the bacteriophages that infect *Bacillus subtitis* contains 5-hydroxymethyluracil instead of uracil, and 5-dioxipentyluracil instead of thymine (Figure 6)

Instead of guanine (Figure 7) as a result of methylation, O - methylguanine is formed (Figure 8)

Figure 4. Fisher formula for Methylated Adenine

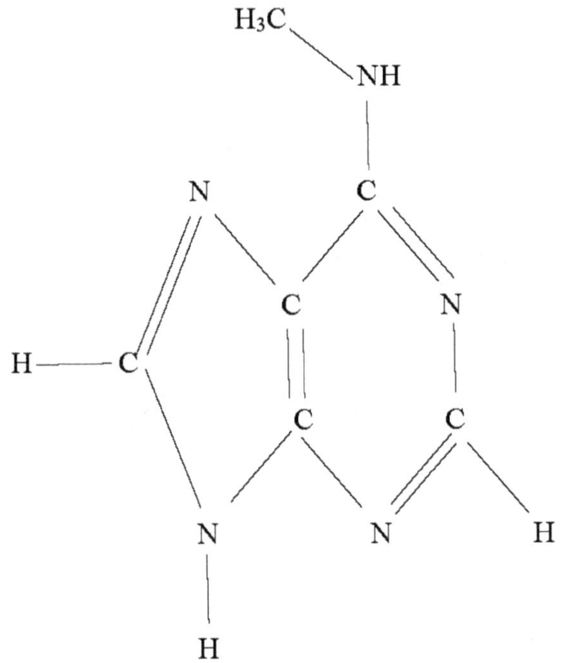

Figure 5. Fisher's formula of Oxymethylcytosine

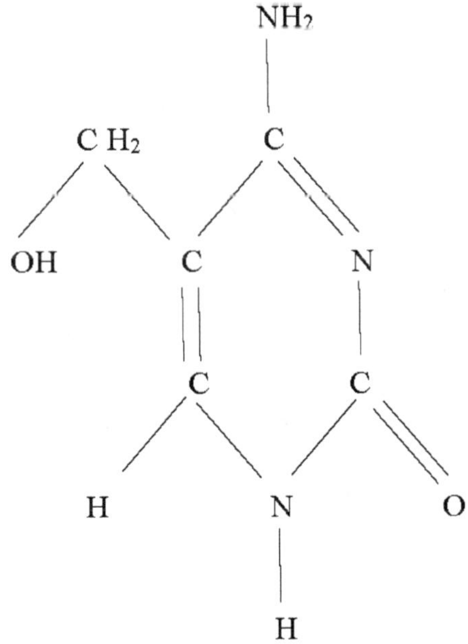

Figure 6. Fisher's formula 5-Dioxipentylluracil

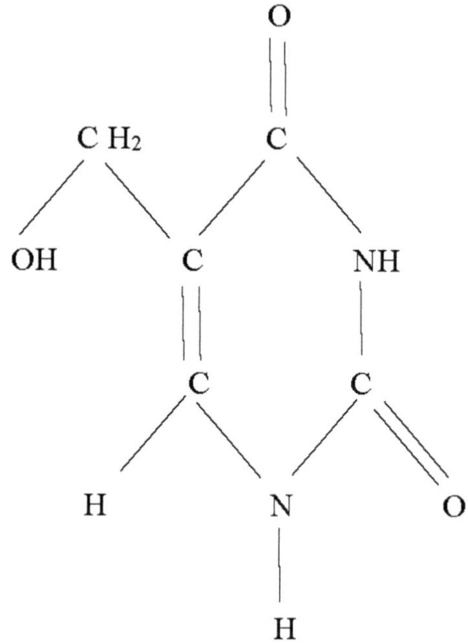

Figure 7. Fisher`s formula Guanine

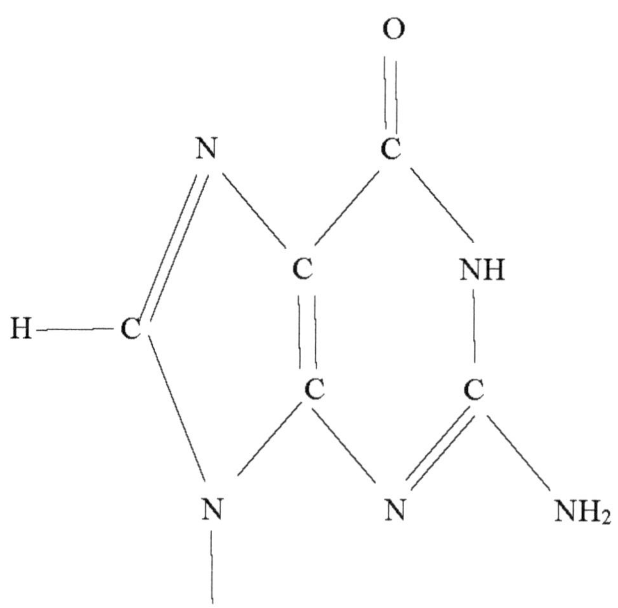

Figure 8. Fisher's formula O^6 - Methylguanine

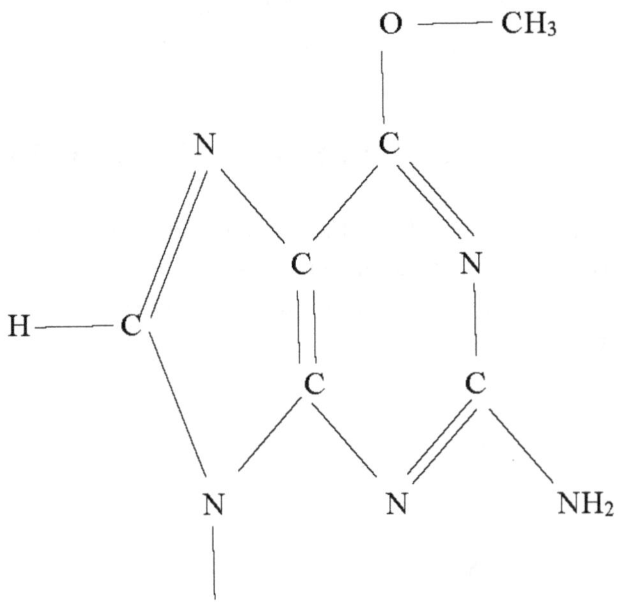

Thus, all classic nitrogenous bases as a result of methylation can take a different form. Moreover, it was established that these changes are very significant for the opposing cells of organisms and viruses.

In eukaryotic DNA, about 5% of the total cytosine content is 5′-methylcytosine. The degree of DNA methylation changes during development, and also depends on the age of the body and hormonal influences. DNA in vertebrates is especially strongly methylated. Methylation can accompany DNA replication; it is inherited and transmitted to offspring. Modified DNA can affect the local structure of the chromosome. Methylation promotes the transition to DNA Z-conformation, controlling the activity of genes (Spirin, 1990). However, the mechanism of action of methylation has not yet been disclosed.

Thus, all nucleotides of the classical type can, under appropriate conditions, be replaced by nature itself with modified nucleotides with methylated nitrogen bases. In this case, the complementarity of the base compound is preserved, and the symmetry of nucleic acids is preserved. What is the secret to the need for such modifications of nitrogen bases? Try to answer this question. In the author's previous works (Zhizhin, 2017, 2018, 2019 e), it was proved that many biomolecules have a higher dimension. In particular, it was proved that the connection of a carbon atom with four other atoms or functional groups has a dimension of 4. A methyl group attached to nitrogen bases is a special case of such a compound. Consequently, methylation changes the dimension of nitrogenous bases. Without a methyl group, the dimension of nitrogen bases is 2, with the addition of a methyl group, the dimension of nitrogen bases becomes 4 or more. Find out what this leads to.

METILATED POLYTOPE OF HEREDITARY INFORMATION

In the author's work (Zhizhin, 2019 a) it was found that between two nucleic acids in each local region of their bond with nitrogen bases there is a polytope of dimension 13 of the cross-polytope type. In a simplified three-dimensional form, this polytope is shown in Figure 9.

Figure 9. The polytope of hereditary information

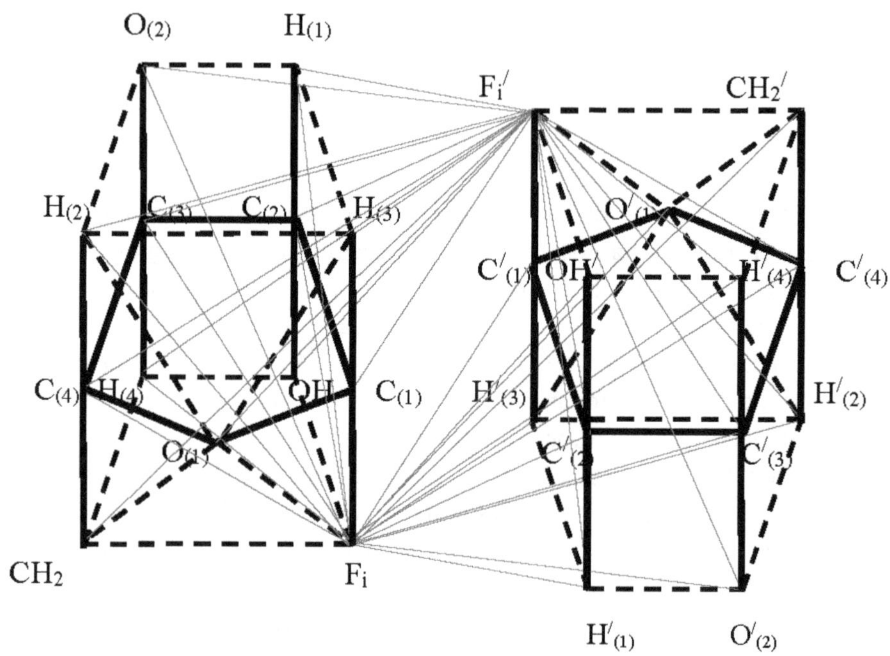

It is formed by two five-carbon sugar molecules in the antiparallel position, each of the sugar molecules has a dimension of 12. This polytope is called a polytope of hereditary information since it transmits hereditary information from one nucleic acid to another nucleic acid. A complete (un simplified) view of the polytope of hereditary information is presented in Figure 7.2 Chapter 7 (Zhizhin, 2019 a, b).

This polytope has 12 flat coordinate planes, i.e. exactly as many different pairs of connected nitrogenous bases exist in molecules DNA (Spirin, 2019).

$$A{:}U,\ U{:}A,\ G{:}C,\ C{:}G,\ G{:}U,\ G{:}A,\ U{:}U,\ U{:}C,\ A{\cdot}A,\ A{\cdot}C,\ U{:}G,\ A{:}G \qquad (10)$$

(the number of points between the bases indicates the number of bonds between the bases). For molecules RNA in (10) need change base *U* to base *T*. The blue color in Figure 8.9 indicates 12 coordinate planes passing through the vertices of the polytope, including through the vertices F_i and $F_i^{/}$ between which flat nitrogenous bases could be located in the coordinate planes. However, as shown in (Zhizhin, 2020), out of 12 coordinate planes (indicated in blue by Figure 9), only 5 planes can include nitrogenous bases. It is planes,

$$\frac{F_i^{/}H_{(1)}}{H_{(1)}^{/}F_i},\ \frac{F_i^{/}C_{(2)}}{C_{(2)}^{/}F_i},\ \frac{F_i^{/}H_{(3)}}{H_{(3)}^{/}F_i},\ \frac{F_i^{/}C_{(1)}}{C_{(1)}^{/}F_i},\ \frac{F_i^{/}OH}{(OH)^{/}F_i} \qquad (11)$$

This leads to an unambiguous correspondence of the number of different nucleotides in each type of nucleic acid (4), taking into account their possible location in a space of higher dimension (5) with the number of different canonical amino acids (20). When nucleic acids are labeled, their nitrogenous bases cease to be flat, since a tetrahedron with a center is formed in the vicinity of one of the carbon atoms of these bases, i.e. four-dimensional polytope. In this case, due to the antiparallel nature of molecular structures formed by antiparallel nucleic acid helices, it should be assumed that such a four-dimensional polytope appears in both linked nitrogen bases (Figure 10).

In Figure 10, the coordinate plane $\frac{F_i^{/}H_{(1)}}{H_{(1)}^{/}F_i}$ is selected as an example. In Figure 8.10, the edges of the tetrahedra with the center, marked in red, correspond to chemical covalent bonds. Thin black ribs are needed to create a closed convex body. The general construction of the polytope of hereditary information synchronizes the type of cross-polytope, but the number of vertices in it is 10 more than that of the non-methylated polytope of hereditary information. The complementarity of the nitrogenous base compound under these conditions is preserved, the symmetry of nucleic acids is preserved. This is facilitated by the mechanisms laid down by nature in nucleic acids that support complementarity and break the chains of nucleic acids when such conditions are violated (Spirin, 1990). According to (Zhizhin, 2019 c), there is a relation between f_i the number of dimension elements i in the cross-polytope and the dimension d of the cross-polytope itself,

Figure 10. The formation of four-dimensional polytopes in nitrogen bases during nucleation of nucleic acids

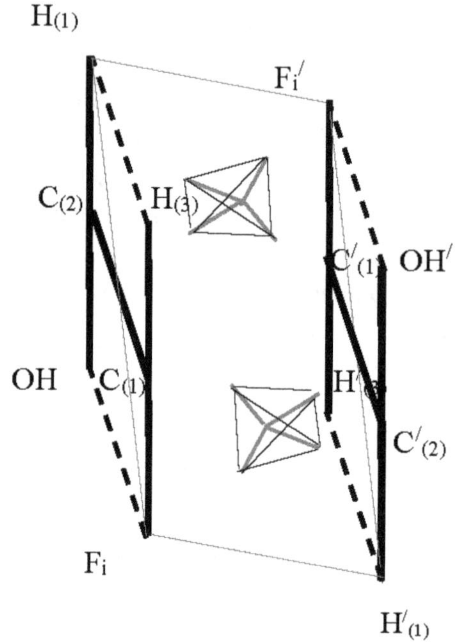

$$f_i(d) = 2^{1+i} C_d^{d-1-i} \quad (12)$$

Using formula (12) from the number of vertices at which the dimension $i = 0$, we can calculate the dimension of the methylated polytope of hereditary information

$$f_0 = 2C_d^{d-1} = 2d = (26+10) = 36, d = f_0 / 2 = 18. \quad (13)$$

Thus, the methylated polytope of hereditary information has dimension 18.

Consequently, the dimension of the methylated polytope of hereditary information increased by 5 units compared with the dimension of the unmethylated polytope of hereditary information (13). By the ratio (8.13), it is possible to determine the number of elements of different dimensions and in the methylated polytope of hereditary information.

The number of edges is $f_1(18) = 2^2 C_{18}^{16} = 612$,
The number of triangles is $f_2(18) = 2^3 C_{18}^{15} = 6528$,

The number of tetrahedrons is $f_3(18) = 2^4 C_{18}^{14} = 48960$,
The number of fourth-dimensional faces (simplexes) is $f_4(18) = 2^5 C_{18}^{13} = 274176$,
The number of fives-dimensional faces (simplexes) is $f_5(18) = 2^6 C_{18}^{12} = 1188096$,
The number of six-dimensional faces (simplexes) is $f_6(18) = 2^7 C_{18}^{11} = 4073472$,
The number of seven-dimensional faces (simplexes) is $f_7(18) = 2^8 C_{18}^{10} = 11202048$,
The number of eight–dimensional faces (simplexes) is $f_8(18) = 2^9 C_{18}^{9} = 24893440$,
The number of nine-dimensional faces (simplexes) is $f_9(18) = 2^{10} C_{18}^{10} = 44808192$,
The number of ten-dimensional faces (simplexes) is $f_{10}(18) = 2^{11} C_{18}^{11} = 65175552$,
The number of eleven-dimensional faces (simplexes) is $f_{11}(18) = 2^{12} C_{18}^{12} = 76038144$,
The number of twelve-dimensional faces (simplexes) is $f_{12}(18) = 2^{13} C_{18}^{13} = 70189056$,
The number of thirteenth-dimensional faces (simplexes) is $f_{13}(18) = 2^{14} C_{18}^{14} = 50135040$,
The number of fourteenth-dimensional faces (simplexes) is $f_{14}(18) = 2^{15} C_{18}^{15} = 26738688$,
The number of fifteenth-dimensional faces (simplexes) is $f_{15}(18) = 2^{16} C_{18}^{16} = 10027008$,
The number of sixteenth-dimensional faces (simplexes) is $f_{16}(18) = 2^{17} C_{18}^{17} = 2359296$,
The number of eighteenth-dimensional faces (simplexes) is $f_{17}(18) = 2^{18} C_{18}^{18} = 262144$.

The obtained numbers determine the structure of the methylated polytope of hereditary information. To prove the existence of a methylated polytope of hereditary information, we use the Euler-Poincaré equation (Poincaré, 1895)

$$\sum_{k=0}^{d} (-1)^i f_i(d) = (-1)^i + 1, \tag{14}$$

f_i is the number of elements of dimension i in polytope of dimension d.
Substitute the obtained values f_i in the Euler-Poincaré equation (14)

$36 - 612 + 6528 - 48960 + 274176 - 1188096 + 4073472 - 11202048 + 24893440 - 44808192 + 65175552 - 76038144 + 70189056 - 50135040 + 26738688 - 10027008 + 2359296 - 262144 = 0.$

Equality to zero of the familiar variables proves that the methylated polytope of hereditary information exists, its dimension is even and equal to 18.

THE LAW OF CONSERVATION OF INCIDENTS IN THE METHYLATED POLYTOPE OF HEREDITARY INFORMATION

The monograph (Zhizhin, 2019 c) introduced the concept of the incidence coefficients of elements of lower dimension with respect to elements of the higher dimension and elements of higher dimension with respect to elements of the lower dimension. The first characterizes the number of elements of a certain higher dimension to which the given element of a lower dimension belongs. The second characterizes the number of elements of a given lower dimension that are included in a particular element of a higher dimension. Here we must remember that the vertices of geometric elements of various dimensions are atoms, molecules or functional groups. Therefore, the incidence of geometric elements to a friend means contacts between particles of the matter, including living matter. The contact between particles of matter can be interpreted as the transfer of information on material structures, including biological structures. We introduce the notation: $k_{e_j^n e_u}$ is the number of elements of dimension *u*, which include an element of dimension *j* ($u > j$) with number *n*. Thus, $k_{e_j^n e_u}$ is the incidence factor of element *n* with dimension *j* relative to elements with dimension *u*. We introduce the notation also: $k_{e_j e_u^n}$ is the number of elements of dimension *j*, which included in element *n* with a dimension *u* ($u > j$). Thus, $k_{e_j e_u^n}$ is the incidence factor of element *n* with dimension *u* relatively to elements with dimension *j*.

In work (Zhizhin, 2019 c) was proved state,

Theorem 8.1. In any cross - polytope of dimension *d*, the sum of all incidents of elements of a lower dimension with respect to all elements of a higher dimension is equal to the sum of all incidents of elements of a higher dimension with respect to all elements of a lower dimension and equals the sum of the series,

$$\begin{aligned} & f_0(d)3^{d-1} + f_1(d)3^{d-2} + f_2(d)3^{d-3} + \dots + f_{d-1}(d)3^0 \\ & = \sum_{i=0}^{d-1} f_i(d)3^{d-1-i}, f_i(d) = C_d^{d-1-i} 2^{1+i}, i = 0 \div (d-1). \end{aligned} \tag{15}$$

Substituting in the formula (15) the values of $f_i(d)$ for $d = 18$ calculated in the previous section, we find that the sum of the incidents (information flow) for the methylated polytope of hereditary information takes the value

$$\sum_{i=0}^{d-1} f_i(d)3^{d-1-i} = 3.8143 \cdot 10^{12}. \tag{16}$$

Recall that the value of the information flow for the unmethylated polytope of hereditary information (see Zhizhin, 2019 b) was $1.78 \bullet 10^9$. Thus, the methylation of nucleic acids led to an increase in the flow of information between elements of the polytope of hereditary information by more than 2000 times.

CONCLUSION

From experimental studies of the chemistry of nucleic acids, it follows that mytilization of nucleic acids to a certain extent contributes to the protection of nucleic acids from various damaging factors. However, their role in protecting nucleic acids is not always understood. On the other hand, it follows from studies of viral infections that the concentration of methyl groups in the DNA of viruses is significant. Apparently, this allows the DNA of viruses to suppress the activity of the DNA of living cells and force them to work for the production of viruses in viral infections. Impact investigated of nucleic acid methylation on the geometric characteristics of nucleic acids. Such methylation is characteristic of viral nucleic acids when infected with viruses of organisms. It was shown that methylation of nitrogenous bases leads to an increase in their dimension, creating a higher dimension of the bases. This, in accordance with the recently established law for the conservation of incidents (Zhizhin, 2019 d), sharply increases the intensity of information transfer between elements of different dimensions of nucleic acids. At the same time, the intensity of transmission of hereditary information in the polytopic of hereditary information increases significantly. It has been established that the intensity of information transfer in the methylated polytope of hereditary information increases by more than 2000 times in comparison with the unmethylated polytope of hereditary information. It should be assumed that it is this property that activates viral nucleic acids when they infect organisms, weakening organisms, including due to the use of methyl groups of organism cells by viruses. To counteract this activation of viral nucleic acids it is necessary to increase the concentration of methyl groups in the nitrogenous bases of the nucleic acids of the organism. This will enhance the transfer of information in the nucleic acids of the organism, activating them in the process of confrontation with the nucleic acids of viruses. It should be emphasized that the dimension of the methylated polytope of hereditary information is 18, on 5 units higher than the dimension of unmethylated polytope of hereditary information. This indicates that the interaction of nucleic acids during infection of the organism with viruses occurs in space of a very high dimension.

To counteract this activation of viral nucleic acids, apparently, it is necessary to increase the concentration of methyl groups in the nitrogenous bases of the nucleic acids of the organism. This will enhance the transfer of information in the nucleic acids of the organism, activating them in the process of confrontation with the nucleic acids of viruses. It should be emphasized that the dimension of the methylated polytope of hereditary information is 18, i.e. 5 units higher than the dimension of unmethylated polytope of hereditary information. This indicates that the interaction of nucleic acids during infection of the organism with viruses occurs in space of a very high dimension, which must be taken into account when analyzing infection of organisms.

The use of methylation for the experimental study of possible practical counteraction to viral infections will require further development of the methylation technique, measurement of the methylation degree, and determination of ways to change this degree. These issues are currently practically not investigated and have not been considered in this work.

REFERENCES

Agol, V. I., Bogdanov, A. A., & Gvozdev, V. A. (1990). *Molecular biology: Structure and biosynthesis of nucleic acids*. Higher School.

Belov, O.V. (2011). Modeling of excisional repair of damaged stems in bacterial cells. *Physics of elementary particles and the atomic nucleus, 8,* 237 - 248.

Berlin, A. P., Wolfson, S. A., & Enikolopyan, N. S. (1978). *Kinetics ofpolymerization processes*. Chemistry.

Caly, L., Wagstaff, K. M., & Jans, D. A. (2012). Nuclear trafficking of proteins from RNA viruses: Potenial targetfor anti-viral? *Antiviral Research, 95*(3), 202–206. doi:10.1016/j.antiviral.2012.06.008 PMID:22750233

Carlin, S. (1971). *Fundamentals of the theory of random processes*. World.

Crotty, S., Cameron, C., & Audino, R. (2002). Rubaviriu's antiviral mechanism of action: Lethal mutagenesis? *Journal of Molecular Medicine, 80*(2), 86–95. doi:10.1007/s00109-001-0308-0 PMID:11907645

De Clerey, E. (2004). Antiviral drugs in current clinical use. *Journal of Clinical Virology*, *30*, 2. PMID:15125867

De Maio, D., & Coen, D.M. (2001). Replia Yion Strategies of DNA Viruses. *Fields Virology*, 1.

Dong, L., Hu, S., & Gao, J. (2020). Discovering drugs to treat coronavirus disease 2019(COVID -19). *Drug Discoveries & Therapeutics*, *14*(1), 58–60. doi:10.5582/ddt.2020.01012 PMID:32147628

Gale, M., Tan Seng-Lai, J. R., & Katze, M. G. (2000). Translational Control of Viral Gene Expression in Eukaryotes. *Microbiology and Molecular Biology Reviews*, *64*(2), 239–280. doi:10.1128/MMBR.64.2.239-280.2000 PMID:10839817

Gillespie, D. N. (1977). Stochastic simulations of coupled chemical reactions. *Journal of Physical Chemistry*, *81*(25), 2340–2361. doi:10.1021/j100540a008

Gillespie, D. N. (2001). Approximate accelerated stochastic stimulation of chemically reacting systems. *The Journal of Chemical Physics*, *115*(4), 1716–1733. doi:10.1063/1.1378322

Golubev, A. G. (2009). *Biology of life expectancy and aging*. N-L Publishing House.

Gonzalez Canga, A., Sahagún Prieto, A. M., Diez Liébana, M. J., Fernández Martínez, N., Sierra Vega, M., & García Vieitez, J. J. (2008). The pharmacokinetics and interactions of ivermectin in humans-a mini-review. *The AAPS Journal*, *10*(1), 42–46. doi:10.1208/s12248-007-9000-9 PMID:18446504

Harris, T. (1966). *Theory of branching random processes*. World.

Hawkes, E. J., Hennellu, S. P., Novikova, I. V., Irwin, J. A., Dean, C., & Sanbonmatsu, K. Y. (2016). COOLAIR antisense RNAs from evolutionarily conserved elaborate secondary structures. *Cell Reports*, *16*(12), 3087–3096. doi:10.1016/j.celrep.2016.08.045 PMID:27653675

Hayor, F., Jayaprakash, C.A. (2006). A tutorial on cellular stochastisity and Gillespies algorithm.

Lindquist. (2016). Luminidependens (LD) is an Arabidopsis protein with prion behavior. (2016). *Proceedings of the National Academy of Sciences of the United States of America*, *113*(21), 6065–6070. doi:10.1073/pnas.1604478113 PMID:27114519

Lundberg, L., Pinkham, C., Baer, A., Amaya, M., Narayanan, A., Wagstaff, K. M., Jans, D. A., & Kehn-Hall, K. (2013). Nuclear import and export inhibitors alter capsid protein distribution in mammalian cells and reduce Venezuelan Equine Encephalitis Virus replication. *Antiviral Research, 100*(3), 662–672. doi:10.1016/j.antiviral.2013.10.004 PMID:24161512

Lyles, D.S. (2000). *Cytopathogenesis and Inhibition of Host Gene Expression by RNA Viruses.*

Lyles, D. S. (2000, December). Cytopathogenesis and Inhibition of Host Gene Expression by RNA Viruses. *Microbiology and Molecular Biology Reviews, 64*(4), 709–724. doi:10.1128/MMBR.64.4.709-724.2000

Mancuso, S. (2017). *PLANT REVOLUTION. Le piante hanno già inventato il nostro futuro.* GUIUTI EDITORE.

Melekhova, O. P. (2010). *Free radical processes in epigenomic regulation of development.* Nauka.

Novikova, N. A. (2007). *Storage and sale of virus genetic information.* Lower Novgorod State University named N.I. Lobachevsky: Lower Novgorod.

Novikova, N.A., Novikov, V.V., Dobrotina, N.A., & Mazepa, V.N. (2002). *Virology.* – Lower.

Poincaré, A. (1895). Analysis situs. *J de e'Ecole Polytechnique., 1*, 1–121.

Rasonable, R. R. (2011). Antiviral Drugs for Viruses Otber Than Humanimmunodeficiency virus. *Mayo Clinic Proceedings, 86*(10), 1009–1026. doi:10.4065/mcp.2011.0309 PMID:21964179

Schaich, K. M. (2005). In F. Shahidi (Ed.), *Lipid Oxidation: Theoretical Aspects. Bailey's Industrial Oil and Fat Products* (1st ed., pp. 269–355).

Semenov, N. N. (1934). *Chain reactions.* Goshimtekhizdat.

Spirin, A. S. (Ed.). (1990). *Molecular biology. The structure and biosynthesis of nucleic acids.* High School.

Watson, J. D., & Crick, F. H. C. (1953a). Molecular structure of nucleic acids. *Nature, 171*(4356), 738–740. doi:10.1038/171737a0 PMID:13054693

Watson, J. D., & Crick, F. H. C. (1953b). Genetical implication of the structure of deoxyribose nucleic acid. *Nature, 171*(4361), 964–967. doi:10.1038/171964b0 PMID:13063483

Wulan, W. N., Heydet, D., Walker, E. J., Gahan, M. E., & Ghildyal, R. (2015). Nucleocytoplasmic transport of nucleocapsid proteins of enveloped RNA viruses. *Frontiers in Microbiology, 6*, 553. doi:10.3389/fmicb.2015.00553 PMID:26082769

Wurm, T., Chen, H., Hodgson, T., Britton, P., Brooks, G., & Hiscox, J. A. (2001). Localization to the nucleolus is a common feature of coronavirus nucleoproteins, and the protein may disrupt host cell division. *Journal of Virology, 75*(19), 9345–9356. doi:10.1128/JVI.75.19.9345-9356.2001 PMID:11533198

Zhizhin, G. V. (2017). The structure, topological and functional dimension of biomolecules. *J. Chemoinformatics and Chemical Engineering, 5*(2), 44–58.

Zhizhin, G. V. (2018). *Chemical Compound Structures and the Higher Dimension of Molecules: Emerging Research and Opportunities*. IGI Global. doi:10.4018/978-1-5225-4108-0

Zhizhin, G. V. (2019). Methylated Polytope of Hereditary Information. *Int. Journal of Applied Research on Public Health Management, 6*(2), 49–60.

Zhizhin, G. V. (2019a). The Polytope of Hereditary Information the Structure, Location, Signification. *Biochemistry and Modern Applications, 2*(1), 56–62. doi:10.33805/2638-7735.124

Zhizhin, G. V. (2019b). The Structure of the Polytope of Hereditary Information. *International Journal of Chemoinformatics and Chemical Engineering, 8*(2), 7–22. doi:10.4018/IJCCE.2019070102

Zhizhin, G. V. (2019 c). *The Geometry of Higher – Dimensional Polytopes*. IGI Global. doi:10.4018/978-1-5225-6968-8

Zhizhin, G. V. (2019d). The Law of Conservation of Incidents of the Space of Nanoworld. *International Journal of Chemoinformatics and Chemical Engineering, 8*(1), 25–46. doi:10.4018/IJCCE.2019010103

Zhizhin, G. V. (2019 e). *Attractors and Higher Dimensions in Population and Molecular Biology: Emerging Research and Opportunities*. IGI Global. doi:10.4018/978-1-5225-9651-6

Zhizhin, G. V. (2020). Hidden Nucleic Acid Bond Order? *Acta Scientific Biotechnology, 1*(4), 34–36.

Chapter 9
New Concept of the Genetic Code (in the Space of Higher Dimension)

ABSTRACT

Spatial models of the β - structures of protein molecules, forming layers of amino acids, in principle, of unlimited length for different conformations have been constructed. It is shown that the simplified flat Pauling models do not reflect the spatial structure of these layers. Using the recently developed theory of higher - dimensional polytopic prismahedrons, models of the volumetric filling of space with amino acid molecules are constructed. The constructed models for the first time mathematically describe the native structures of globular proteins. A model that simultaneously contains β - structures in parallel and anti - parallel positions and α - spirals. The turn of the polypeptide chain, which is often found in globular proteins, has been studied in higher dimensional space.

INTRODUCTION

At present, in numerous works of the author (Zhizhin, 2016, 2018, 2019 c, 2021 a, b, c), the highest dimension of biomolecules, including sugar molecules, phosphoric acid residue and nucleic acids, has been shown. The established ideas about the genetic code are based on the assumption that biomolecules are three-dimensional (Gamov, Rich, Yeas, 1956; Yeas, 1969; Crich, 1966) which contradicts the real geometry of biomolecules. Currently, the accepted genetic code is that each amino acid residue is encoded by three nucleotides in nucleic acids. It is believed that this is the only way to match nucleotides and amino acid residues due to the different number of types of nucleotides and types of amino acid residues. A list of this correspondence has been proposed (see, for example, Spirin, 2019), which is considered the universal

DOI: 10.4018/979-8-3693-0588-1.ch009

genetic code for all living organisms. However, experimental studies in recent years (Spirin, 2019) have discovered many deviations from the proposed genetic code in animals, protozoa, bacteria and fungi, algae, and plants. It is expected that new examples of deviations from the universal genetic code will be discovered in the future. Many amino acid residues are degenerate (corresponding to several triplets); for some amino acid residues this degeneracy reaches significant values. Therefore, ideas about the genetic code must be corrected in connection with the latest studies of the geometry of biomolecules. This chapter is dedicated to this. The first part of the chapter deals with the established ideas about the genetic code with critical remarks about its justification. The second part of the chapter presents a new concept of the genetic code, which takes into account the higher dimension of biomolecules, takes into account the geometric properties and features of polytopes of higher dimension (Zhizhin, 2019). This concept requires further research and experimental confirmation.

ESTABLLISHED IDEAS ABOUT THE GENETIC CODE

After the discovery of mRNA (1956 - 1961), the first step was to establish the order by which the amino acid sequence of proteins is written in the structure nucleotide of mRNA (Ramanathan, Brett, Siu – Hong Chan, 2016) and, accordingly, in the structure of nucleotide of one of the two DNA strands (Watson, Crick, 1953 a, b; Spirin, 1969; Spirin, Nemer, 1965; Crick, 1958, 1959, 1966). At the same time, the ideas were spread that the nucleotide chain in DNA can be considered linear, although there is no evidence in favor of such ideas (neither experimental nor theoretical). They were based only on the convenience of reasoning. Even before the discovery of mRNA, these ideas led to the conclusion that each amino acid must be encoded by at least three nucleotides (Gamov, Rich, Ycas, 1956; Ycas, 1969; Spirin, 1996, 2019). By this time it was known that proteins consist of 20 types of natural amino acids: Phenylonine (Phe, F), Leucyl (Leu, L), Isoleucyl (Ile, I), Methionyl (Met, M), Valyl (Val, V), Tyrosyl (Tyr, Y), Histidyl (His, H), Glutaminyl (Glu, Q), Asparginyl (Asn, N), Lysyl (Lys, K), Klutomil (Glu, Q), Seryl (Ser, S), Prolyl (Pro, P), Threonyl (Thr, T), Alanyl (Ala, A), Cysteinyl (Cys, C), Tryptophanyl (Trp, W), Arginyl (ARG, R), Glycyl (Gly, G), Aspartyl (Asp, D). While nucleic acids contain four types of nucleotides (A, C, G, T or U). It is obvious that only one nucleotide can correspond (encode) to each amino acid, since there are many more types of amino acids than types of nucleotides. There are 16 combinations of two nucleotides (AA, AC, AG, AT, CC, CA, CG, CT, GG, GA, GC, GT, TT, TA, TC, TG). However, if we assume that each amino acid corresponds to some two nucleotides, then the possible number of 11 base pairs to encode all 20 amino acids is also not enough.

From this it was concluded that the minimum number of nucleotides in a combination corresponding to one amino acid should be at least three. However, here the logic is clearly lame, since purely logically it can be assumed that one nucleotide corresponds to some amino acids, some pairs of nucleotides correspond to some amino acids, and some triple of nucleotides corresponds to some amino acids. After all, there are no conditions for the number of nucleotides corresponding to an amino acid, except for the simplest arithmetic calculation. Nevertheless, it was concluded that amino acids must be encoded by nucleotide triplets. This conclusion entailed many consequences necessary to reconcile the various consequences of this assumption. For example, the number of possible triplexes (as you can easily see) is 64. This number is more than enough to encode 20 amino acids. There can be two possible explanations for the excess number of triplets: either only 20 of some triplets are valid, i.e. can encode one or another amino acid, while the remaining 44 triplets have no meaning; or amino acids can be encoded by more than three nucleotides, in particular by more than three nucleotides. In the latter case, we will assume that the code is degenerate. It is easy to see that these variants are a consequence of previously accepted unfounded assumptions. Further, in principle, a triplet code, if we continue to insist on its tripletness, can be overlapping. In this case, the same nucleotide belongs to two overlapping triplets (weak overlap) or it can simultaneously belong to three overlapping triplets (strong overlap). In addition, coding triplets in a template nucleic acid may be adjacent to each other or separated by non-coding nucleotides. Numerous observations show that point mutations (substitutions of a single nucleotide in a nucleic acid molecule) usually lead to a change in only one amino acid in the corresponding protein.

This argued against the idea of an overlapping code and against the idea of an arrangement between triplets of non-coding nucleotides. Moreover, an overlapping code would inevitably lead to restrictions in the possible neighbors of a given amino acid residue, i.e. to a situation that has never been observed in real protein amino acid sequences. Thus, it was considered that a non-overlapping code is a more likely code variant. In addition, it was necessary to decide whether the coding triplets are separated by non-coding inserts, i.e. commas. If the code does not have commas, there is a problem of reading frame of the matrix nucleic acid. Only precise, triple-by-triplet reading, starting from a fixed point in the nucleotide chain, could lead to the formation of the corresponding amino acid sequence. Experimental work (Crick, Barnett, Brenner, Watts-Tobin, 1961; Benzer, 1961) established, according to the authors, that the code is triplet, degenerate, non-overlapping and does not contain commas. In these experiments, chemicals were used that led to the insertion or deletion (deletion) of one nucleotide during DNA replication and numerous gene B mutants were obtained in the field r II of the bacteriophage T4 (Lehninger, 1982; Nelson, Cox, 2020). For this purpose, proflamin and acridine dyes were used. Insertions or deletions of nucleotides near the start of a gene resulted in loss of gene expression. By recombination of various phage mutants in *E. coli*, phenotypic reverents were obtained, in which normal gene expression again occurred.

Revelent analysis showed that gene expression was restored if the region with the deletion was located close to the region with the insert, or vice versa. Gene expression could also be restored if two additional inserts (in mutants with an insert) or two additional deletions (for deletion mutants) were introduced in the vicinity of the region with the initial insertion or, respectively, the deletion.

Based on these experiments, the following conclusions were drawn.

1. Insertion or deletion of a single nucleotide at the beginning of a coding region results in the loss of the entire coding potential of the corresponding gene, rather than a simple point mutation.
2. A deletion (or insertion) located close to the initial insertion (or deletion, respectively) restored the initial coding potential of the sequence.
3. Three insertions or deletions located in the vicinity of the insertion (or deletion) also restored the initial coding potential sequence.

It remains not completely clear how it is possible to unambiguously introduce nucleotides and exclude them from the nucleotide chain with the help of a chemical agent with sufficient accuracy and reliability. Nevertheless, according to the authors, the experimental results prove that the code is triplet, the triplets are read sequentially without commas, starting from a strictly fixed point in one reading frame. These experiments, according to the authors, provided additional evidence that the code is degenerate: if many of the 64 possible triplets were meaningless, then there would be a high probability that at least one triplet appeared in the region between the insertion and deletion or between three inserts, where frameshift reading occurs. This would lead to interruption of the synthesis of the polypetide chain.

The deciphering of which amino acid a particular nucleotide triplet encodes began in 1961, when Nirenberg and Mystery-coding properties of synthetic polyribonucleotides in a cell-free protein synthesis system (Nirenberg, Mattei, 1961). The possibility of synthesizing polyribonucleotides of various compositions using the enzyme polynucleotide phosphorylase was first demonstrated earlier. The composition of polyribonucleotides synthesized in the system described by the authors depended only on the set of substrates. Using polyuridil acid as a polyribonucleotide template for ribosomes *E.coli*, Nirenberg and Mattei showed that this template provides the synthesis of polyphenylalanine. It was concluded that the triplet UUU codes for phenylalanine. Similarly, experiments with polyacidyl and polycytidylic acids showed that AAA codes for lysine and CCC codes for proline.

Further decoding of the genetic code was based on the use of synthetic heteropolyribonucleotides of various compositions, which were obtained by changing the amount and ratio of nucleoside diphosphate in the polynucleotide phosphorylase reaction (Grunberg – Manago, Ochoa, 1955; Nirenberg, Jones, Leder, Clarck, Sly, Pestka, 1963; Nirenberg, Leder, 1964). Thus, it was shown that the random polymer ensures the inclusion of four amino acids in the polynucleotide chain: phenylalanine, leucine, serine and proline. When the ratio in the polyribonucleotide was 1: 1, then all four amino acids were included in the chain in equal proportions (Speyer, Lengyel, Basilio, Wahba, Gardner, Ochoa, 1963).

Note that the statistical approach cannot give the exact composition in the chain. Soon the nucleotide sequences for all amino acids were determined.

Phe: UUU, UUC; Leu: UUA, UUG, CUU, CUC, CUA, CUG; Ile: AUU, AUC, AUA;
Met: AUG; Val: GUU, GUC, GUA, GUG; Ala: GCU, GCC, GCA, GCG;
Thr: ACU, ACC, ACA, ACG; Pro: CCU, CCC, CCA, CCG;
Ser: UCU, UCC, UCA, UCG, AGU, AGC; Tyr: UAU, UAC; OCHRE: UAA; (9.1)
Amber: UAG; His: CAU, CAC; Gin: CAA, CAG; Asn: AAU, AAC; Lys: AAA, AAG;
Asp: GAU, GAC; Glu: GAA, GAG; Gly: GGU, GGC, GGA, GGG;
Arg: AGA, AGG, CGU, CGC, CGA, CGG; Trp: UGG; Opal: UGA; Trp: UGG;
Cys: UGU, UGC.

Of the 64 triplets, three triplets UAG, UAA, UGA do not code for amino acids. They ensure the termination of the synthesis of the polypeptide chain and therefore are called stop codons.

From the list of the genetic triplet code (9.1) it follows that the degeneracy of the code does not extend to all 20 amino acids. Two amino acids - methionine and tryptophan are encoded by single triplets AUG, UGG and respectively, and three amino acids, namely leucine, serine and argenine are encoded by six triplets each. The remaining amino acids, with the exception of isoylicin, are encoded by two or four triplets, while isoylecin alone is encoded by three triplets. The difference in the degeneracy of amino acids remains unclear, it is not clear how the degeneracy of the code is related to the chemical structure of amino acids that have different types of radicals. It is easy to see that the universality of the presented triplet code is not absolute. It is suitable for most bacteria and eukaryotes. But later deviations from this code were discovered (Watanabe, Osawa, 1995). It turned out that eubacteria of the genus *Mycoplasma* and *Spiroplasma* have not one, but two coding triplets for tryptophan UGG, UGA. In one genus of ciliated protozoa (*Euplotes)*, UGA codes for cysteine, and not as in the genetic code (9.1). In other ciliate genera (*Tetrhymena, Paramecium, Stylonichia, Oxytricha*), and in at least one genus of unicellular green stop algae (*Acetabularia*), codons do not code for glutamine. In some yeasts

(*Candida*), the triplet codes for serine, not leucine. The protein-synthesizing systems of animals and fungal mitochondria are also characterized by a number of significant deviations from the triplet code (9.1). Tryptophan in such mitochondria is encoded by two triplets UGG, UGA, not one. In the mitochondria of all vertebrates, most invertebrates, and some fungi, the triplet AUA codes for methionine, not isoylicine. Thus, methionine is encoded by two triplets rather than one.

Triplets AGA and AGG in the mitochondria of most animals do not code for argenine. They are stop codons in vertebrate mitochondria and code for serine in the mitochondria of many invertebrates. In yeast mitochondria (*Saccharomyces Torulopsis*), a whole family of triplets CUU,CUA, CUC, CUG encodes threonine, not leucine.

Perhaps later additional examples of deviations from the presented triplet code will be discovered, especially among unicell ulnar eukaryotes (protozoa, algae, fungi).

Thus, the presented triplet genetic code is not universal. Numerous deviations from it have been proven. It should be recalled that the formulation of the triplet code arose due to a discrepancy between the number of nucleotide types and the number of amino acid types, and the need to restore this correspondence by introducing nucleotide triplets. In addition, the proof of the correctness of the triplet code was based on complex experiments related to the removal and insertion of nucleotides into the nucleotide chain, using synthetic substances, using statistical methods for processing the results of experiments, excluding their strict accuracy.

NEW CONCEPT OF THE GENETIC CODE

Now let us recall that in the previous chapters (Zhizhin, 2014, 2016, 2018, 2019 a; Zhizhin, Diudea, 2016) it was proved (Poincare, 1895) that the transmission of hereditary information takes place in a polytope of hereditary information of dimension 13. This polytope has 12 two-dimensional flat coordinate planes (Zhizhin, 2019 b, c, d). In each of these planes, in principle, pairs of nitrogenous bases connected by a hydrogen bond could be located. However, as follows from the previous geometric analysis, of these 12 coordinate planes, only five planes are free for the location of nitrogenous bases. It is on them that the canonical pairs of nitrogenous bases can be located. Now let us recall that in the previous chapters it was proved that the transmission of hereditary information takes place in a polytope of hereditary information of dimension 13. This polytope has 12 two-dimensional flat coordinate planes. In each of these planes, in principle, pairs of nitrogenous bases connected by a hydrogen bond could be located (Zhizhin, 2020, 2021 a, b, c; 2020). However, as follows from the previous geometric analysis, of these 12 coordinate planes, only five planes are free for the location of nitrogenous bases. It is on them that the canonical pairs of nitrogenous bases can be located, that is, any of the four pairs of nitrogenous bases can be located on each of these coordinate planes. When moving along the nucleotide chain (transition from one polytope of hereditary information to the neighboring polytope of hereditary information), 20 coordinate planes are formed with a characteristic pair of nitrogenous bases. Thus, each of these 20 coordinate planes is characterized by its location in the space of higher dimension and its own pair of nitrogenous bases. There are exactly as many of these coordinate planes as there are types of amino acids. Let us designate the types of amino acids a_i, 1÷20. Let us designate the selected coordinate planes in the polytope of hereditary information s_i, 1÷20 . Each amino acid a_i corresponds to a certain coordinate plane s_i. Let's say we have a four nucleotide type A, T, C, G. Each nucleotide is paired with another nucleotide from this quadruple $A \leftrightarrow T$, $G \leftrightarrow C$. The genetic code in this case consists in determining the correspondence of each of the 20 coordinate planes to one of the 20 amino acids. Suppose one of the chains has a nucleotide A with a certain coordinate plane s_i plane in the polytope of hereditary information. Let's denote this nucleotide A_i. It connects on this plane to the nucleotide T. Let's denote this nucleotide T_i This connection leads to the synthesis of amino acids a_i. It is easy to understand that in order to form all 20 amino acids, it is necessary to have five A nucleotides on the selected chain on five coordinate planes, five T nucleotides on the other five coordinate planes, five G nucleotides on five more coordinate planes, and five more C nucleotides on other coordinate planes. planes. In this case, all 20 amino acids are formed as a result of the connection of the corresponding nucleotides on 20 coordinate planes. Suppose one of the chains has a nucleotide A with a certain coordinate plane in the polytope of hereditary information. Let's denote this nucleotide F. It connects on this plane to the nucleotide E. This connection leads to the synthesis of amino acids. It is easy to understand that in order to form all 20 amino acids, it is necessary to have five A nucleotides on the selected chain on five coordinate planes, five T nucleotides on the other five coordinate planes, five P nucleotides on five more coordinate planes, and five more C nucleotides on other coordinate planes. planes. In this case, all 20 amino acids are formed as a result of the connection of the corresponding nucleotides on 20 coordinate planes in the polytope of hereditary information. The introduction of codons (triplets) is not required. This eliminates all the complexities associated with triplets. It is required to experimentally determine the correspondence between coordinate planes and types of amino acids. This match will also include matching a given location in the chain to one of the nucleotides A, C, T, G. Thus, five coordinate planes on the selected strand are associated with five A nucleotides, five coordinate planes on the selected strand are associated with nucleotides T, five coordinate planes are associated with nucleotides C, five coordinate planes are associated with nucleotides G. Naturally, when moving along the chain, the number of bonds on the planes increases. The order of the nucleotides will determine the order in which the amino acids are joined. The genetic code in this case consists in determining the correspondence of each of the 20 coordinate planes to one of the 20 amino acids.

CONCLUSION

An analysis of the scientific literature on the studies of the three -fluid genetic code showed that despite the many serious scientific works performed by outstanding researchers in many countries of the world, the proposed genetic code has many shortcomings, and it is universal to seize it. Experimental studies in recent years have found many deviations from the proposed genetic code. This applies to animals and protozoa, bacteria and fungi, algae, and plants. The triplets encoding the amino acid residues of many of them differ from the triplets recommended by the triplet genetic code. More examples of deviations from the universal genetic code are expected to be discovered in the future. The heterogeneous degeneracy of the genetic code is amazing. Some amino acid residues do not have degeneracy, and some amino acid residues have degeneracy equal to 6. There is no explanation for this situation within the framework of the universal genetic code. No relationship was found between the degree of degeneracy and the chemical structure of amino acid residues. The reason for many contradictions in the genetic code is the original position of its triplet nature. This position was based on an arithmetic discrepancy between the number of types of nucleotides and the number of types of amino acid residues. To coordinate these numbers, triplet codons were introduced, and this required even more complicating the genetic code. Determining the content of triplets required complex experiments related to changing the number of nucleotides in the nucleotide chain using chemical reagents. Synthetic statistical heteropolyribonucleotides of various compositions were used in the studies. Statistical methods cannot lead to unambiguous correct results.

These studies relied on old geometric ideas about the three-dimensional shape of biomolecules. Although almost 10 years ago it was proved that biomolecules have a higher dimension. This completely changes the idea of the interaction of molecules. In this regard, this chapter proposes a new concept of the genetic code, taking into account the higher dimension of nucleic acids and their components. Within the framework of this approach, it was proved that the area in which the transmission of hereditary information occurs is a polytope of dimension 13. This area has 12 two-dimensional flat coordinate planes, in which pairs of nitrogenous bases connecting nucleotides could be located. The geometrical analysis of polytopes of hereditary information showed that five pairs of nitrogenous bases can be located only in five of these 12 coordinate planes, since other coordinate planes are filled with elements of the polytope. Considering that any of the four pairs of nitrogenous bases can, in principle, be located in each of these planes, then 20 variants of coordinate planes arise with the corresponding variants of pairs of nitrogenous bases. Thus, the number of these variants of coordinate planes is exactly equal to the number of types of amino acids. Each amino acid corresponds to one of the coordinate planes with the corresponding pair of nitrogenous bases. The genetic code within the framework of these ideas consists in establishing a correspondence between the selected amino acid of a certain coordinate plane and a certain pair of nitrogenous bases. Establishing this correspondence should be the task of new experimental studies. Within the framework of these representations, it is not necessary to introduce triplets and the complexities and uncertainties associated with them.

REFERENCES

Benzer, S. (1961). On the topography of the genetic fine structure. *Proceedings of the National Academy of Sciences of the United States of America*, *47*(3), 403–415. doi:10.1073/pnas.47.3.403 PMID:16590840

Crick, F. H. C. (1958). On the protein synthesis. *Symposia of the Society for Experimental Biology*, *12*, 138–163. PMID:13580867

Crick, F. H. C. (1959). The present position of the coding problem. *Brookhaven Symposia in Biology*, *12*, 35–39. PMID:13812855

Crick, F. H. C. (1966). The genetic code: Yesterday, today and tomorrow. *Cold Spring Harbor Symposia on Quantitative Biology*, *31*(0), 1–9. doi:10.1101/SQB.1966.031.01.007 PMID:5237190

Crick, F. H. C., Barnet, L., & Brenner, S. (1961). General nature of the genetic code for proteins. *Nature*, *253*(4809), 374–375. doi:10.1038/1921227a0 PMID:13882203

Gamow, G., Rich, A., & Ycas, M. (1956). The problem of information transfer from the nucleic acids to proteins. *Advances in Biological and Medical Physics*, *4*, 23–68. doi:10.1016/B978-1-4832-3110-5.50006-6 PMID:13354508

Grunberg-Manago, M., & Ochoa, S. (1955). Enzymatic synthesis and breakdown of polynucleotides: Polynucleotide phosphorylase. *Journal of the American Chemical Society*, *77*(11), 3165–3166. doi:10.1021/ja01616a093

Lehninger, A. L. (1982). *Principles of Biochemistry*. The Jons Hopkins University: Worth Publishers, Inc.

Nelson, D. L., & Cox, M. M. (2022). *Lehninger Principles of Biochemistry*. W.H. Freeman.

Nirenberg, M. W., Jones, O. W., Leder, P., Clarck, B. F. C., Sly, W. S., & Pestka, S. (1963). On the coding of genetic information. *Cold Spring Harbor Symposia on Quantitative Biology*, *28*(0), 549–557. doi:10.1101/SQB.1963.028.01.074

Nirenberg, M. W., & Leder, P. (1964). RNA codeword and protein synthesis: The effect of trinucleotides upon the binding of sRNA to ribosomes. *Science*, *145*, 549–557. doi:10.1126/science.145.3639.1399 PMID:14172630

Nirenberg, M. W., & Mattei, J. H. (1961). The dependence of cell-free protein synthesis in *E. coli* upon naturally occurring or synthetic polyribonucleotides. *Proceedings of the National Academy of Sciences of the United States of America*, *47*(10), 1588–1602. doi:10.1073/pnas.47.10.1588 PMID:14479932

Poincare, A. (1895). Analysis situs. *J de é Ecole Polytechnique, 1*, 1 – 121.

Ramanathan, A., Brett, R.G., Siu – Hong, Chan. (2016). mRNA capping: biological functions and applications. *Nucleic Acids Res., 19*, 7511 – 7526.

Speyer, J. F., Lengyel, P., Basilio, C., Wahba, A. J., Gardner, R. S., & Ochoa, S. (1963). Synthetic polynucleotides and the amino acid code. *Cold Spring Harbor Symposia on Quantitative Biology*, *28*(0), 559–567. doi:10.1101/SQB.1963.028.01.075

Spirin, A. S. (1969). Informosomes. *European Journal of Biochemistry*, *10*, 20–35. doi:10.1111/j.1432-1033.1969.tb00651.x PMID:4899460

Spirin, A. S. (1996). Masked and translatable messenger ribonucleoproteins in higher eukaryotes. In J. W. B. Hershey, M. B. Mathews, & N. Sonenberg (Eds.), *Translational Control* (pp. 319–334). Cold Spring Harbor Laboratory Press.

Spirin, A. S. (2019). *Molecular biology. The ribosomes and protein biosynthesis*. Knowledge Lab.

Spirin, A. S., & Nrer, M. (1965). Messenger RNA in early see – urchin embryos: Cytoplasmic particles. *Science*, *150*(3693), 214–217. doi:10.1126/science.150.3693.214 PMID:5837337

Watanabe, K., & Osawa, S. (1955). tRNA sequences and variations in the genetic code. In: tRNA: Structure, Biosynthesis, and Function (eds. Soell D. and RajBhandary, U.L.): 225 – 250. ASM Press, Washington, DC.

Watson, J. D., & Crick, F. H. C. (1953a). Molecular structure of nucleic acids. *Nature*, *171*(4356), 738–740. doi:10.1038/171737a0 PMID:13054693

Watson, J. D., & Crick, F. H. C. (1953b). Genetical implications of the structure of deoxyribose nucleic acid. *Nature*, *171*(4361), 964–967. doi:10.1038/171964b0 PMID:13063483

Ycas, M. (1969). *The Biological Code*. North – Holland Publ. Co.

Zhizhin, G. V. (2014). On the higher dimension in nature. *Biosphere*, *6*(4), 313–318.

Zhizhin, G. V. (2016). The structure, topological and functional dimension of biomolecules. *J. International Journal of Chemoinformatics and Chemical Engineering*, *5*(2), 44–58. doi:10.4018/IJCCE.2016070104

Zhizhin, G. V. (2018). The Metric Structure of Nucleic Acids and the Higher Dimension of Their Constituents. *International Journal of Chemoinformatics and Chemical Engineering*, *7*(2), 1–15. doi:10.4018/IJCCE.2018070101

Zhizhin, G. V. (2019a). The Structure of the Polytope of Hereditary Information. *International Journal of Chemoinformatics and Chemical Engineering*, *8*(2), 7–22. doi:10.4018/IJCCE.2019070102

Zhizhin, G. V. (2019 b). *Attractors and Higher Dimensions in Population and Molecular Biology: Emerging Research and Opportunities*. IGI Global. doi:10.4018/978-1-5225-9651-6

Zhizhin, G. V. (2019d). The Law of Conservation of Incidents of the Space of Nanoworld. *International Journal of Chemoinformatics and Chemical Engineering*, *8*(1), 25–46. doi:10.4018/IJCCE.2019010103

Zhizhin, G. V. (2020). Hidden Nucleic Acid Bond Order? *Acta Scientific Biotechnology*, *1*(4), 34–36.

Zhizhin, G. V. (2021a). Interaction of Nucleic Acids: Hidden Order of Interaction. *International Journal of Applied Research in Bioinformatics*, *11*(2), 1–8. doi:10.4018/IJARB.2021070101

Zhizhin, G. V. (2021 b). The Structures of DNA taking into account the Higher Dimension of its Components. In Kosrapohr (Ed.). Encyclopedia of Organizational Knowledge, Administration, and Technologies. Hershey, PA: IGI Global. 730 - 745. doi:10.4018/978-1-7998-3479-3.ch050

Zhizhin, G. V. (2021c). Methylated Polytope of Hereditary Information. *International Journal of Applied Research on Public Health Management*, *6*(2), 49–60. doi:10.4018/IJARPHM.2021070104

Zhizhin, G. V. (2022). The Properties of Biomaterials Hidden in the Space of Higher Dimension. *Annals of Advanced Biomedical Sciences*, *5*(2), 1–9. doi:10.23880/aabsc-16000189

Zhizhin, G. V., & Diudea, M. V. (2016). Space of Nanoworld. In M. V. Putz & M. C. Mirica (Eds.), *Sustainable Nanosystems, Development, Properties, and Applications* (pp. 214–236). IGI Global.

Chapter 10
Synthesis and Structure of the Protein

ABSTRACT

The structure of the adamantane molecule, which coincides with the unit cell of diamond, is constructed. It is proved that the dimension of the adamantane molecule is nine. This is more than twice the dimension of the adamantane molecule previously considered. The higher dimension of adamantane molecules indicates the higher dimension of the molecules of substances containing adamantane molecules in their composition, in particular drugs based on adamantane. Examples of calculating the dimensions of known medicines are given. In this work, it is shown that the main part of adamantane itself, containing ten active carbon atoms, can be a means of impeding the formation of beta-amyloids that play the main role in various diseases of nerve cells.

INTRODUCTION

In 1950, it was found that the synthesis of proteins in living cells perform ribosomes and that RNA present in ribosomes (ribosomal RNA) is the overwhelming part of the RNA of the cell (Spirin, 2019). RNA chains are copies of certain functional segments of the DNA chain (genes). They serve as matrices for the synthesis of polypeptide protein circuits. Experiments were conducted on the definition of nucleotide composition of cellular RNA (Spirin, Belozersky, et. al., 1957; Belozersky, Spirin, 1958). It turned out that, contrary to the "central dogma of biology", the composition of the cellular RNA does not repeat the composition of the DNA, a copy of which should be cellular RNA. Thus, the existence of non - dodging RNA in living cells was open. Genetic control of protein synthesis was established in research by Francois Jacob and Jaco Monod (Jacob, Monod, 1961 a, b). Their

DOI: 10.4018/979-8-3693-0588-1.ch010

classical work on the induction of β - galactosidasnos activity in the cells *E.coli* led the authors to the formation of an operon hypothesis to explain the genetic control of protein synthesis in prokaryotes. This hypothesis later received confirmation in direct biochemical experiments. The type of regulation of the protein biosynthesis, considered in the Operon hypothesis, is control at the transcription stage, since regulation here is carried out mainly due to changes in the speed of genes transcription, i.e. on stadia formation of matrix RNA. The presented model of Operon first made it possible to consider the regulation of genes at the molecular level. The operator and a group of interconnected genes make up an Operon. The transcription of the group of metabolically interconnected genes is regulated by joining (or separation) of a special protein - repressor to the DNA operator section.

Each organism contains thousands of proteins that perform various functions. Structural proteins are responsible for maintaining the shape and structure of cells and tissues. Transport bands ensure the transfer of oxygen and carbon dioxide, ions and metabolites. Regulatory proteins are responsible for signaling. The largest group of proteins includes enzymes that control all processes in the body associated with chemical reactions (Dixon, Webb, 1979). In their form, proteins can represent a linear chain of amino acids linked by a peptide bond (primary structure). The secondary structure includes sections of the peptide bond with a certain conformation, stabilized by hydrogen bonds. The conformation of a protein, consisting of elements of the secondary structure and unstructured fragments, is called the tertiary structure. Due to the absence of non - covalent interactions, many proteins form symmetrical complexes with each other, which are called quaternary structures (Metzler, 1977; Lehnindger, 1982; Koolman, Roehm, 2013). The most common variants of the secondary structure of proteins are α - helices and β - layers. With such conformations, hydrogen bonds arise between adjacent helices and between adjacent layers. If the peptide chains in the β - layer go in the same direction, then the β - layers are called parallel. If the peptide chains in the β - layer run in opposite directions, then the β - layer is called antiparallel. Stacked in even layers, antiparallel β - layers form a conformation called fibroin. Soluble proteins differ from insoluble (fibrillar) proteins by the formation of spherical globules. Soluble proteins differ from insoluble (fibrillar) proteins by the formation of spherical globules. In a biologically active state, globular proteins are characterized by a specific structure - a native conformation. This conformation simultaneously includes both α - helices and β - layers. The linear Pauling model is currently used to describe various protein conformations (Pauling, Corey, 1950, 1953; Pauling, et al., 1951; Corey, Pauling, 1953). However, it has recently (Zhizhin, 2020 b, e; 2021 b, 2022 a) been shown that the Pauling model does not reflect and even contradicts the spatial structure of amino acid polypeptide molecules without taking into account their higher dimension. At present, the question of the correct description of protein conformations is associated, in particular with the need to find ways to combat so far incurable prion diseases Alzheimer`s, Parkinson`s, Kreietzfeld–Jakob`s and others. In all these diseases, insoluble large protein aggregates with β - structure are formed (β – amyloids), disrupting the brain, leading to neurodegeneration and disrupting protein metabolism. The structure of these formations is still unknown, and ways to combat their formation also remain unknown. It has been established that amyloids can be formed not only in the brain, but also in the liver, and enter the brain with lipoproteins. In this chapter, protein of different conformations in the form of fibroin and native structure are considered taking into account their real extent in the higher - dimensional space.

DIMENSIONS OF PROTEIN MOLECULES

Monomer units which are built of proteins are the 20 standard amino acids. These small molecules containing two different chemical functional groups capable of reacting with each other to form a covalent bond. Amino acids are chemical compounds that simultaneously include a carboxyl group (- COOH) and an amine group (-NH_2). These groups are covalently linked to a carbon atom (α - carbon), to which a hydrogen atom H and a side chain R are also covalently attached (Koolman, Roehm, 2013). Connection which determines the formation of protein polymer is called a peptide bond. In the formation of such a connection by joining together - COOH and -NH_2 with secretion a molecule of water. Amino acids have two enantiomorphic forms L and D, which are Fischer planar views shown in Figure 1. From Figure 1 it can be seen that the two enantiomorphic forms are related by mirror image relative to the line passing through the amine and carboxyl groups.

The side chains in Figure 10.1 have 20 variants in classic amino acids (Koolman, Roehm, 2013). They determine the properties of the corresponding amino acids. Fisher's projections do not reflect the spatial structure of amino acids, although it is clear that these are spatial objects. Spatial images of enantiomorphic amino acids (Zhizhin, 2016 a) are presented in Figure 2.

In Figure 2, covalent bonds are indicated in red. The rest of the segments define the spatial shape of the molecule, they are marked in black.

From Figure 2 it follows that the amino acid molecule is a tetrahedron with a center. The dimension of such a polyhedron can be determined by the Euler - Poincaré equation (Poincaré, 1895)

$$\sum_{i=0}^{n-1}(-1)^i f_i(P) = 1 + (-1)^{n-1}. \tag{1}$$

Figure 1. Enantiomorphic forms of the amino acid molecule
a) L - amino acid, b) D - amino acid

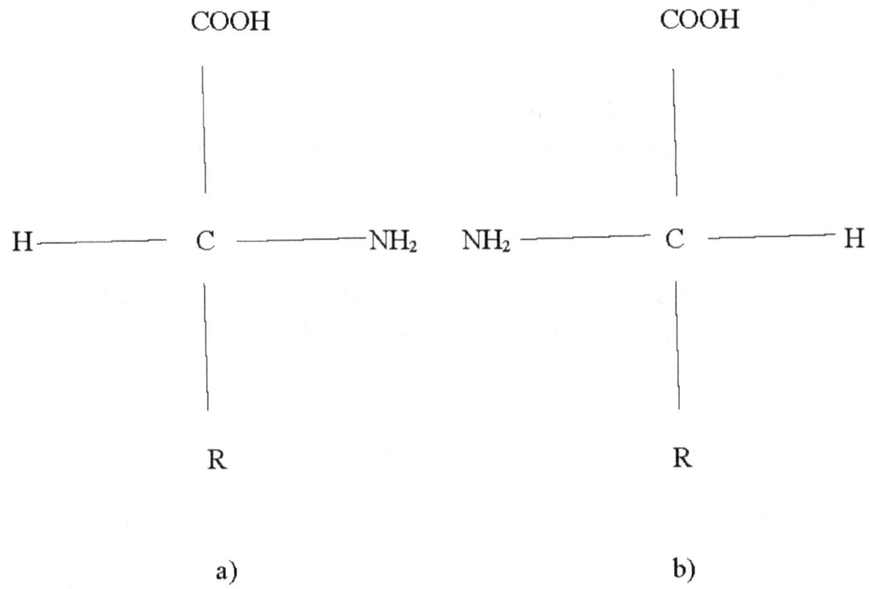

Figure 2. Spatial images of amino acids
a) L - amino acid, b) D - amino acid

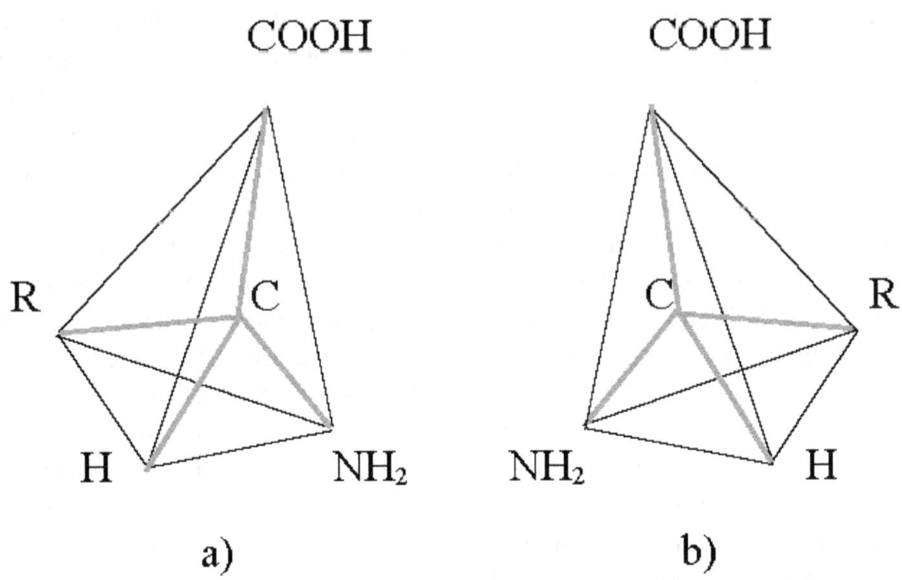

There n is dimension of a polytope P, $f_i(P)$ is the number of elements with dimension i in the polytope P.

From Figure 2 it follows that in any of the two enantiomorphic forms of the amino acid molecule, the number of vertices is 5, the number of edges is 10, the number of two - dimensional faces is 10, the number of three - dimensional faces is 5. In this case $f_0=5, f_1=10, f_2=10, f_3=5$.

Substituting the obtained values f_i, $i=0\div3$ into equation (1), we obtain,

$$5 - 10 + 10 - 5 = 0.$$

Thus, Euler - Poincaré's equation for a tetrahedron with a center holds for $n = 4$. This proves that a tetrahedron with a center (amino acid molecule) has dimension 4.

LINEAR POLYPEPTIDE CHAIN STRUCTURE

The study of diffraction patterns of crystals of amino acids (Pauling, Corey & Branson, 1951; Pauling & Corey, 1950, 1953; Corey & Pauling, 1953) made it possible to establish some values of bond bonds and bond angles in an amino acid. However, the interpretation of the spatial arrangement of atoms in an amino acid requires additional instructions. An example of the possibility of ambiguous interpretation of diffraction patterns can be found in numerous works on the description of quasicrystals. In particular, the apparent icosahedrical symmetry of the diffraction patterns of quasicrystals was proposed to be explained (Pauling, 1987) by multiple twinning of cubic crystals. Pauling based his arguments on the radial arrangement of the spots in the diffractograms. However, high - resolution photomicrographs refuted this assumption (Gratias, Cahn, 1986). The statement about the absence of translational symmetry in quasicrystals (Shechtman, et al., 1984) was constructed using the representations of the three-dimensional geometry of Euclid. Later this statement was refuted (Zhizhin, 2013 a, b; Zhizhin, 2014 b; Zhizhin, Diudea, 2016). It was proved that translational symmetry in the diffractograms of quasicrystals appears if the diffractograms are considered as projections of structures from the space of higher dimensions. When studying diffraction patterns of amino acid crystals in the works of Pauling joint authors the values of a number of parameters were not determined, for example, bond lengths $R - C$ depending on the species R, bond angles between carbon atoms and nitrogen atoms, between hydrogen atoms and R, etc. To compensate for these disadvantages, it was proposed to use a simplifying assumption about the arrangement of a group of atoms of a peptide bond and centers of tetrahedrons in a plane (Zhizhin, 2020 b, e) (Figure 3)

Figure 3. A simplifying assumption about the arrangement of a group of atoms of a peptide bond

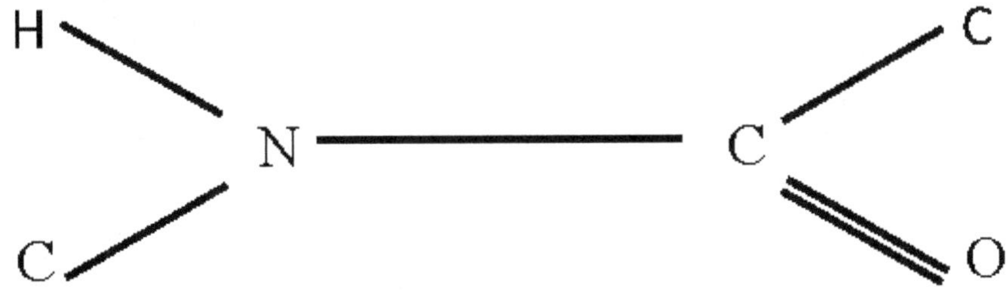

Under this assumption, knowledge of the indicated quantities was not required. It should be noted that in subsequent publications by other authors (Metzler, 1977; Lehninger, 1982; Koolman, Roehm, 2013) with reference to Pauling's work, it was argued that the arrangement of these atoms in the plane was proved in Pauling's works. But this is not so, to be convinced of this it is enough to refer to these works. This unproven statement became the basis for the construction of three - dimensional models of protein molecules of various conformations in Pauling's works. Let us consider this issue in detail when constructing spatial structures of a linear polypeptide chain. In this case, the atoms of amino acid molecules must be located on a set of parallel lines. Then the polypeptide chain of two molecules of *L* - amino acids, taking into account the four - dimensionality of amino acid molecules, has the form shown in Figure 4.

Figure 4. Linear polypeptide chain of L–molecules amino acids

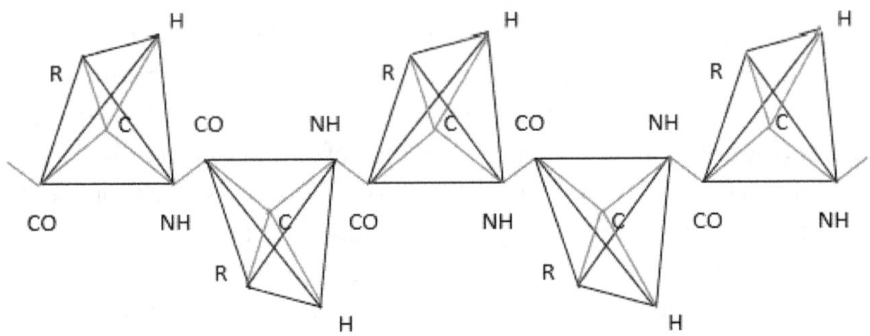

In the case of a *D* - amino acid, such a polypeptide chain is shown in Figure 5.

Figure 5 Linear polypeptide chain of D–molecules amino acids

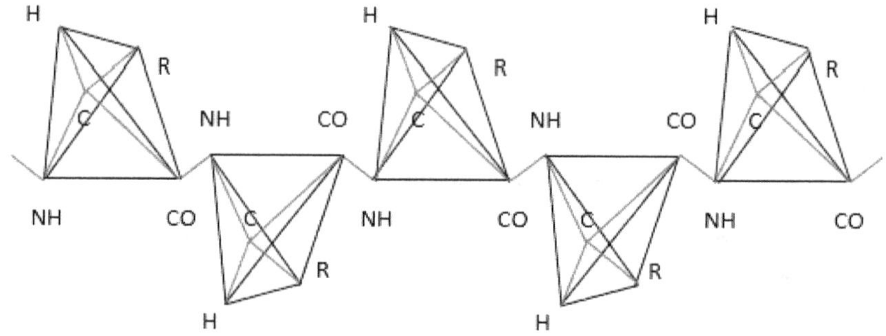

From Figure 4 and Figure 5 it can be seen that in order for the corresponding vertices of tetrahedrons with the center (i.e., specific atoms or functional groups) could lie on a system of parallel lines (i.e., form a linear polypeptide chain), it is necessary that the amino acid molecules in the chain alternate with their mirror image relative to the edge in the tetrahedron connecting the bond centers of the polypeptide chain (i.e., relative to the segment CO – NH). The chains in Figure 4 and Figure 5 differ by the mirror reflection of the molecules relative to the perpendicular to the segment *CO – NH* (*L* - and *D* - molecules).

In Figures 4 and 5, you can see the translational symmetry of the circuits. Moreover, the elementary translational element in this case is a group of two linked amino acid molecules, one of which is a mirror image of another amino acid molecule relative to the segment. From these two linked amino acid molecules, you can create a convex polytope by connecting edges of each vertex of any of the molecules with all other vertices in the group. This is how a simplex polytope of dimension 9 is formed, since the dimension of a simplex polytope is one less than the number of vertices in a simplex (Zhizhin, 2019 a), and the number of vertices in two tetrahedra with a center is 10. Therefore, a polypeptide linear amino acid chain has translation symmetry, in which an elementary translation element is a simplex polytope of dimension 9. In Figures 4 and 5, as well as in Fischer's projections (Figure 1) and in spatial images of molecules (Figure 2), in accordance with the definition of functional dimension, functional groups R, CO, NH were used. Analysis of Pauling's works on the structure of protein molecules (Pauling, Corey, 1950; Pauling, et al., 1951; Pauling, Corey, 1953; Corey, Pauling, 1953) shows that Pauling's model is nothing more than an assumption. Consider three amino acid residues linked by a peptide bond (Figure 6).

Figure 10.6 shows three amino acid residues in the form of tetrahedra, in the center of which there are three carbon atoms, from which four valence bonds proceed to functional groups located at the vertices of the tetrahedra. These valence bonds are marked with red ribs. Three amino acid residues are linked by two peptide bonds located within two contours, indicated by blue dashed lines. According to Pauling's model, the atoms located on the dotted contour and inside the contour are on the same plane. Here we must take into account that the lengths of chemical bonds connecting different pairs of atoms are different. Calculations of bond angles in Pauling's works are not given for any of the bonds. The plan is determined, as is known, by three points. The statement that some other point belongs to a given plane requires geometric proof. There is no such evidence in Pauling's work with coworkers. Therefore, Pauling's model remains unproven. From the point of view of chemistry and geometry, the tetrahedral coordination of functional groups relative to the central carbon atom in the amino acid molecule is much more important. This fact should be the basis for the analysis of protein structures.

Thus, it is essential (see Figure 6) that the α - carbon atoms do not lie in the same plane with the peptide bond atoms. From a comparison of Figures 4 – 6 it follows that α - carbon atoms lie either above the plane in which the atoms of the peptide bond are located, or below this plane, and can never lie on this plane. Therefore, Pauling's idea that α - carbon atoms lie in the same plane with functional groups CO, NH is erroneous. Therefore, Pauling's approximation cannot be used both in the construction of linear polypeptide chains and in the construction of more complex conformations of amino acid molecules.

Figure 6. Spatial image of three amino acid residues linked by a peptide bond

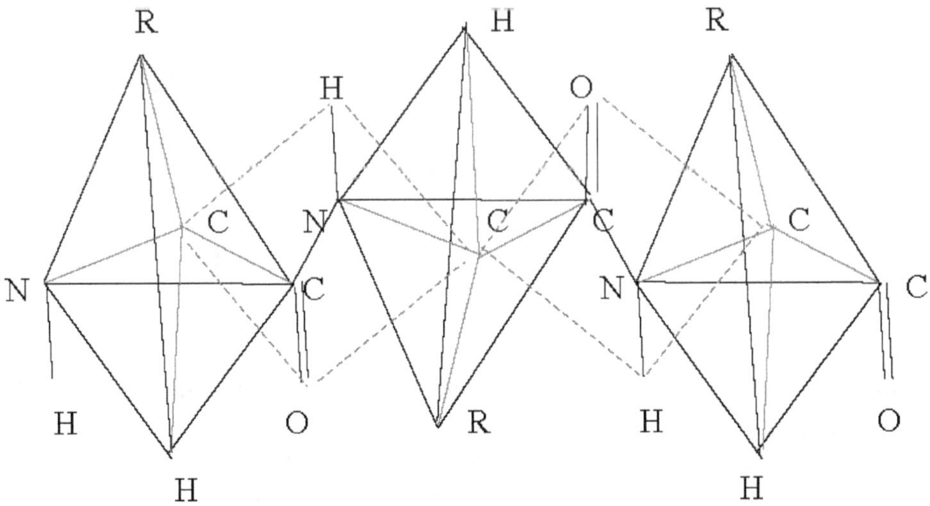

TURNS OF POLYPEPTIE CHAINS

The polypeptide chain can change its direction. This, in particular, is required in the formation of globular proteins. In this case, loops of the polypeptide chain are formed. Since the polypeptide chain, as we have seen, is a chain of polytopes of higher dimension, it is necessary to verify the possibility of such turns in the chain of polytopes of higher dimension, which are a chain of amino acid molecules. Figure 7 shows a possible implementation of such a rotation. It is interesting to note that when the polypeptide chain is rotated, the enantiomorphic shape of the amino acid molecules changes.

Figure 7. The turns of polypeptide chains

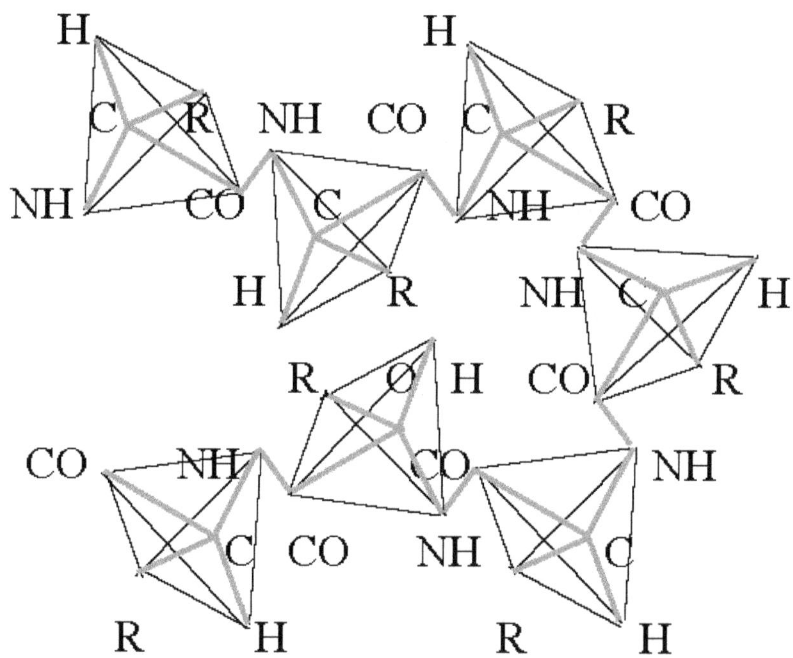

This follows from the geometric analysis of the rotation presented in Figure 7. If the upper part of the polypeptide chain in Figure 7 is considered as the initial one, then it is a sequence of *D* -molecules of the amino acid from left to right. When the chain is rotated, a polypeptide chain is formed, going from right to left (right turn), while the amino acid molecules look like *L* -molecules. If the lower part of the polypeptide chain in Figure 7 is considered the initial one, then the chain is a sequence of L - molecules going from left to right. When turning in this case (left turn), the molecules take the form of D - molecules going in sequence from right to left. The change in the shape of amino acid molecules during the rotation of the polypeptide chain can be considered, in a sense, the justification for the existence of enantiomorphic forms of amino acid molecules.

Changes in the enantiomorphic forms of amino acid molecules during chain rotations have not previously been paid attention to.

SPIRAL POLYPEPTIDE CHAINS

In the previous section, it was proved that the centers of tetrahedrons cannot lie on a flat sheet in which the peptide bond is located. Therefore, Pauling's idea of the connection of such flat leaves at the centers of tetrahedrons to form a helical peptide chain, as Pauling did, cannot be used. There is a relatively simple way of forming a helical polypeptide chain based on the concept of the four - dimensionality of amino acid molecules. This method was proposed in 2016 (Zhizhin, 2016 a, b). Suppose another amino acid molecule (not in a mirror image) is attached to an amino acid molecule (tetrahedron with a center) by a peptide bond C – N (Figure 8) so that in the general case between the valence bond C – C of the first molecule and the

Figure 8. Spiral polypeptide chains

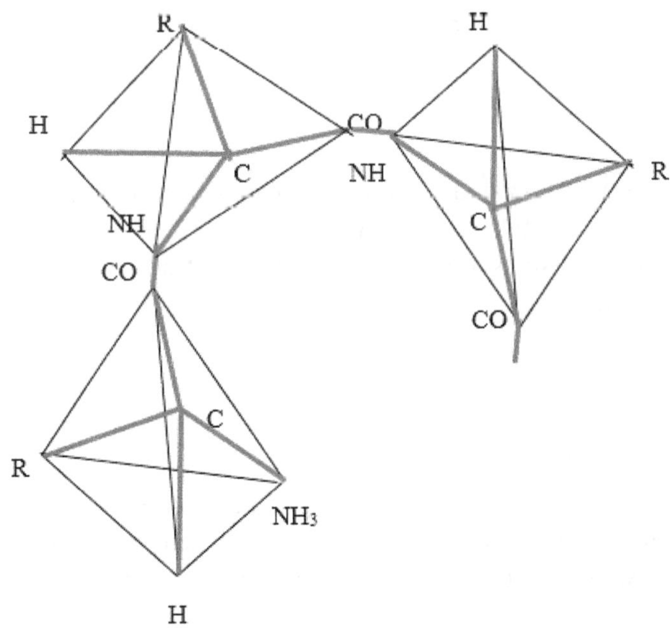

peptide bond C – N there is a certain angle less than 180 degrees (since there is no reason to assert that one of these connections is an exact continuation of the other).

Therefore, the second amino acid molecule is rotated by some angle relative to the first amino acid molecule. At the same time, while maintaining the shape of the second amino acid molecule, the second peptide bond, with the help of which the third amino acid molecule can be attached, turns out to be significantly (almost 90 degrees) rotated with respect to the first peptide bond. The third peptide bond is thus almost antiparallel to the first peptide bond. There is a rotation of the attached amino acid molecules relative to some center O (Zhizhin, 2016 a, 2018). If we designate the length of the peptide bond "*a*", the length of the edge of the tetrahedron between the atoms C and N "*b*", then in the projection onto the plane perpendicular to the axis of rotation, we get a polygon (Figure 9), which corresponds to the particular with the period of rotation of the molecules equal to 4.

Figure 9. Rotation of peptide bonds

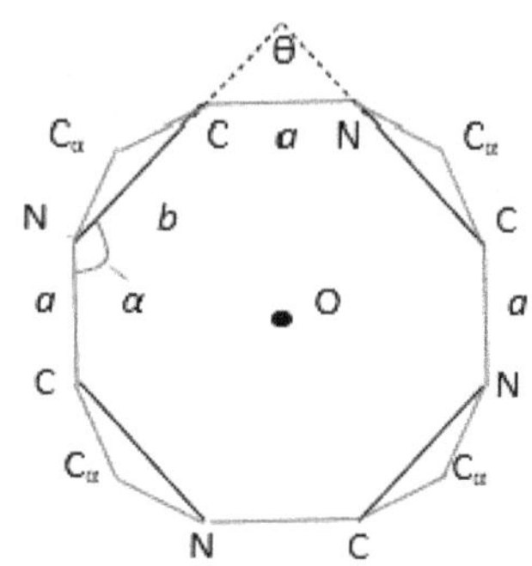

If we continue the nearest sides of the polygon *b* until they intersect with each other, we find the angle between these extensions. This is the angle θ=180–2(180–α), α *is* the angle between sides a, *and* b. *F*rom here we find the connection between the period n a*n*d the angle α

$$n\theta = 360, \alpha = \frac{1}{2}(\frac{360}{n} + 180).$$

If the period n = 4 so α = 135^0. Pauling had a minimum period of 3.6, but it should be noted that the rotation period of an amino acid molecule can only be an integer. If we now assume that the lines connecting the centers of the tetrahedrons do not lie in the plane of rotation but have a certain angle with respect to this plane, then we get a helicoid of rotation of the amino acid molecule (i.e., α - helix, since α -carbon atoms are located at the centers of the tetrahedrons). It can be shown that if we use the values of the valence bond lengths and the bond angle determined in Pauling's works, then to achieve the distance between the turns of 0.54 nm in the spiral (adopted in Pauling's works), the required value of the angle between the line connecting the centers of the tetrahedra and the plane of rotation, is equal to 1.5 degrees. It should be noted that when constructing a helix, a fundamentally different method of connecting amino acid molecules is used in comparison with a linear peptide chain. This method is more natural since it does not require the use of mirror - opposite amino acid molecules. Between adjacent turns of the spiral of amino acid molecules there is a stabilizing hydrogen bond between the *H* atoms of the peptide bond and the *O* atoms of the peptide bond with a double bond.

FOLDER STRUCTURES OF THE AMINO ACIDS

One of the important principles of protein structure formation is the formation of as many hydrogen bonds between groups as possible. This is especially pronounced in the so - called β - structures (folder structures) in which elongated chains of amino acid molecules are hydrogen bonded to each other to form a continuous layer. In these structures, the hydrogen bond plays a decisive role, and when calculating the dimensions of the compounds, it must be taken into account on an equal footing with the covalent bond. The repeating element in these β - structures is a polyhedron composed of two amino acid molecules. Figure 10 shows one of two variants of such polyhedrons in a fairly arbitrary position.

Figure 10. Polyhedron composed of two amino acid molecules

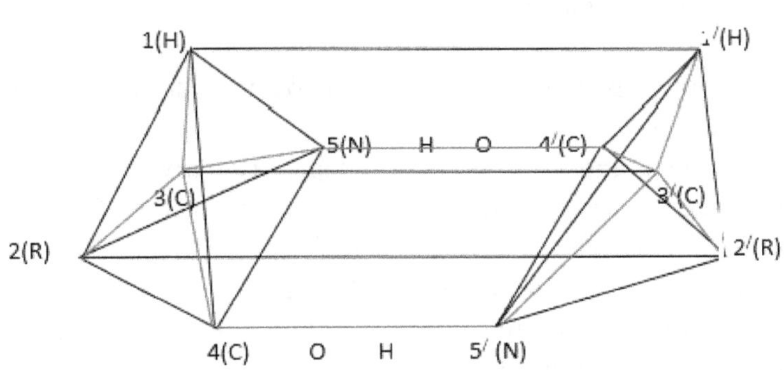

It should be noted that the H and O atoms are not the tops of the figure. Therefore, the polyhedron in Figure 10 has 10 vertices (f_0=10), 24 edges (f_1=24), 29 two - dimensional faces (f_2=29), 20 three - dimensional faces (f_3=20), 7 four -dimensional faces (f_4=7). Four - dimensional faces are faces 12345, 1′2′3′4′5′, like tetrahedrons with a center. In addition, tetrahedrons are four - dimensional faces located on four triangular faces of tetrahedron with centers 13451′3′4′5′, 13251′3′2′5′, 12341′2′ 3′4′, 32543′2′5′4′. These figures are topologically equivalent to the products of tetrahedrons and a one - dimensional segment (Zhizhin, 2019 a). The difference between these figures and the products of tetrahedron and a one - dimensional segment is only that in this case the one - dimensional segment as a factor in these figures discretely changes its length. But this does not change the number of elements of different dimensions included in the figure. Therefore, as well as the product of a tetrahedron and a one - dimensional segment, these figures include 8 vertices (f_0=8), 16 edges each (f_1=16), 14 flat faces (f_2=14), 6 three - dimensional faces (f_3=6). Substituting these numbers into the Euler - Poincaré equation (10.1), we obtain

$$8 - 16 + 14 - 6 = 0.$$

This proves that for all the listed figures with vertices 3 and 3′ the Euler - Poincaré equation is fulfilled for n = 4, i. e. all of them have dimension 4. Another four - dimensional figure is formed by two tetrahedrons 1254 and 1′2′5′4′ connected by parallel lines. They are also topologically equivalent to the product of a tetrahedron and a one - dimensional segment. Thus, the total number of four - dimensional figures included in Figure 10 is 7. Substituting the values obtained for Figure 10 into the Euler - Poincaré equation (10.1), we find,

$$10 – 24 + 29 – 20 + 7 = 2.$$

This proves that in this case the Euler - Poincaré equation is fulfilled for n = 5, i. e. Figure 10.10 has dimension 5.

In a continuous layer of amino acid molecules, the polyhedron in Figure 10 alternates with another polyhedron formed by two tetrahedrons with center at other positions (Figure 11).

It can be seen that Figure 10.11 is also topologically equivalent to the product of a tetrahedron with center on a one - dimensional segment and therefore has dimension 5. The polyhedrons in Figure 10.10 and Figure 10.11 cannot directly adjoin each other. Between them are other polyhedron formed from two three - dimensional pyramids attached to each other along flat faces. One of these topologically equivalent polyhedrons is shown in Figure 12.

Figure 11. Second polyhedron formed by two tetrahedrons with center

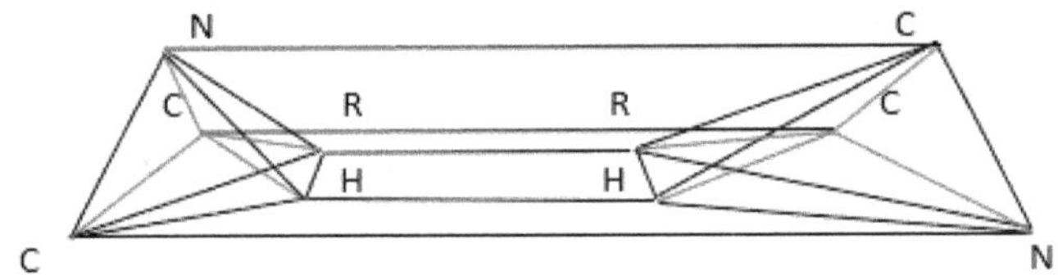

Figure 12. Polyhedron of two three-dimension pyramids

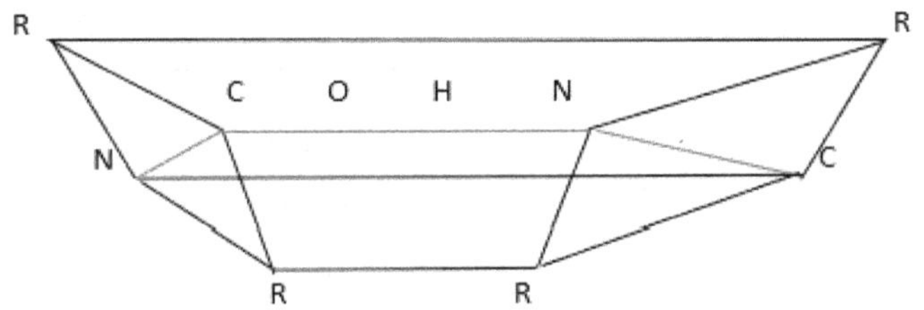

You can see that it has 8 vertices (f_0=8), 14 edges (f_1=14), 9 two - dimensional faces (f_2=9), 3 three - dimensional faces (f_3=3). Substituting these numbers into the Euler - Poincaré equation (11), we find

8 - 14 + 9 - 3 = 0

Therefore, Figure 10.12 satisfies the Euler - Poincaré equations for n = 4, i. e. it has dimension 4. A topologically equivalent Figure 13 also has dimension 4.

Figure 13. Polyhedron of two three-dimension pyramids equivalent figure 12

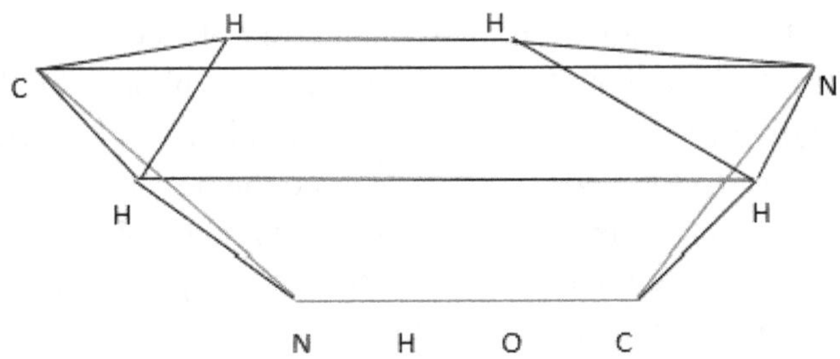

A part of the common layer composed of periodically repeating polyhedrons 8, 9, 10, 11 is shown in Figure 14.

Figure 14. Layer composed of periodically repeating polyhedrons 8,9,10, 11

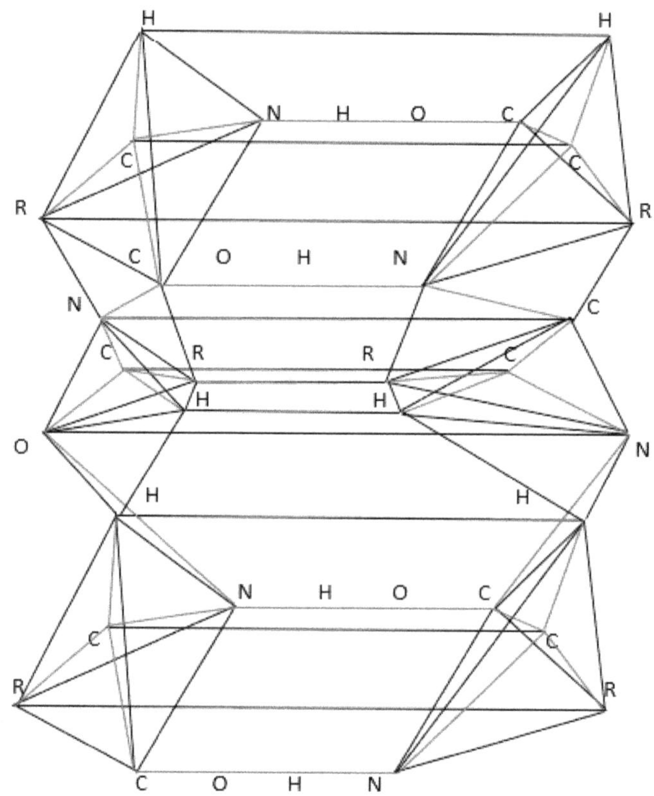

Figure 14 shows that the β - structure has translational symmetry. An elementary element of translational symmetry is the sequence of polyhedrons shown in Figures 10 – 13. If we denote the polyhedron in Figure 10 by symbol 5 (1), the polyhedron in Figure 11 by symbol 5 (2), and the polyhedron in Figure 12 by symbol 4 (1), the polyhedron in Figure 13 by the symbol 4 (2), then the translational element of the β - structure in Figure 14 is the polyhedron 5 (1)-4 (1)-5 (2)-4 (2). This polyhedron contains 24 vertices. If we connect each vertex of this polyhedron with all other vertices by edges, we get a single convex simplex polyhedron of dimension 23. Thus, the translational element of the β - structure in Figure 14 is a simplex of dimension 23.

In the literature, the β - structures traditionally described are called antiparallel and are depicted as alternating flat two - dimensional stripes (sheets).

However, it is clear that the name anti - parallel in this case does not reflect much of the essence of the structure in Figure 14. More precisely, this structure could be called non - parallel. In addition, and this is essential, the β - structures are fundamentally not flat and represent periodically repeating 5 -dimensional and 4 - dimensional regions, and each of these regions has two topologically equivalent images.

There are β - structures that include parallel 4 - dimensional amino acid molecules with two different orientations. In these structures, alternation of 5 - dimensional and 4 - dimensional regions (polyhedrons) also occurs. In continuous form, it is presented in Figure 15. Figure 15 shows that the β - structure has translational symmetry too.

Figure 15. Folder structure with parallel molecules amino acid in two different orientations

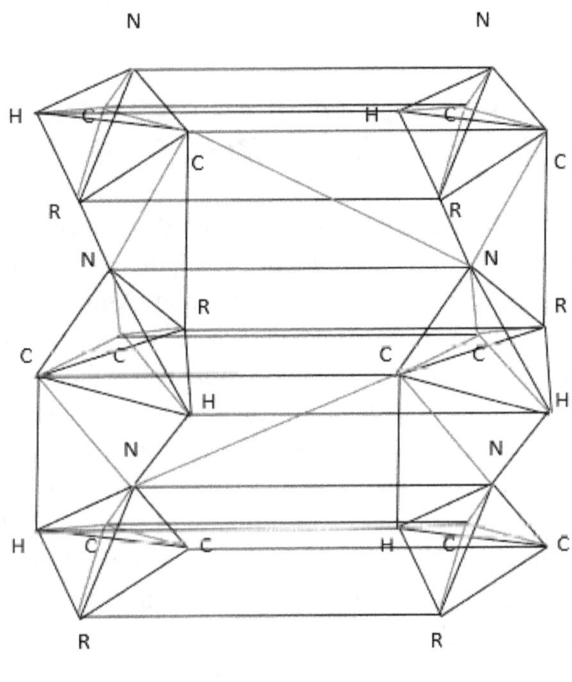

Similar this the β - structure in Figure 14, it is also possible to define the sequence of polyhedrons of dimensions 5 and 4, which make up the translational element of the β - structure in Figure 15. Given the difference in the form of polyhedrons in Figures 14 and 15, the sequence of polyhedrons that make up the translational element of the structure in Figure 15 can be written as 5 (3)-4 (3)-5 (4)-4 (4). The particular type of polyhedrons of dimensions 5 and 4 is easy to determine from Figure 15. The translational element of β - structure in Figure 15 also includes 24 vertices and, therefore, is a simplex polyhedron of dimension 23.

Most proteins in a living cell are characterized by a more complex arrangement of amino acid molecules associated with a significant influence of hydrogen bonds and the formation of quaternary structures based on secondary and tertiary structures (i.e., secondary structures with the addition of unstructured fragments). The dimension of such structures (hemoglobin, collagen, etc.) should be yet higher.

Spatial models of the β - structures of protein molecules, forming layers of amino acids, in principle, of unlimited length for both antiparallel and parallel conformation have been constructed. It is shown that the simplified flat Pauling models do not reflect the spatial structure of these layers. Using the recently developed theory of higher - dimensional polytopic prismahedrons, models of the volumetric filling of space with amino acid molecules are constructed. The constructed models for the first time mathematically describe the native structures of globular proteins.

NATIVE STRUCTURE OF GLOBULAR PROTEINS WITH PARALLEL ARRANGEMENT OF AMINO ACID RESIDUES

A polytopic prismahedron, according to the theory developed in (Zhizhin, 2019 a, b, 2020 b, 2021 b, c, 2022 b), is a product of two polytopes on top of each other. Moreover, each polytope can have its own dimension. As a result of the product of polytopes, each of them is evenly distributed over the space of their product as a polytope. Therefore, using the theory of polytopic prismahedrons, one can obtain various options for a uniform distribution of polytopes over space in a parallel position. If we take a polytope in an antiparallel position as the initial state, then as a result of the product of polytopes, one can obtain various options for a uniform distribution over space of polytopes in an antiparallel position. In both cases, if the remainder of the amino acid molecule is used as one of the factors, we will obtain uniform spatial distributions of the residues of the amino acid molecule in a strictly parallel or strictly antiparallel position. These distributions will differ from the fairly random distributions discussed earlier in this chapter.

It is known that the residues of amino acid molecules (after the cleavage of water molecules) can be located in a protein in a parallel position (Metzler, 1980; Lehninger, 1982; Koolman, Roehem, 2013). Two such residues, considered together, form a polytopic prismahedron (Figure 16).

Figure 16. The polytopic prismahedron from two residues of amino acid molecules in parallel position

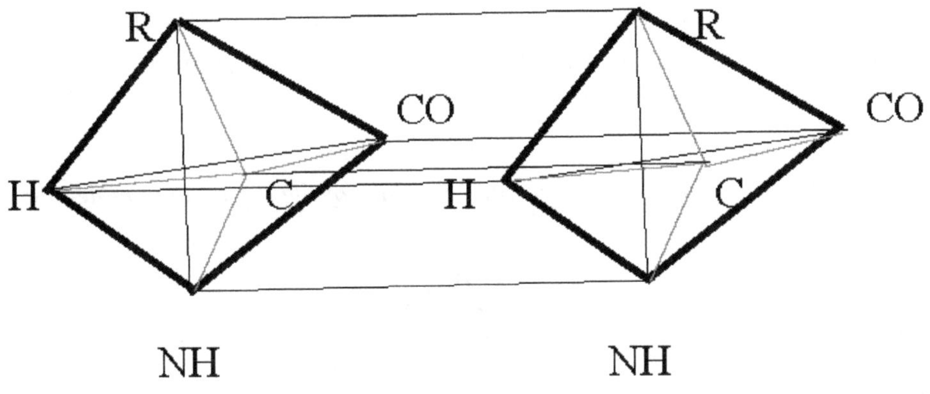

Its dimension can be determined by equation (1). In this prismahedron the number of vertices is 10, the number of edges is 24, the number of two - dimensional faces is 29, the number of three - dimensional faces is 20, the number of four - dimensional faces is 7. In this case f_0=10, f_1=24, f_2=29, f_3=20, f_4=7.

Substituting the obtained values f_i, i=0÷4 into equation (1), we obtain,

$$10 - 24 + 29 - 20 + 7 = 2.$$

Thus, Euler - Poincaré's equation in this case holds for $n = 5$. This proves that a prismahedron on Figure 16 has dimension 5. The polytopic prismahedron in Figure 16 is the product of a tetrahedron centered on a one - dimensional segment. The bond between the residues of the amino acid molecule is facilitated by a covalent bond and a hydrogen bond. An example of the resulting structure is shown in Figure 17, in which the covalent bond between the residues of the amino acid molecule is indicated by red solid lines, and the hydrogen bond is indicated by red dashed lines.

This is how the so - called β - structure in the form of a sheet is formed. However, these sheets are high -dimensional spatial formations. The highest dimension is possessed not only by polytopic prismahedrons with bases in the form of tetrahedra with a center, but also by the figures located between these polytopic prismahedrons. They are polytopic prisms of dimension 4, attached to each other along whole quadrangular faces.

To describe the distribution of residues of amino acid molecules, taking into account their exit from the specified layer, it is necessary to go from the product of a tetrahedron with a center to a one - dimensional segment to the product of a tetrahedron with a center by geometric elements of higher dimension. Consider the product of a tetrahedron centered on a triangle. The process of obtaining the product of a tetrahedron and a triangle was considered earlier (Zhizhin, 2019 a). The product of a tetrahedron with a center and a triangle is obtained in the same way, with the difference that in this case the product result is a polytope of dimension 6. Its image is shown in Figure 18.

Figure 17. The β – structure with parallel residues of the amino acid molecule

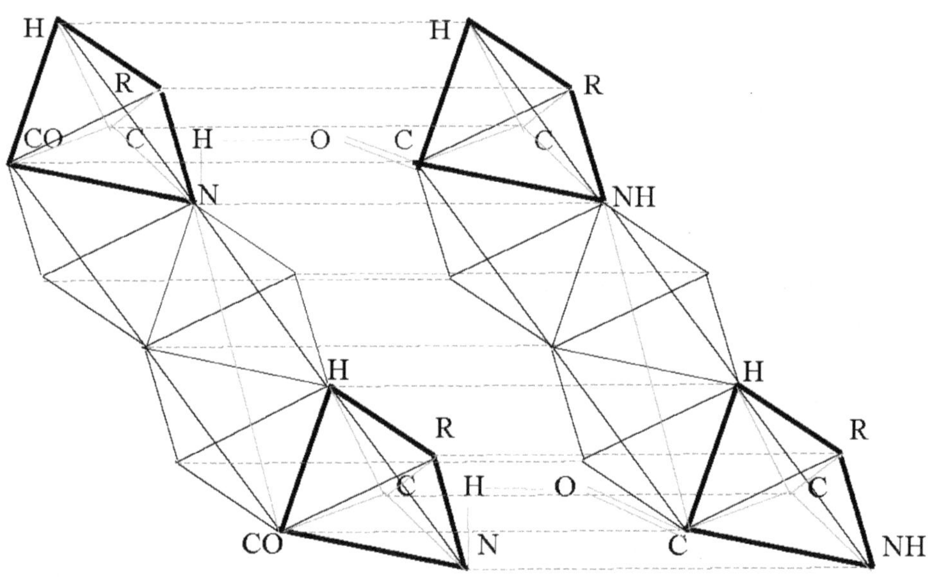

Each of the three pairs of parallel tetrahedrons with a center can lead to the formation of a β -layer of amino acid molecules perpendicular to the forming tetrahedral prism. The space between the residues of amino acid molecules is filled with tetrahedral prisms with tetrahedrons without a central carbon atom. Figure 18 shows that the residues of amino acid molecules are distributed in space, moving away from the center of the structure in three directions, forming a globule. (The chemical covalent and hydrogen bonds connecting amino acid residues are not depicted in Figure 18 for ease of reference.) Multiplying the tetrahedron centered (the remainder of the amino acid molecule) by the tetrahedron creates a space of dimension 7 (Figure 19).

Figure 18. The product of the tetrahedron with center by a triangle

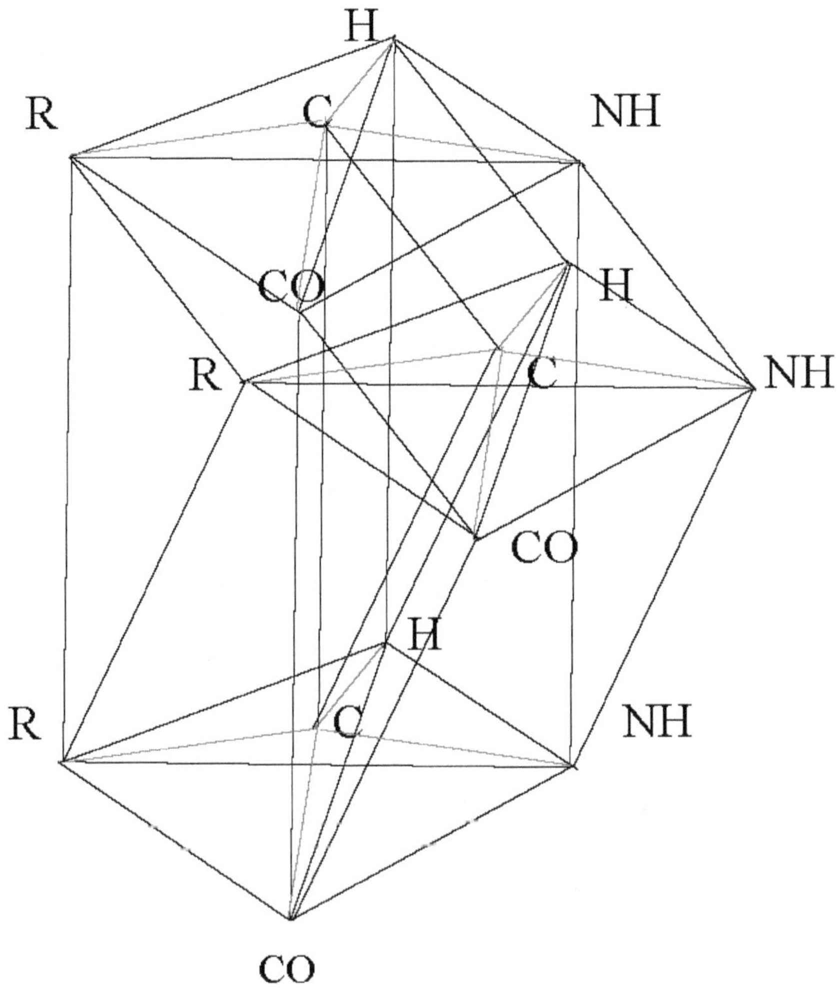

There are 6 pairs of parallel tetrahedrons with a center, i. e. 6 polytopic prisms, each of which forms a β - structure. An even more densely packed globule is thus formed.

Multiplying the tetrahedron with the center (the remainder of the amino acid molecule) by the simplex (dimension 4), we get a polytope of dimension 8 with five residues of the amino acid molecule. This process can be continued, obtaining polytopes of increasing dimensions and with an increasing number of residues of the amino acid molecule. In each case, they will generate periodic β - structures with the formation of ever - increasing globules.

Figure 19. The product of the tetrahedron with center by a tetrahedron

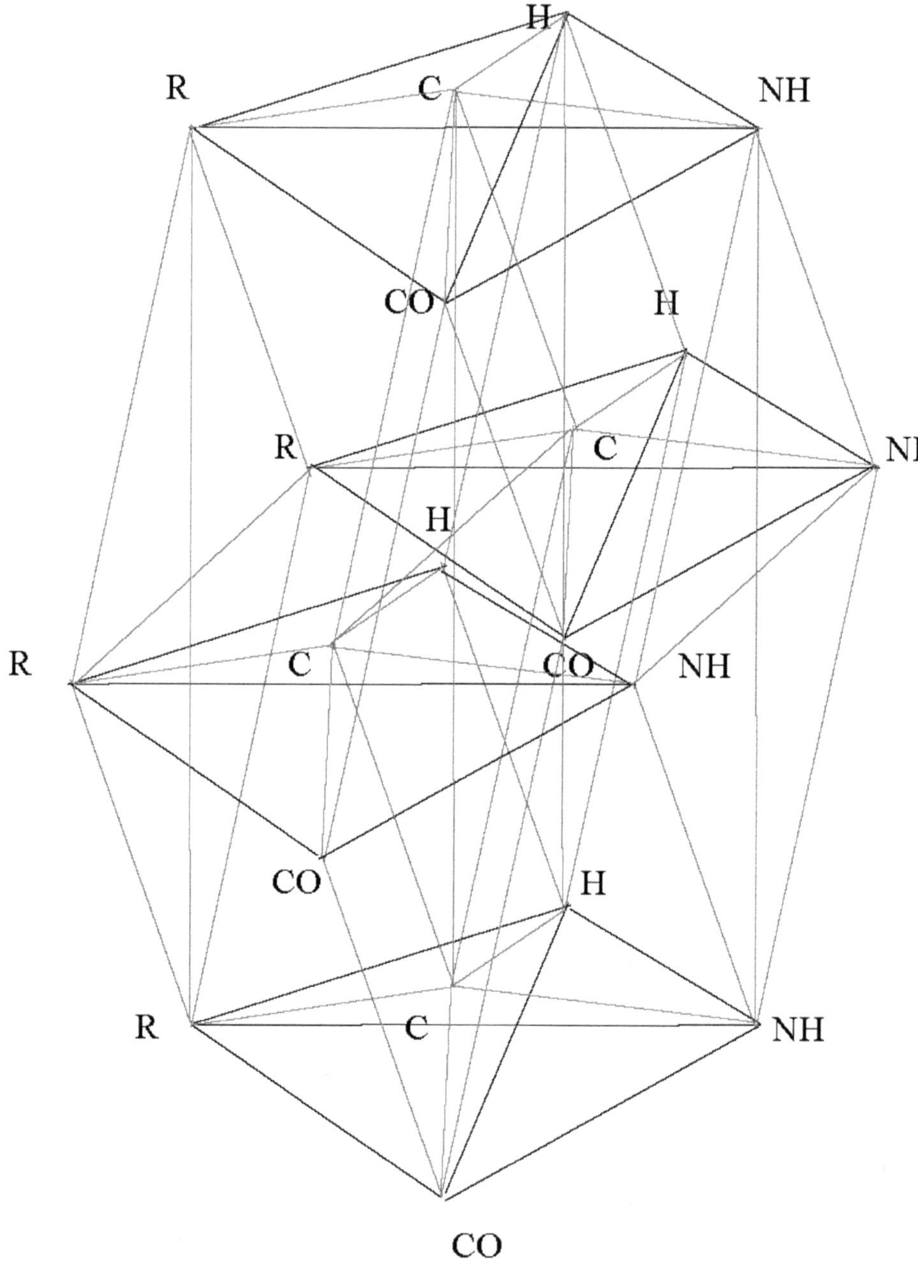

NATIVE STRUCTURE OF GLOBULAR PROTEINS WITH ANTIPARALLEL ARANGEMENT OF AMINOACIDE RESIDUES

Two tetrahedrons centered in the antiparallel position form a cross - polytope of dimensions 5 (Zhizhin, 2019 a, b). Considering the arrangement of atoms and functional groups at the vertices of the tetrahedron centered on the amino acid molecule, Figure 20 shows an image of two residues of the amino acid molecule in the antiparallel position.

Figure 20. Protein structure with antiparallel position of amino acid residues

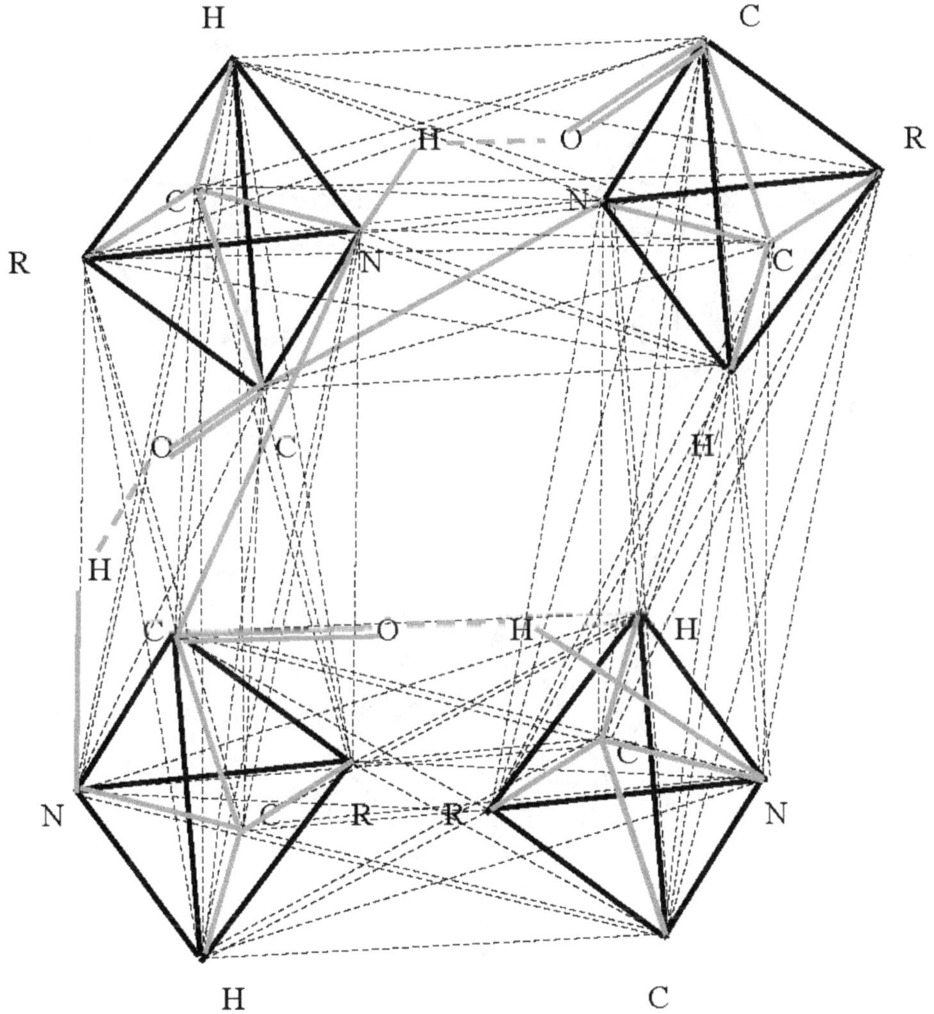

It should be noted here that antiparallelism of figures means rotation of one figure relative to another along a helicoidal curve, i. e. in a spiral. Therefore, when depicting the second pair of tetrahedra with centers antiparallel to the first pair, each tetrahedron from this pair also forms a 5 – cross - polytope with the corresponding tetrahedron of the first pair of tetrahedra (Figure 20).

A closed chain of 5 - cross - polytopes with intersection along tetrahedrons with the center is formed. Thus, each tetrahedron with a center participates simultaneously in motion along two helicoidal curves.

Figure 21. Combination of parallel and antiparallel tetrahedrons with a center

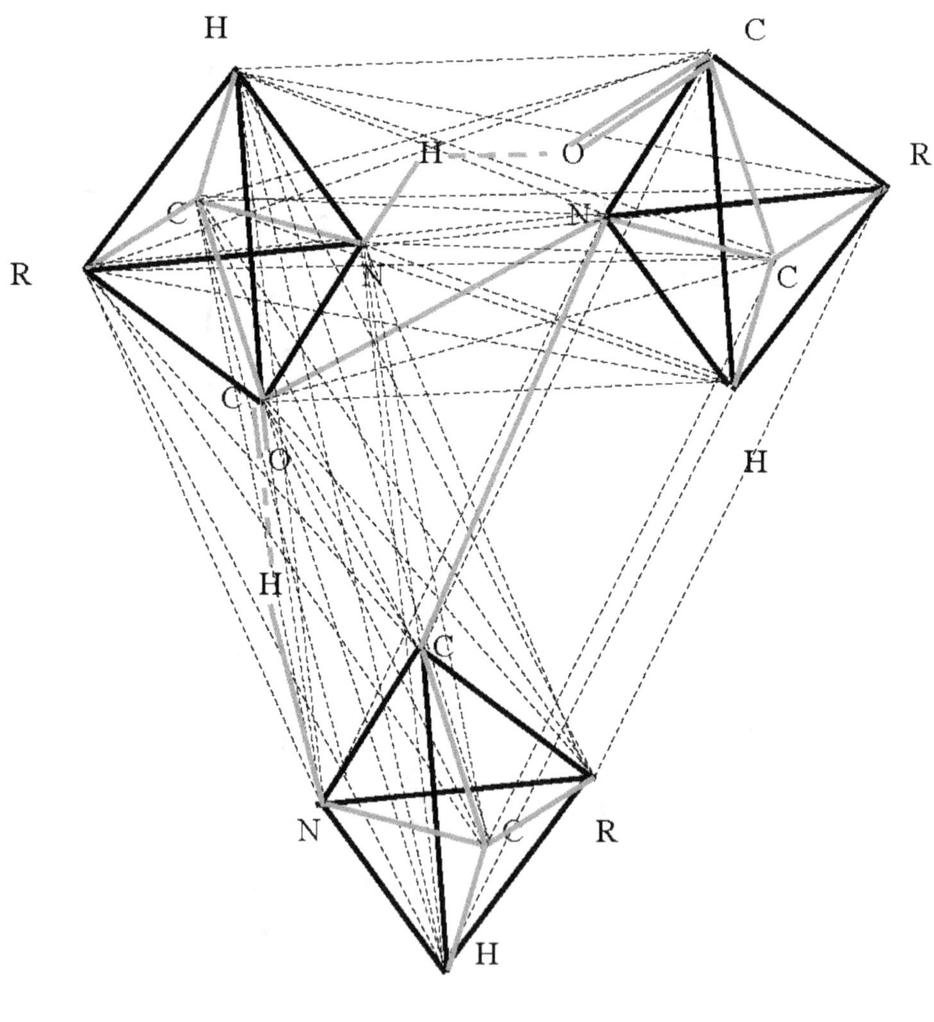

If we multiply a construction of four 5 – cross - polytopes by a one - dimensional segment, then one more construction will appear, located parallel to the first construction of 5 – cross - polytopes. The multiplication process can be continued by increasing the number of tetrahedra with a center (residues of amine acid molecules), the number of 5 – cross - polytopes united in chains.

The process of constructing the structure of proteins with antiparallel residues of amino acid molecules naturally leads to the appearance of parallel tetrahedrons in the distribution (see Figure 10.20), since double antiparallelism is parallelism.

However, it is possible to construct a structure of tetrahedrons with center, in which, along with antiparallel tetrahedrons with a center, parallel tetrahedrons with a center from a continuous region (Figure 21).

In Figure 21, two intersecting 5 – cross – polytopes end in two tetrahedrons centered in a parallel position, which form a polytopic prismahedron of dimension 5. In can be continued to form a region of tetrahedrons centered in a parallel position.

NATIVE STRUCTURE OF GLOBULAR PROTEINS WITH PARALLEL AND ANTIPARALLEL ARRAGEMENT OF AMINO ACID RESIDIE, AND α – SPIRALS

In order to describe the α - helices existing in the native structure of proteins with the help of tetrahedron with a center, in addition to parallel and antiparallel tetrahedrons with a center, it is necessary to have an intermediate state of a tetrahedron with a center between the two indicated states. It can be obtained by rotating each edge of the tetrahedron centered at 90 degrees, since the antiparallel state to this state is obtained by rotating each edge of the tetrahedron centered at 180 degrees. Each of the polytopes, consisting of two tetrahedra with a center, rotated relative to each other by 90 degrees, have dimension 5. Indeed, let us denote by integers the vertices of any of these polytopes (Figure 22).

Figure 22. Polytope, consisting of two tetrahedra with the center, rotated relative to each other by 90 degrees

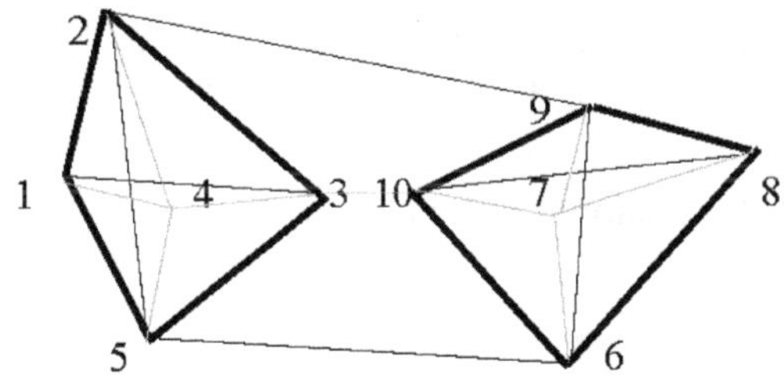

The polytope in Figure 22 is not a prismahedron, since the lines connecting the vertices of the two tetrahedra in it are not parallel to each other. This polytope can be called a self - orthogonal polytope. Its dimension can be determined by equation (11). In this polytope the number of vertices is 10,

The number of edges is 22: 1 - 2, 1 – 4, 1 – 3, 1 – 5, 2 – 5, 2 – 4, 2 – 3, 3 – 4, 3 – 5, 3 – 10, 4 – 5, 5 – 6, 6 – 10, 6 – 9, 6 – 8, 6 – 7, 7 – 8, 7 – 9, 7 – 10, 8 – 9, 8 – 10, 10 – 9;

The number of two - dimensional faces is 22: 1 – 5 – 3, 1 – 2 – 3, 1 – 2 – 5, 1 – 4 – 5, 1 – 2 – 4, 1 – 3 – 5, 2 – 4 – 5, 2 – 3 – 5, 2 – 3 – 4, 3 – 4 – 5, 3 – 5 – 10 – 6, 2 – 3 – 9 – 10, 7 – 9 – 10, 7 – 10 – 8, 7 – 6 – 8, 7 – 6 – 10, 7 – 9 – 6, 6 – 8 – 10, 6 – 8 – 9, 8 – 9 – 10, 6 – 9 – 10, 2 – 5 – 9 – 6;

The number of three - dimensional faces is 11: 1 – 2 – 3 – 5, 1 – 4 – 3 – 5, 1 – 2 – 4 – 3, 1 – 2 – 4 – 5, 2 – 4 – 3 – 5, 6 – 8 – 9 – 10, 7 – 9 – 8 – 6, 7 – 8 – 9 – 10, 7 – 8 – 6 – 10, 7 – 9 – 6 – 10, 1 – 2 – 3 – 5 – 6 – 8 – 9 – 10 (polyhedron bounded by the outer surface of the polytope);

The number of four - dimensional faces is 3: 1 – 2 – 3 – 4 – 5, 6 – 7 – 8 – 9 – 10, 2 – 4 – 3 – 5 – 10 – 9 – 7 – 6. In this case f_0=10, f_1=22, f_2=22, f_3=11, f_4=3.

Substituting the obtained values f_i, $i=0 \div 4$ into equation (1), we obtain,

10 – 22 + 22 – 11 + 3 = 2.

Thus, Euler - Poincaré's equation in this case holds for $n = 5$. This proves that a polytope on Figure 10.22 has dimension 5. The four - dimensional polytopes include two tetrahedrons with a center and a polytope 2 – 4 – 3 – 5 – 10 – 9 – 7 – 6. Let us prove that this polytope has dimension 4. In this polytope the number of vertices is 8, the number of edges is

15: 2 – 4, 2 – 5, 2 – 3, 4 – 3, 5 – 3, 3 – 10, 4 – 7, 5 – 6, 2 – 9, 10 – 9, 10 – 7, 10 – 6, 7 – 6, 7 – 9, 9 – 6;

the number of two - dimensional faces is

13: 2 – 4 – 5, 2 – 4 – 3, 5 – 4 – 3, 2 – 3 – 5, 10 - 7 – 9, 7 – 9 – 6, 10 – 7 – 6, 10 – 9 – 6, 2 – 3 – 10 – 9, 2 – 4 – 7 – 9, 2 – 5 – 9 – 6, 5 – 4 – 7 – 6, 5 – 3 – 10 – 6;

the number of three - dimensional faces is

6: 2 – 3 – 5 – 10 – 9 – 6, 2 – 3 – 4 – 5, 7 – 9 – 10 – 6, 2 – 4 – 5 – 7 – 9 – 6, 5 – 4 – 3 – 10 – 7 – 6, 2 – 3 – 4 – 10 – 7 - 9.

In this case f_0=8, f_1=15, f_2=13, f_3=6.
Substituting the obtained values f_i, i=0÷3 into equation (1), we obtain,

$$8 - 15 + 13 - 6 = 0.$$

Thus, Euler - Poincaré's equation in this case holds for n = 4. This proves that a polytope 2 – 4 – 3 – 5 – 10 – 9 – 7 – 6 has dimension 4.

Consequently, all possible polytopes that can be created from two tetrahedrons with a center somehow, polytopic prisms, cross - polytopes and self - orthogonal polytopes, have dimension 5. Each such polytope leads to polytopic prismahedrons, with the smallest dimension 6, with the help of which, in accordance with the theory of polytopic prismahedrons, it is possible to construct a space of the highest dimension.

When this intermediate state and the state antiparallel to it are included in the set of possible states of a tetrahedron with a center, a picture is obtained that describes all the variety of states observed in the native structure of the protein (Figure 23).

If we multiply the picture in Figure 10.23 by geometric elements of different dimensions (segment, triangle, tetrahedron, etc.), we can see how different spatial β - sheets are formed in the native protein structure with parallel and antiparallel arrangement of residues of acid amine molecules, and various α - helices. The presence or absence of certain secondary conformations of protein molecules will be limited by the presence or absence of certain chemical bonds between tetrahedrons centered on a common structure.

GLOBULAR PROTEINS AS MOLECULAR MACHINES

This class of proteins known as globular proteins that perform complex biological functions. Stabilized by the metal ion contained in them, they control the work of many biological organs, performing the most complex functions, for example, in enzymes and ribosomes. For example, the protein is a globular myoglobin – oxygen - binding protein present in the muscles. In the center of myoglobin globule is hem - group containing Fe - porphyrin (iron atom surrounded by five nitrogen atoms).

Figure 23. Native structure of globular proteins

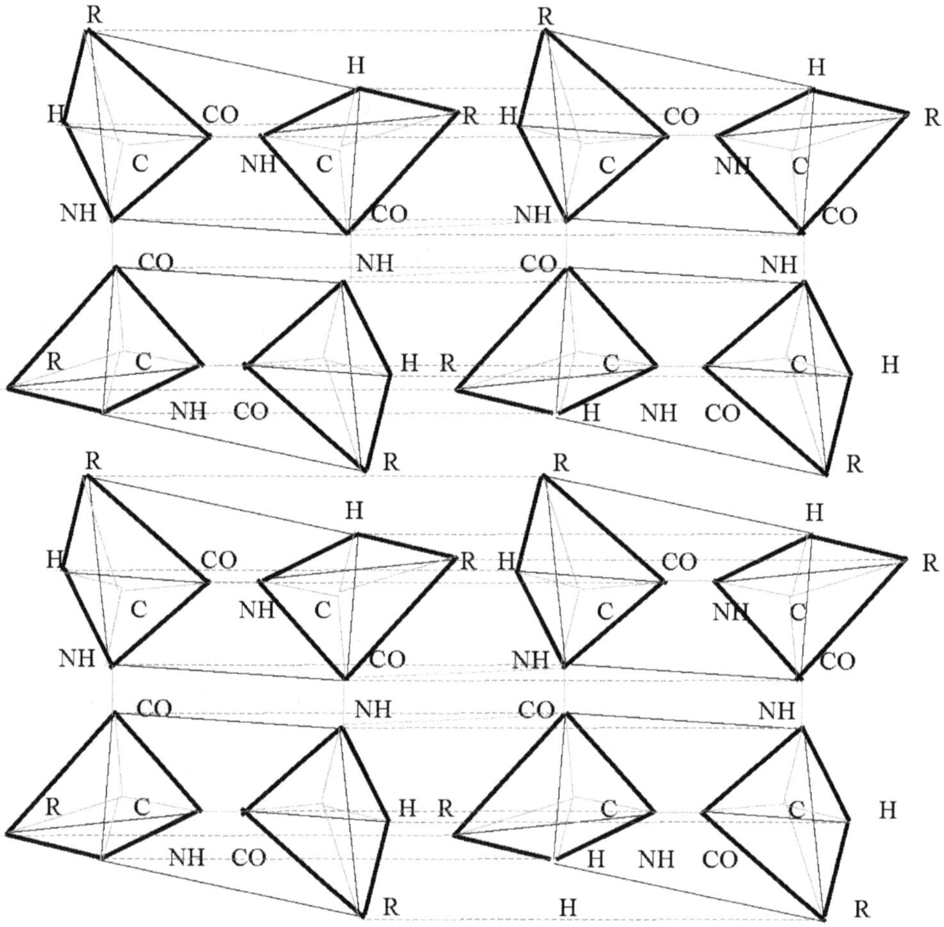

Theorem 10.1

The dimension of the Fe - porphyrin before joining the oxygen atom is equal to 5.

Proof.

Consider the first coordination sphere of the iron atom in the center of the porphyrin (Zhizhin, 2015 b, c), since only in the first coordination sphere of atoms are linked by a covalent bond, and in the following focal areas of intermolecular bonds between atoms. Before joining of the oxygen atom the first coordination sphere of Fe - porphyrin may be represented as a plane projection (Figure 24), at the vertices *a, c, d, f* of which the nitrogen atoms of the porphyrin are located, an iron atom is located at the vertex *g*, and the nitrogen atom of the nearest histidine residue is located at the vertex *b*. The deflection of vertex *g* from the center of the rectangle *acdf* corresponds to a certain "dome" character of porphyrin (Steed, Atwwod, 2007; Lehn, 1998). The projection in Figure 24 represents some polytope (let`s denote *A* - polytope).

Figure 24. The first coordination sphere of Fe - porphyrin before binding oxygen

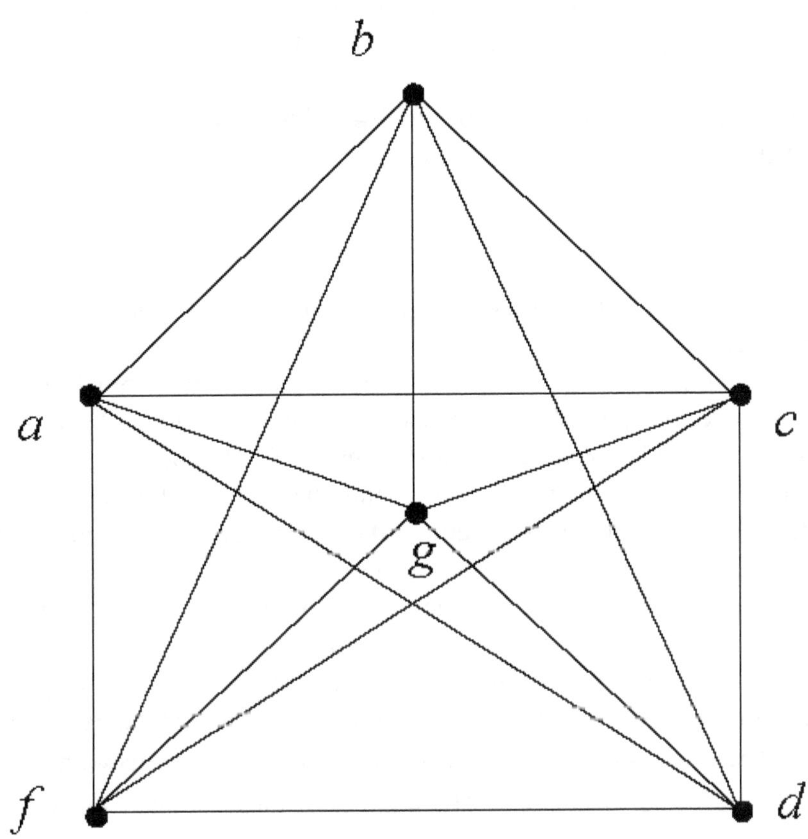

The *A* – polytope has six elements with dimension 0, $f_0(A)=6$. There are vertices *a, c, d, f, g, b*. The number of elements with dimension 1 is $f_1(A) = C_6^2 = 15$. It are edges *ab, bc, bd, bf, bg, ac, cd, fd, af, fc, ad, ag, gc, fg*. The number of elements with dimension 2 is $f_2(A) = C_6^3 = 20$. It are triangles *abf, bfg, bgd, dbc, bga, bgc,*

agc, dfg, adc, acf, fcd, bgd, fbg, agd, fgc, fbc, abd, afg, gcd, afd). The number of elements with dimension 3 is $f_3(A) = C_6^4 = 15$. It are tetrahedons *abgf, bsgd, abfc, abcd, bfcg, abdg, acfg, abdf, acdg, bfdg, abgc, fbcd, fgcd, afgd, afcd*). The number of elements with dimension 4 is $f_4(A) = C_6^4 = 6$. It are simplexes *abcdf, adcdg, abdfg, abcfg, bcdfg, acdfg*. Substituting the received numbers of elements of different dimensions in the equation (1) at a value of $n = 5$, we obtain

$$6 - 15 + 20 - 15 + 2 = 2,$$

Thus, the Euler - Poincare equation is satisfied for A - polytope with $n = 5$. This is a simplex of dimension 5. This proves theorem 10.2.

Theorem 10.2. The Dimension of Fe - Porphyrin After Joining of Oxygen Atom is 6

Proof

The first coordination sphere after joining oxygen atoms is complemented by one vertex *e* (Figure 25).

"Dome" character of Fe - porphyrin after joining of oxygen atom decreases, but it is not possible to affirm that it disappears completely (Steed, Atwood, 2007). Therefore, the deflection of vertex *g* from the center of rectangle in Figure 25 quality is maintained qualitatively. Taking into account the significant difference between the geometry and mass of the groups attached to the iron atom at the top and bottom, it is shown in Figure 25 that the vertices *e* and *b* do not lie on the same line. In the polytope in Figure 25 (B - polytope) the number of elements of zero dimension is increased compared with to the A-polytope by one vertex *e*, $f_0(B)$=7. This leads to the increase in the dimension of the polytope by 1, as the number of edges issuing from each top also increases by 1. In the polytope B the number of elements of dimension 1 is $f_1(B) = C_7^2 = 21$ (edges). The number of elements with dimension 2 is $f_2(B) = C_7^3 = 35$ (triangles). The number of three - dimensional figures is $f_3(B) = C_7^4 = 35$ (tetrahedrons). The number of elements with dimension 4 is $f_4(B) = C_7^5 = 21$, (simplexes of dimension 4). The number of elements with dimension 5 is $f_5(B) = C_7^6 = 7$, (simplexes of dimension 5). Substituting the numbers of the elements of different dimensions in equation (10.1) with $n = 6$, we get

$$7 - 21 + 35 - 35 + 21 - 7 = 0,$$

i.e. the Euler - Poincare equation for B - polytope is satisfied when $n = 6$. Therefore, B - polytope is a simplex of dimension 6. This proves theorem 10.2.

The dimensions of molecules increase with an increase of their energy again. It is shown that myoglobin is associated coil circuit elements of higher dimension (4) and, moreover, in the center of the coil is a group of atoms even greater dimension.

Figure 25. The first coordination sphere of Fe - porphyrin after joining of oxygen atom

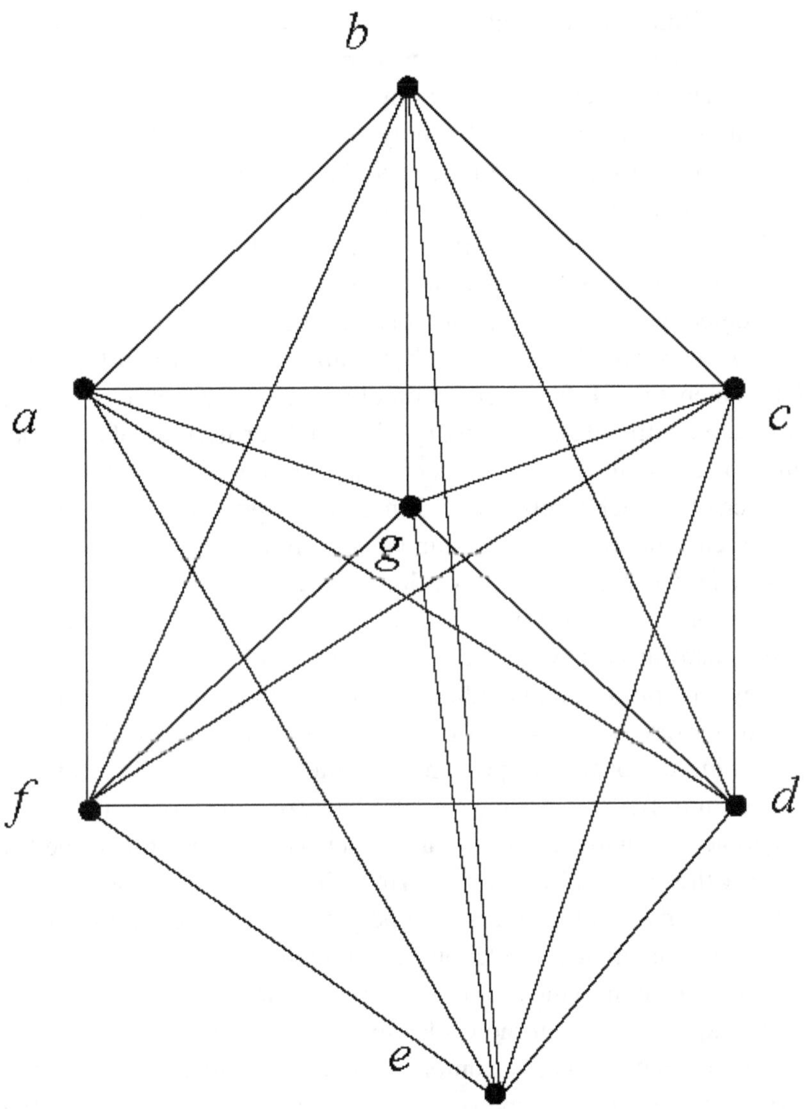

CONCLUSION

It is shown that the widespread quasi - plane Pauling models do not reflect and even contradict the spatial structure of polypeptide amino acid molecules, taking into account their highest dimension. This applies to all types of polypeptide chain conformation. In this regard, new models of linear polypeptide chains were construct and it was found that these chains have translational symmetry in the space of higher dimensions. The elementary element of the translational symmetry of these chains is the 9 - dimensional polyhedron, which consists of two mirror - opposite amino acid molecules. Helical conformations of polypeptide chains are also composed of higher - dimensional elements. The rotation of monomeric amino acid units in a helical conformation is determined only by the alternation of active centers in an asymmetric amino acid molecule. The study of the amino acid β - structures showed that they also have translational symmetry in the space of the highest dimension. It was found that the β - structures can have various forms, including not containing amino acid molecules in either parallel or antiparallel positions. Moreover, β - structures have elements of translation symmetry with dimension 23, consisting of two topologically equivalent polyhedrons of dimension 5 and two topologically equivalent polyhedrons of dimension 4. The resulting polyhedrons in high dimension, as translation elements of the construction of spaces protein molecules, should be further compared with the general theory of normal partitioning of n - dimensional spaces and corresponding high - dimensional parallelohedrons. The turn of the polypeptide chain, which is often found in globular proteins, has been studied in the higher dimensional space. It was found that the rotation of the polypeptide chain is accompanied by a change in the enantiomorphic form of the protein molecules.

The deposition of protein molecules in the brain with the formation of β - amyloid is the cause of many diseases of the nervous system. Until now, the possible structure of the β - amyloid complex and methods of treating these diseases have been little studied. In this work, with the aim of constructing mathematical models of the volumetric filling of space with protein molecules, this possibility is investigated without using simplified flat Pauling models. Analyzing the process of attachment of amino acid molecules under these conditions, it was found that the formed layers of molecules have the highest dimension both in antiparallel and parallel conformation. Based on the recently developed theory of polytopic prismahedrons (Zhizhin, 2019 a, 2021 c, d), a mathematical model of the volumetric filling of space with amino acid molecules was built, which for the first time makes it possible to describe the structure of plaques of protein molecules in the brain. This model can serve as a mathematical model of the native conformation of a globular protein in the general case. Obviously, the specific type of the native structure of the globular protein depends on the type of protein molecules and the conditions for the formation of the globule.

REFERENCES

Belozersky, A. N., & Spirin, A. S. (1958). A correlation between the compositions of deoxyribonucleic and ribonucleic acids. *Nature*, *182*(4628), 111–112. doi:10.1038/182111a0 PMID:13566202

Corey, R. B., & Pauling, L. (1953). Fundamental dimensions of polypeptid chains. *Proc R Soc Lord B Biol Sci, 141*(902), 10-20.

Dixon, M., & Webb, E. C. (1979). *Ensymes*. Longman Group Ltd.

Gratias, D., & Cahn, J. W. (1986). Periodic and qusiperiodic crystals. *Scripta Metallurgica*, *20*(9), 1193–1197. doi:10.1016/0036-9748(86)90030-X

Grünbaum, B. (1967). *Convex Polytopes*. Springer.

Jacob, F., & Monod, J. (1961a). On regulation of gene activity. *Cold Spring Harbor Symposia on Quantitative Biology*, *26*(0), 193–209. doi:10.1101/SQB.1961.026.01.024

Jacob, F., & Monod, J. (1961b). Genetic regulatory mechanisms in the syntesis of proteins. *Journal of Molecular Biology*, *3*(3), 318–356. doi:10.1016/S0022-2836(61)80072-7 PMID:13718526

Koolman, J., & Roehm, K. (2013). *Color Atlas of Biochemistry*. Thieme.

Lehn, J. M. (1998). *Supramolecular chemistry. Concept and perspectives*. Science.

Lehninger, A. L. (1982). *Principes of Biochemistry*. Worth Publishers, Inc.

Metzler, D. E. (1977). *Biochemistry. The Chemical Reactions of Living Cells*. Academic Press.

Pauling, L. (1987). So-called Icosahedral Decagonal Quasicrystals are Twins of an 820-atom Cubic Cristal. *Physical Review Letters*, *58*(4), 365–368. doi:10.1103/PhysRevLett.58.365 PMID:10034915

Pauling, L., & Corey, R. B. (1950). Two Hydrogen-bonded Spiral Configurations of the Polypeptide Chain. *Journal of the American Chemical Society*, *72*(11), 5349. doi:10.1021/ja01167a545

Pauling, L., & Corey, R. B. (1953). Compound Helical Configurations of the Polypeptide Chain:Structure of Proteins of the α - keratin Type. *Nature*, *171*(4341), 59–61. doi:10.1038/171059a0 PMID:13025480

Pauling, L., Corey, R. B., & Branson, H. R. (1951). The Structure of Proteins: Two Hydrogen-bonded Helical Configurations of the Polypeptide Chain. *Proceedings of the National Academy of Sciences of the United States of America*, *37*(4), 205–211. doi:10.1073/pnas.37.4.205 PMID:14816373

Poincare, A. (1895). Analysis situs. *J. de e` Ecole Polytechnique, 1,* 1 – 121.

Shechtman, D., Blech, I., Gratias, D., & Cahn, J. W. (1984). Metallic phase with longerange orientational order and no translational symmetry. *Physical Review Letters*, *53*(20), 1951–1953. doi:10.1103/PhysRevLett.53.1951

Spirin, A. S. (2019). *Molecular biology. The ribosomes and protein biosynthesis*. Knowledge Lab.

Spirin, A. S., Belozersky, A. N., Shugaeva, A. N., & Vanyushin, B. F. (1957). Study of the species specificity of nucleic acids in bacteria. *Biochemistry (Moscow)*, *22*, 744–754. PMID:13479533

Steed, J. V., & Atwood, J. L. (2000). *Supramolecular chemistry*. John Wiley and Sons.

Zhizhin, G. V. (2013 a). Relations for the number of faces of different dimensions in the tower of n - dimensional convex polytopes. Materials 9th All-Russian Scientific School "Mathematical research in the natural sciences. Geological institute KSC RAS, Apatity.

Zhizhin, G. V. (2013b). Images of convex regular and semiregular n-dimensional polytopes. Materials 9th All-Russian Scientific School Mathematical research in the natural sciences. Geological institute KSC RAS, Apatity.

Zhizhin, G. V. (2014). *World – 4D*. Polytechnic Service.

Zhizhin, G. V. (2015). November. *Polytopic prismahedrons – fundamental regions of the n -dimension nanostructures*. Paper presented at The International conference Nanoscience in Chemistry, Physics, Biology and Mathematics, Cluj – Napoca, Romania.

Zhizhin, G. V. (2016). The structures, topological and functional dimension of biomolecules. *International Journal of Chemoinformatics and Chemical Engineering*, *5*(2), 44–58. doi:10.4018/IJCCE.2016070104

Zhizhin, G. V. (2018). *Chemical Compound Structures and the Higher Dimension of Molecules: Emerging Research and Opportunities*. IGI Global. doi:10.4018/978-1-5225-4108-0

Zhizhin, G. V. (2019 a). *The Geometry of Higher-Dimensional Polytopes*. IGI Global. doi:10.4018/978-1-5225-6968-8

Zhizhin, G. V. (2019 b). *Attractors and higher dimensions in population and molecular biology: emerging research and opportunities*. IGI Global. doi:10.4018/978-1-5225-9651-6

Zhizhin, G. V. (2020a). The Dimension and Life. *Acta Scientific Biotechnology, 1*(11), 35–40.

Zhizhin, G.V. (2020 c). Higher Dimensions in Linear, Helicoidal and Folder Structures of Protein Molecules. *Acta Scientific Biotechnology, 1*(12), 02-09.

Zhizhin, G. V. (2021 a). *Normal partitions and hierarchical fillings of n - dimensional spaces*. IGI Global. doi:10.4018/978-1-7998-6768-5

Zhizhin, G. V. (2021b). Spatial Models of Sugars and their Compounds. *International Journal of Applied Research in Bioinformatics, 11*(2), 9–23. doi:10.4018/IJARB.2021070102

Zhizhin, G. V. (2021 c). *Nanotechnology and Clusters in the Space of Higher Dimension: Emerging Research and Opportunities*. IGI Global. doi:10.4018/978-1-7998-3784-8

Zhizhin, G. V. (2022a). On the Possible Spatial Structures of the β – Amyloid. Towards the theory of the native structure of proteins. *Int. J. of Applied Research on Public Health Management, 20*(Jan).

Zhizhin, G. V. (2022 b). *The Classes of Polytopes Higher Dimensions in Physics, Chemistry and Biology*. IGI Global.

Zhizhin, G. V., & Diudea, M. V. (2016). Space of Nanoworld. In M. V. Putz & C. M. Marius (Eds.), *Sustainable nanosystems, development, properties, and applications* (pp. 221–245). IGI Global.

Chapter 11
Complementary Condition in Higher Dimensional Space

ABSTRACT

Peculiarities of the conditions of correspondence of biomolecules of higher dimension, necessary for the passage of chemical reactions between them, are considered. As specific examples, the interactions of amino acids with each other and nucleic acids with each other, and monosaccharide molecules are considered. It is shown that in the case of chemical interaction of biomolecules of higher dimension, the conditions of correspondence (complementarity) are associated with a mirror image of the geometric elements of the corresponding polytopes of higher dimension and opposite electric charges.

INTRODUCTION

The spatial structure (conformation) of biomolecules plays an extremely important role in many biochemical processes: with the interaction of catalytic centers of enzymes with substrates, when connecting the hormone molecules with its receptor on the surface of the cell, during DNA replication, when transmitting inhuman information and many processes in the cell. Therefore, the study of the spatial structure of biomolecules is an important part of modern studies aimed at clarifying the relationship between the structure of the cells, the structure of their constituent elements and their biochemical functions. To ensure normal biological functions in the interaction of biomolecules, they must sterically accurately correspond to each other (Metzler, 1980; Lehninger, 1985). However, this issue requires additional research. A well-known formulation of the principle of complementarity (Metzler, 1980) states: "two molecules whose surfaces are complementary tend to interact

DOI: 10.4018/979-8-3693-0588-1.ch011

and connect with each other, while molecules that do not contain complementary surfaces do not interact." Watson called this the principle of selective "stickiness" of molecules (Watson, 1976). However, in this definition, it is not known what should be called the surface of the molecule, and which surfaces can be complementary. It is explained that "if there is a protrusion on the surface of one molecule, then there must be a recess on the complementary surface of the other molecule; opposite the positive charge there must be a negative charge." But if it is not defined what to call the surface of a molecule, then it is not clear what is a protrusion on the surface, and what is a depression on the surface. This is especially not clear if the molecule has a higher dimension, where the concept of the surface of the molecule loses its meaning altogether.

The fact is that recent studies of the structure of biomolecules (Zhizhin, 2016, 2018 a, b, 2019 a, b, c, d, 2020 b, c, d, e, 2021 b, c, d, e, 2022 a, b, 2023 a, b) show that biomolecules have the highest dimension, i.e. A description of the form of biomolecules in the form of a geometric closed figure leads to a multi -party in the dimension space of more than 3, that is, to a polytope of the highest dimension. In this case, the question of the steric accordance of the interacting biomolecules requires the consideration of the correspondence of biomolecules in the space of the highest dimension. At the same time, it cannot be limited to indicating the proximity of the values of some characteristic linear dimensions of the interacting molecules, since the orientation of linear areas in space also plays an important role, especially if this space has the highest dimension.

In this chapter, the question of the interaction of molecules in a higher-dimensional space is considered using three examples already studied in the previous chapters of the monograph: the formation of an amino acid chain, the formation of polysaccharides, and the interaction of nucleic acids in the polytope of hereditary information. Some details of these interactions will be skipped (they can be found in previous chapters), and the main attention will be paid to the features of the interaction in higher-dimensional spaces.

THE INTERACTION OF AMINO ACID MOLECULES IN THE FORMATION OF PROTEIN

Dimensions of Protein Molecules

Monomer units which are built of proteins are the 20 standard amino acids. These small molecules contain two different chemical functional groups capable of reacting with each other to form a covalent bond. Amino acids are chemical compounds that simultaneously include a carboxyl group (- COOH) and an amine group (-NH_2).

These groups are covalently linked to a carbon atom (α - carbon), to which a hydrogen atom H and a side chain R are also covalently attached (Koolman, Roehm, 2013). Connection which determines the formation of protein polymer is called a peptide bond. In the formation of such a connection by joining together - COOH and -NH_2 with secretion a molecule of water. Amino acids have two enantiomorphic forms L and D, which are Fischer planar views shown in Figure 9.1 (Chapter 9). From Figure 1 it can be seen that the two enantiomorphic forms (L and D) are related by mirror image relative to the line passing through the amine and carboxyl groups.

Spatial images of enantiomorphic amino acids (Zhizhin, 2016) are presented in Figure 2 (Chapter 9).

In Figure 2, covalent bonds are indicated in red. The rest of the segments define the spatial shape of the molecule, they are marked in black.

From Figure 2 it follows that the amino acid molecule is a tetrahedron with a center. The dimension of such a polyhedron can be determined by the Euler - Poincaré equation (Poincaré, 1895)

$$\sum_{i=0}^{n-1}(-1)^i f_i(P) = 1 + (-1)^{n-1}. \quad (1)$$

There n is dimension of a polytope P, $f_i(P)$ is the number of elements with dimension i in the polytope P.

From Figure 9.2 it follows that in any of the two enantiomorphic forms of the amino acid molecule, the number of vertices is 5, the number of edges is 10, the number of two - dimensional faces is 10, the number of three - dimensional faces is 5. In this case f_0=5, f_1=10, f_2=10, f_3=5.

Substituting the obtained values f_i, i=0÷3 into equation (1), we obtain,

$$5 - 10 + 10 - 5 = 0.$$

Thus, Euler - Poincaré's equation for a tetrahedron with a center holds for $n = 4$. This proves that a tetrahedron with a center (amino acid molecule) has dimension 4.

LINEAR POLYPEPTIDE CHAIN STRUCTURE

The study of diffraction patterns of crystals of amino acids (Pauling, Corey & Branson, 1951; Pauling & Corey, 1950, 1953; Corey & Pauling, 1953) made it possible to establish some values of bond bonds and bond angles in an amino acid. However, the interpretation of the spatial arrangement of atoms in an amino acid requires additional instructions. An example of the possibility of ambiguous interpretation of diffraction patterns can be found in numerous works on the description of quasicrystals. In particular, the apparent icosahedrical symmetry of the diffraction patterns of quasicrystals was proposed to be explained (Pauling, 1987) by multiple twinning of cubic crystals. Pauling based his arguments on the radial arrangement of the spots in the diffractograms. However, high - resolution photomicrographs refuted this assumption (Gratias, Cahn, 1986). The statement about the absence of translational symmetry in quasicrystals (Shechtman, et al., 1984) was constructed using the representations of the three-dimensional geometry of Euclid. Later this statement was refuted (Zhizhin, 2013 a, b; Zhizhin, 2014; Zhizhin, Diudea, 2016). It was proved that translational symmetry in the diffractograms of quasicrystals appears if the diffractograms are considered as projections of structures from the space of higher dimensions. When studying diffraction patterns of amino acid crystals in the works of Pauling joint authors the values of a number of parameters were not determined, for example, bond lengths $R - C$ depending on the species R, bond angles between carbon atoms and nitrogen atoms, between hydrogen atoms and R, etc. To compensate for these disadvantages, it was proposed to use a simplifying assumption about the arrangement of a group of atoms of a peptide bond and centers of tetrahedrons in a plane (Zhizhin, 2020 b, e) (Figure 3m, chapter 9)

Under this assumption, knowledge of the indicated quantities was not required. It should be noted that in subsequent publications by other authors (Metzler, 1980; Lehninger, 1985; Koolman, Roehm, 2013) with a reference to Pauling's work, it was argued that the arrangement of these atoms in the plane was proved in Pauling's works. But this is not so, to be convinced of this it is enough to refer to these works. This unproven statement became the basis for the construction of three - dimensional models of protein molecules of various conformations in Pauling's works. Let us consider this issue in detail when constructing spatial structures of a linear polypeptide chain. In this case, the atoms of amino acid molecules must be located on a set of parallel lines. Then the polypeptide chain of two molecules of L - amino acids, taking into account the four - dimensionality of amino acid molecules, has the form shown in Figure 1.

In the case of a D - amino acid, such a polypeptide chain is shown in Figure 2.

From Figure 1 and Figure 2 it can be seen that in order for the corresponding vertices of tetrahedrons with the center (i.e., specific atoms or functional groups) could lie on a system of parallel lines (i.e., form a linear polypeptide chain), it is necessary that the amino acid molecules in the chain alternate with their mirror image relative to the edge in the tetrahedron connecting the bond centers of the polypeptide chain (i.e., relative to the segment CO – NH). The chains in Figure 1 and Figure 2 differ by the mirror reflection of the molecules relative to the perpendicular to the segment $CO - NH$ (L - and D - molecules). It is essential that during the formation of a peptide bond, two particles with different signs are combined: the positive charge at the proton (+ H) combines with the negative charge at the hydroxyl group (- OH) with the release of a neutral water molecule. Thus, the opposite of two charges, noted in the definition of a complementary interaction (Metzler, 1980), is preserved when particles interact in a higher-dimensional space.

Figure 1. Linear polypeptide chain of L – molecules amino acids

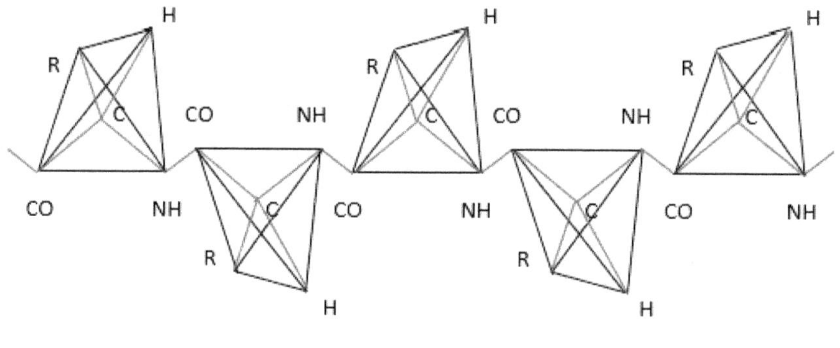

Figure 2. Linear polypeptide chain of D – molecules amino acids

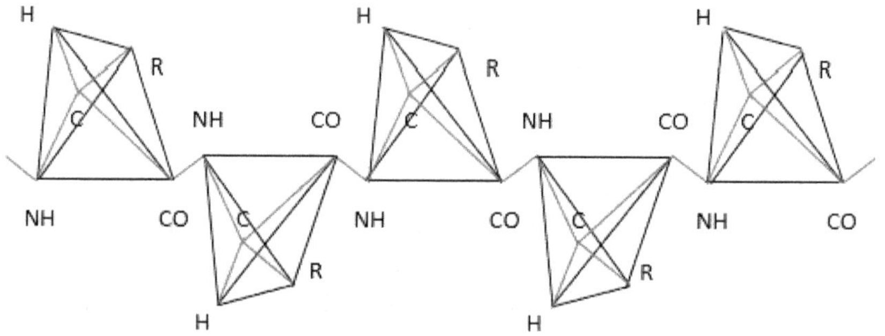

In Figures 1 and 2, you can see the translational symmetry of the circuits. Moreover, the elementary translational element in this case is a group of two linked amino acid molecules, one of which is a mirror image of another amino acid molecule relative to the segment. From these two linked amino acid molecules, you can create a convex polytope by connecting edges of each vertex of any of the molecules with all other vertices in the group. This is how a simplex polytope of dimension 9 is formed, since the dimension of a simplex polytope is one less than the number of vertices in a simplex (Zhizhin, 2019 a), and the number of vertices in two tetrahedrons with a center is 10. Therefore, a polypeptide linear amino acid chain has translation symmetry, in which an elementary translation element is a simplex polytope of dimension 9. In Figures 1 and 2, as well as in Fischer's projections (Figure 1, chapter 9) and in spatial images of molecules (Figure 2, chapter 9), in accordance with the definition of functional dimension, functional groups R, CO, NH were used. Analysis of Pauling's works on the structure of protein molecules (Pauling, Corey, 1950; Pauling, et al., 1951; Pauling, Corey, 1953; Corey, Pauling, 1953) shows that Pauling's model is nothing more than an assumption. Consider three amino acid residues linked by a peptide bond (Figure 3).

Figure 3. Spatial image of three amino acid residues linked by a peptide bond

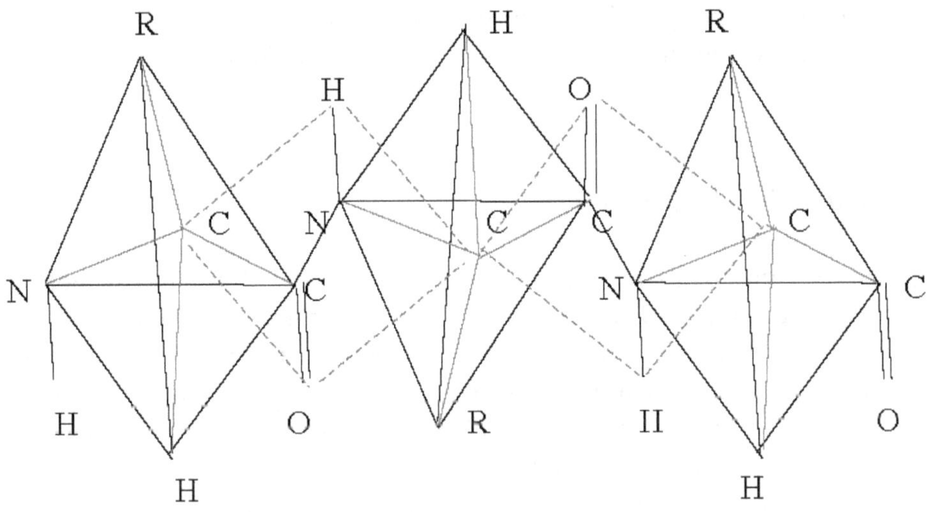

Figure 3 shows three amino acid residues in the form of tetrahedra, in the center of which there are three carbon atoms, from which four valence bonds proceed to functional groups located at the vertices of the tetrahedra. These valence bonds are marked with red ribs. Three amino acid residues are linked by two peptide bonds located within two contours, indicated by blue dashed lines. According to Pauling's model, the atoms located on the dotted contour and inside the contour are on the same plane. Here we must take into account that the lengths of chemical bonds connecting different pairs of atoms are different. Calculations of bond angles in Pauling's works are not given for any of the bonds. The plane is determined, as is known, by three points. The statement that some other point belongs to a given plane requires a geometric proof. There is no such evidence in Pauling's work with coworkers. Therefore, Pauling's model remains unproven. From the point of view of chemistry and geometry, the tetrahedral coordination of functional groups relative to the central carbon atom in the amino acid molecule is much more important. This fact should be the basis for the analysis of protein structures.

Thus, it is essential (see Figure 3) that the α - carbon atoms do not lie in the same plane with the peptide bond atoms. From a comparison of Figures 1 – 3 it follows that α - carbon atoms lie either above the plane in which the atoms of the peptide bond are located, or below this plane, and can never lie on this plane. Therefore, Pauling's idea that α - carbon atoms lie in the same plane with functional groups CO, NH is erroneous. Therefore, Pauling's approximation cannot be used both in the construction of linear polypeptide chains and in the construction of more complex conformations of amino acid molecules.

Accounting for the highest dimension of amino acid molecules is also important in the analysis of other possible conformations of protein (spirals, layers, globes). Models of these conformations of the protein are built in the works (Zhizhin, 2020 b, e, 2023).

MONOSACCHARIDE CHAINS

In Chapter 3 of this monograph, based on the Fisher model of the glucose molecule (Figure 3), a spatial model (in particular) of the a-D glucose molecule (Figure 4) was built. It has been proven that the dimension of this molecule is 15. In addition, for ease of use, a simplified three-dimensional model of this molecule has been constructed, taking into account the fundamental features of the spatial model of this molecule with a dimension of 15 (c - anomer Figure 7, a - anomer Figure 8). In this section, we focus on the formation of chains from each of these molecules.

Representations of the saccharide molecules in the form of polytope simplifies the understanding of the formation of polysaccharides.

Monosaccharide molecules combine with each other thanks to the combination of two hydroxyl groups with the release of a water molecule. Here, as well as in the previous section, the positively charged particle proton (+ H) combines with the negatively charged particle hydroxyl group (- OH), forming a neutral particle, the water molecule. The remaining oxygen atom connects the remaining part of the monosaccharides. The most common chains of α -*D* - glucose monosaccharide residues. Here, residual α - *D* - glucose molecules can be joined through an oxygen atom, connecting carbon atoms 1 - 1, 1 - 4, 4 - 4, 4 - 1. In some cases, the carbon atom $C_{(2)}$ in the cycle will be involved in the compound. Branching chains of glucose molecules occurs at the sixth carbon atom in the functional group CH_2OH. Monosaccharide chains have different forms depending on the type of these molecules,

i. e. depending on whether these molecules are *β*- anomers or α - anomers. Using the obtained simplified three - dimensional models of α - *D* - glucose molecules, we consider in more detail these compounds in a metric image, taking into account the angles between the valence bonds. Let two *β*- anomers of the α - *D* - glucose molecule join together, linking the carbon atom $C_{(1)}$ of the first α - *D* - molecule to the carbon atom $C_{(4)}$ of the second α -*D* - molecule (lactose). Based on the geometric image in Figure 4, you can imagine a top view of this connection (Figure 4).

Figure 4. Projections of β- anomer disaccharides

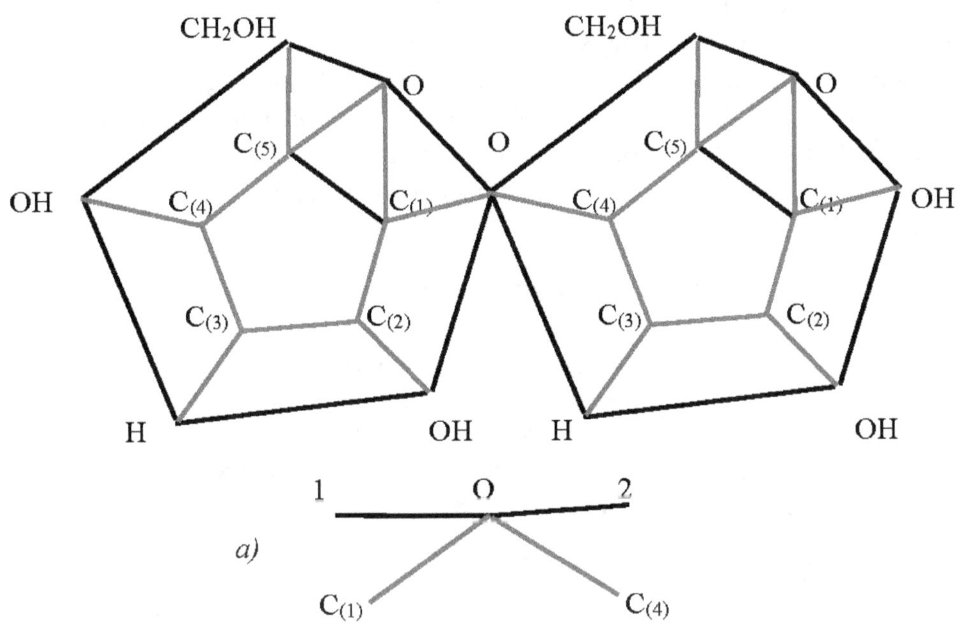

It is easy to see that the same attachment of the third *α* - *D* - glucose molecule leads to a linear arrangement of molecules when viewed from above on this compound. Linearity will not change with the following similar connections. However, given the three - dimensionality of the image, you should look at this connection from a different view. Figure 4 a) shows an image of the junction of two *α* - *D* - glucose molecules when viewed from the front. The black segments in Figure 4 a) are the traces of the intersection of the upper bases of the prism with a plane passing through the valence bonds of the oxygen atom and carbon atoms. According to the obtained solution of equation (3.4) (Chapter 3), the angle between the base of the prism and the valence bond is 36 degrees. both the first *α* -*D* - glucose molecule and the second α -*D* - glucose molecule. The valence angle at the oxygen atom in compounds with two carbon atoms in the chain of α -*D* - glucose molecules, as you know, is slightly larger than the normal valence angle and is about 112 degrees. Thus, the angle 1O2 in Figure 11.4 a) is 176 degrees. This means that the base of the second α -*D* - glucose molecule has a slope to the base of the first α - *D* - glucose molecule other than zero. More precisely, this slope is 4 degrees, therefore the second α - *D* - glucose molecule, when viewed in full view, turns to the left relative to the first α - *D* - glucose molecule. It is clear that such an addition of a third α - *D* - glucose molecule will increase the angle of inclination of this molecule relative to the first molecule. Thus, the statement about the linearity of the chain of α - *D* - glucose molecules when connecting them in the case of anomers is not entirely accurate. One can say that this statement is somewhat one - sided.

Let two α - anomers of the α - *D* - glucose molecule join together, linking the carbon atom $C_{(1)}$ of the first α -*D* - molecule to the carbon atom $C_{(4)}$ of the second α - *D* - molecule (maltose). Based on the geometric image in Figure 3.8 (Chapter 3), you can imagine a top view of this connection (Figure 5).

Figure 5. Projections of α - anomer disaccharides

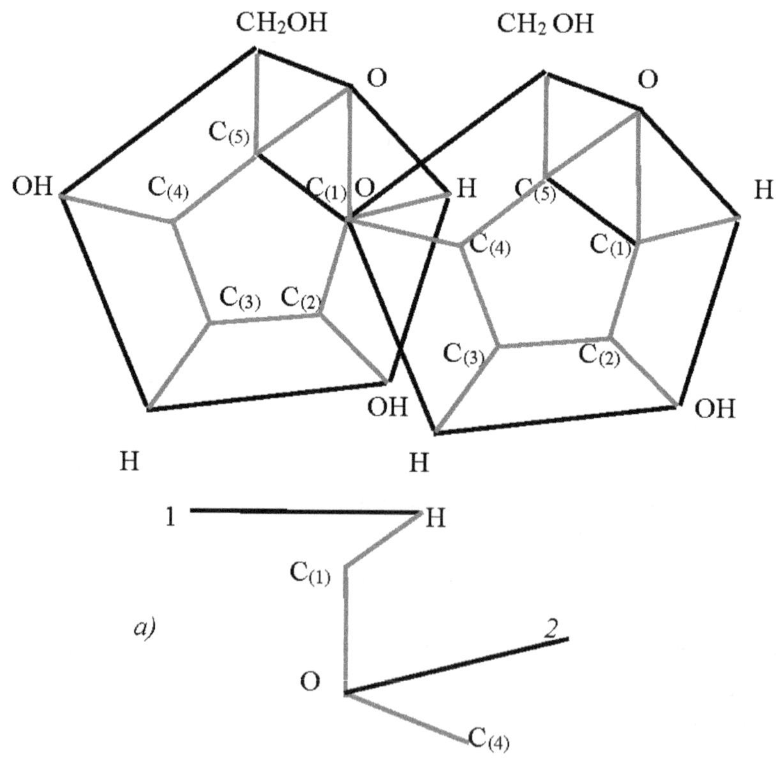

It is easy to see that the same attachment of the third α - D - glucose molecule leads to a nonlinear arrangement of molecules when viewed from above on this compound. Nonlinearity will save with the following similar connections. Thus, the chain of molecules turns to the right. However, given the three - dimensionality of the image, you should look at this connection from a different view. Figure 5 a) shows an image of the junction of two α - D - glucose molecules when viewed from the front. The black segments in Figure 11.5 a) are the traces of the intersection of the upper bases of the prism with a plane passing through the valence bonds of the oxygen atom and carbon atoms. According to the obtained solution of equation (3.4) (Chapter 3), the angle between the base of the prism and the valence bond is 36 degrees both the first α - D - glucose molecule and the second α - D - glucose molecule. Thus, on the Figure 11.5 a the angle 1H $C_{(1)}$ is 36 degrees and the angle $C_{(1)}OC_{(4)}$ is 112 degrees. Since the angle $C_{(4)}O2$ is 36 degrees too, so the angle between bases 1H and O2 is 14 degrees. This means that the base of the second α - D - glucose molecule has a consequent slope to the base of the first α - D - glucose molecule to the left. It is clear that such an addition of a third α - D - glucose molecule will increase the angle of inclination of this molecule relative to the first molecule. In the chains of β- anomers, one of the projections is linearly transmitted along the chain, and the projection perpendicular to it along the chain along a curved line. In chains of α - anomers (spirals), it is observed that when moving along a chain of projections in mutually perpendicular planes, the pyranose molecules rotate in opposite directions (Figure 6).

Figure 6. The chain of α - anomers (spirals)

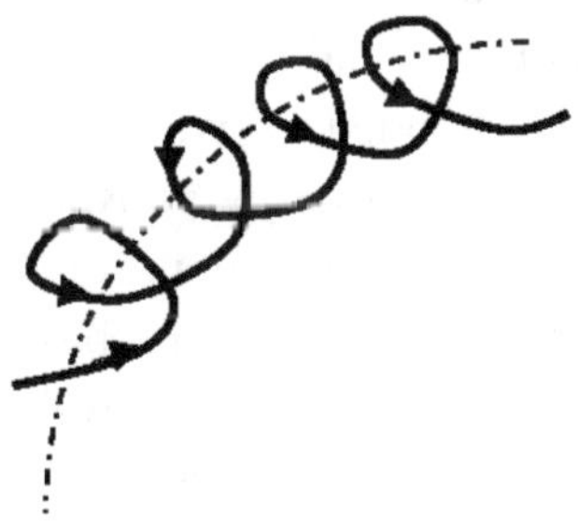

Thus, chains of α - anomers of α - D - glucose molecules rotate simultaneously in opposite directions in perpendicular planes.

The study of furanose chains using simplified three - dimensional models was carried out using nucleic acids as an example, in which furanose models were connected to each other using phosphoric acid residues. These studies showed that, when nucleic acids bound by nitrogen bases interact, there is a latent degree of freedom of arrangement of nitrogen bases in higher dimensional space, leading to an unambiguous corresponding number of nucleotides taking into account their possible location in space and the number of amino acids (Zhizhin, 2019 b, d, 2020 b, c, d, 2021 d, e).

It is interesting to note that during the formation of a molecule b - the disaccharide anomer, the valence bonds connecting two saccharide molecules are mirror symmetrical with respect to each other. While in the formation of a molecule a-anomer of a disaccharide, due to a more complex geometry, a spiral movement occurs in perpendicular planes simultaneously in two opposite directions.

INTERACTION BETWEEN NUCLIEC ACIDS IN THE SPACE OF THE HIGHEST DIMENSION WHEN TRANSMITTING HEREDITARY INFORMATION

Nucleic Acids

There are two known nucleic acids: deoxyribonucleic acid (DNA) and ribonucleic acid (RNA). Direct experiments by Avery, McLeod and McCarthy (1944) proved that DNA can be the carrier of the hereditary traits of an organism, i.e. carrier of genetic information. At the same time, RNA is an intermediary that receives this information from DNA and implements it in the process of protein biosynthesis. Watson and Crick (1953 a, b) proposed a model of the macromolecular structure of DNA. This is a double helix, where two DNA polymer strands are twisted relative to each other around a common axis and are held together due to pair interactions of nitrogenous bases covalently bonded to five - carbon sugar molecules. In turn, sugar molecules are linked to each other by phosphoric acid residues, forming a chain. Bases are large circular molecules with nitrogen atoms. DNA nucleotides (a sugar molecule, a phosphoric acid residue and a base) have one of the four bases adenine (A), guanine (G), cytosine (C) and thymine (T). In RNA, thymine is replaced by uracil (U). Cytosine, thymine and uracil have one ring of atoms and are called pyrimidine bases. Adenine and guanine each have two rings and are called purine bases. It is significant that nitrogenous bases are practically flat. The reproduction (reduplication) of the DNA structure is based on the principle of complementarity: in a double helix, two DNA polymer chains are linked to each other by hydrogen bonds in the formation of pairs G: C, C: G, A: T, T: A (the colon denotes a hydrogen bond). If two chains of the double helix diverge, then a new complementary chain can be built on each of them. At the same time, opposite the G of the original chain,

C of the new chain is established, opposite C of the old chain - G of the new chain, opposite A - T, and opposite T - A. As a result, two child double helixes are obtained, completely identical to the original parent. The same principle of complementarity provides a mechanism for RNA replication on a DNA template. This is another type of interaction between nucleic acids. The only difference is that RNA polymerizes only on one of the two separated strands of the DNA double helix. During RNA synthesis, opposite A of the DNA chain becomes U (uridyl ribonucleide), instead of T during DNA synthesis. The replicating strand of RNA is thus an exact copy of the opposite strand of DNA, with T replaced by Y. As a result of this replication, RNA is formed as a flexible single - stranded polymer.

By the second half of the 1950s, it was established that protein synthesis in living cells is carried out by ribonucleoprotein particles - ribosomes. The main component of ribosomes is ribosomal RNA, it consists of transfer RNA that transports selected amino acids to messenger RNA, which is synthesized and edited as a result of RNA processing. This is the third type of nucleic acid interaction. It was found that due to the interaction of side groups (nitrogenous bases) with each other, a single - stranded RNA polymer folds onto itself, forming secondary and tertiary structures (Spirin, 2009). The formation of these structures also occurs due to complementary interactions result of the interaction of nucleic acids, the central dogma of molecular biology is fulfilled: the flow of hereditary information is transmitted from DNA to RNA and further to protein. At present, in all types of interaction of nucleic acids, it is believed that nitrogenous bases are linearly sequential along the polymeric chain of acids. But since there are four traditional types of nitrogenous bases in each chain, and there are 20 types of amino acids in proteins, it has come to be recognized that each amino acid corresponds to three nitrogenous bases. This assumption is based on the concept of the genetic code, the deciphering of which establishes these correspondences (Gamov, 1954; Crick, 1959). However, in recent works (Zhizhin, 2019, 2020 a, b; 2021) it was found that the region of interaction of nucleic acids is not a linear section of the chain, but a polytope of higher dimension (a polytope of hereditary information), at the tops of which nitrogenous bases are located. In this regard, the question of the interaction of nucleic acids requires additional consideration.

POLYTOPES WITH ANTIPARALLEL EDGES

In single - stranded and double - stranded nucleic acids (RNA, DNA), the constituents of acids (residues of phosphoric acid and sugar molecules) interact with each other (Watson, Crick, 1953a, b; Spirin, Gavrilova, 1971; Frank – Kamenetskiy, 1986, 1988). Phosphoric acid residues are connected by divalent metal ions, mainly magnesium ions, due to the interaction of negative charges of phosphoric acid residues with positive charges ions (Spirin, Gavrilova, 1971). This interaction is essential for the stability of nucleic acid structures, especially in the ribosomes. Sugar molecules interact with each other due to the hydrogen bond between the nitrogenous bases attached to the sugar molecules. The constituents of nucleic acids being geometric forms interact with each other to form new geometric forms - new polytopes. However, it is not known how flat nitrogenous bases are oriented in space, whether their orientation depends on the type of nitrogenous base. Currently there is no information on this. There is also no information on how exactly the metal ions are located, connecting the phosphoric acid residues. It should be remembered here that the adopted three - dimensional model of the components and the nucleic acid molecule itself is only a model for visual perception. As it was shown earlier, the phosphoric acid residue is a polytope of dimension 4, and the sugar molecule has a dimension of 12.

The movement of triangles along a helix leads to the formation of polytopes with antiparallel edges. Consider an arbitrary triangle *ABC* on the plane. Choose some point *O′* on the plane outside the triangle to his left. Let this point be the base of the axis of the helix passing through the triangle. Rotate the *ABC* triangle 180 degrees to the right, moving it up the helix, parallel to the initial plane. In the projection on the plane, both triangles *ABC* and the displaced triangle *A′B′C′* will be located as shown in Figure 7.

Figure 7. Polytopes of dimension three with anti – parallel edges

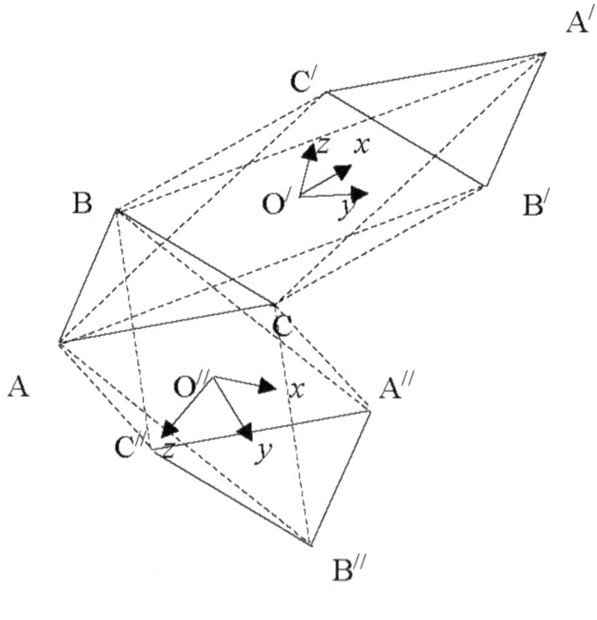

It is easy to see that the edges of the triangle *ABC* and $A'B'C'$ are antiparallel. It can now connect in space the vertices of the triangle *ABC* with the vertices of the triangle $A'B'C'$ so that there are no connections of the vertices with the same letters. In a projection on the plane the connections are represented by dotted segments. It can be seen that the connecting segments also break up into pairs of antiparallel segments. Let us now verify that the image $ABCA'B'C'$ in Figure 1, chapter 2, along with the dotted segments, is a projection of a three - dimensional convex polytope. We use the Euler – Poincaré equation (Poincaré, 1895) (1) for this aim.

The shape $ABCA'B'C'$ in Figure 7 has 6 vertices, 12 edges, 8 triangular faces (rectangles are not faces by construction, since connecting, for example, vertex *A* with vertices B', C' it turns out to be exactly the triangle $A'B'C'$). Substituting these values of elements of different dimensions into equation (1), can find 6 – 12 +8 = 2, i. e. the Euler – Poincaré equation holds in this case for $n = 3$. This proves that the resulting figure is a convex polytope of dimension 3 (if the figure were not convex, the Euler – Poincaré equation would be violated). The point O' in Figure 7 coincides with the center of the three - dimensional figure $ABCA'B'C'$ as diagonal figures pass through it and point O' located on the axis of the helix. Point O' can be considered as the origin of three - dimensional space. Coordinate axes *x, y, z* in this direction emanates from directions AA', BB', CC' respectively. Three pairs of these axes define the coordinate planes of the space of this shape. Choose some point O'' on the plane outside the triangle below it. Let this point be the base of the axis of the helix passing through the triangle. Rotate the *ABC* triangle 180 degrees to the left, moving it up the helix, parallel to the initial plane. In the projection on the plane, both triangles *ABC* and the displaced triangle $A''B''C''$ will be located as shown in Figure 7.

It is easy to see that the edges of the triangle *ABC* and $A''B''C''$ are antiparallel. It can now connect in space the vertices of the triangle *ABC* with the vertices of the triangle $A''B''C''$ so that there are no connections of the vertices with the same letters. In a projection on the plane the connection are represented by dotted segments. It can be seen that the connecting segments also break up into pairs of antiparallel segments. Let us now verify that the image $ABCA'B'C'$ in Figure11.7, along with the dotted segments, is a projection of a three - dimensional convex polytope. Obviously, a shape $ABCA''B''C''$ has as many vertices, edges, and flat faces as a shape $ABCA'B'C'$. Therefore, it satisfies the Euler – Poincaré equation (1) with dimensionality $n = 3$, i. e. it is a convex three - dimensional polyhedron. The point O'' in Figure 7 coincides with the center of the three - dimensional figure $ABCA''B''C''$ as diagonal figures pass through it and point O'' located on the axis of the helix. Point O' can be considered as the origin of three - dimensional space. Coordinate axes *x, y, z* in this direction emanates from directions AA'', BB'', CC'' respectively. The order of the coordinate axes *x, y, z* in the figures $ABCA'B'C'$ and $ABCA''B''C''$, as can be seen from Figure 7, coincide. This suggests that both figures $ABCA'B'C'$ and $ABCA''B''C''$ are topologically one and same figure - the wrong octahedron.

In the same way, polytopes with antiparallel edges of dimension greater than three can be constructed. Such a polytope of dimension 4 is constructed, for example, in Chapter 5 of this monograph. An important example of a polytope with antiparallel edges is the hereditary information polytope (Chapter 6).

The Polytope of Hereditary Information

Let us consider in detail the formation of a polytope of two sugar molecules with anti - parallel edges. Here, as in the case of the tetrahedron, to form a polytope with antiparallel edges from two sugar molecules, you must have one sugar molecule on one helix to turn this helix together with the sugar molecule and move the sugar molecule along this reversed helix in opposite direction to the original helix direction. When the sugar molecule rotates 180 degrees to the right while moving, then the original sugar molecule and the sugar molecule on the reversed helix are two polytopes with anti - parallel edges. Both of these sugar molecules in a simplified form are represented in Figure 8.

Figure 8. The two sugar molecules in a simplified form with anti – parallel edges

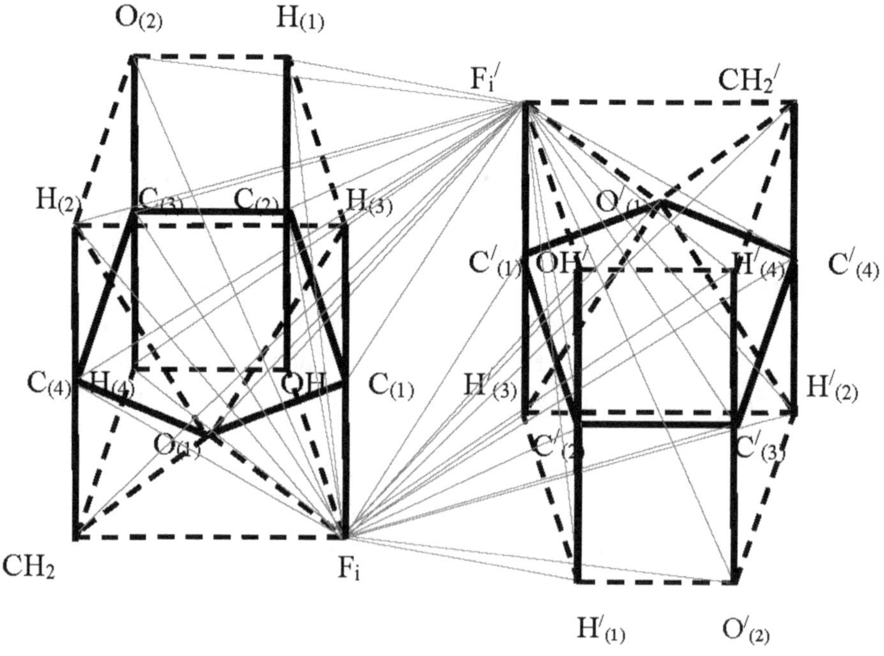

When the five - carbon sugar molecule is rotated to the left by 180 degrees, the nitrogenous bases are on opposite sides of both molecules, so that for their connection it is necessary to cross the entire set of atoms of two molecules. This is unrealistic therefore this option is not considered. When the sugar molecule is rotated 180 degrees to the left, it is possible to connect the sugar molecules through nitrogenous bases only between two different chains of nucleic acids or in the case of turning the chain itself in the opposite direction.

In full, the sugar molecules have a dimension of 12, in the corresponding polytope each vertex must have an edge connection with all the other vertices. Knowledge of this now one need. For the formation of a polytope of dimension 13, it is necessary to connect each vertex of one polytope with the vertices of another polytope so that there are no vertex connections with the same letters. All connecting edges break into pairs of antiparallel edges. At the same time, a set of coordinate a two - dimensional planes emanates from the center of the formed polytope as from the origin of coordinates. Their number is equal to the number of combinations from 13 to 2, i. e. 48 coordinate planes. In order to clarify the possible geometrical circumstances of the connection of helices in double - helix nucleic acid molecules with nitrogenous bases, we are primarily interested in the coordinate planes containing these nitrogenous bases $F_i, F_i^{/}$. There are exactly 12 such coordinate planes in the obtained polytope of dimension 13. They are depicted in Figure 6.2. (Chapter 6) by blue solid lines and are indicated below by the vertices of the polytopes contained in them

$$\begin{matrix} F_i^{/}H_{(1)} & F_i^{/}O_{(2)} & F_i^{/}H_{(2)} & F_i^{/}C_{(3)} & F_i^{/}C_{(2)} & F_i^{/}H_{(3)} & F_i^{/}C_{(4)} & F_i^{/}H_{(4)} & F_i^{/}OH & F_i^{/}C_{(1)} & F_i^{/}O_{(1)} & F_i^{/}CH_2 \\ H_{(1)}^{/}F_i\text{'} & O_{(2)}^{/}F_i\text{'} & H_{(2)}^{/}F_i\text{'} & C_{(3)}^{/}F_i\text{'} & C_{(2)}^{/}F_i\text{'} & H_{(3)}^{/}F_i\text{'} & C_{(4)}^{/}F_i\text{'} & H_{(4)}^{/}F_i\text{'} & (OH)^{/}F_i\text{'} & C_{(1)}^{/}F_i\text{'} & O_{(1)}^{/}F_i\text{'} & (CII_2)^{/}F_i \end{matrix} .$$

In the center of each parallelogram, indicated by its four vertices, is the origin of coordinates and the corresponding pair of coordinate axes (they are not shown). To identify the different hydrogen and oxygen atoms at vertices of polytope, they indicated by numbers in brackets at the lower indices.

It is surprising that the number of coordinate planes containing vertices is exactly equal to the number of possible compounds of nitrogenous bases 12 (Spirin, 2019).

A:U, U:A, G:C, C:G, G:U, G:A, U:U, U:C, A·A, A·C, U:G, A:G.

Since each coordinate plane designated by the vertices of the parallelograms has a specific atomic environment, it can be assumed that each of the 12 possible compounds of nitrogenous bases is located on one particular coordinate plane out of 12 possible. Nitrogenous bases in the coordinate planes are connected to each other using hydrogen bonds. Moreover, there are pairs of nitrogenous bases that are connected by three parallel hydrogen bonds, two parallel hydrogen bonds, two non-parallel hydrogen bonds, and even one hydrogen bond. But in every hydrogen bond there is a division of electric charges into positive and negative, which are located opposite each other. Due to this separation, the combination of nucleic acids occurs. If a nucleic acid consists of several single-stranded nucleic acids, then, as proved in this monograph (Chapter 5, Chapter 6), single-stranded nucleic acids form pairs of interacting nucleic acids, within which the described interaction exists while maintaining the corresponding complementarity. This solves the question of the possible orientation of the bond of nitrogenous flat bases in nucleic acids using ideas about the high dimensionality of the constituent nucleic acids. It is also surprising that in order to create 13 – cross - polytopes, providing the connection with nitrogenous bases $F_i, F_i^{/}$, nature specially created double - stranded nucleic acids with oppositely directed spirals. This is realized in DNA and RNA when creating regions with inverted helices.

In a variety of nucleic acid molecules, the issue of chain interaction is important.

In ribosomes, RNA interacts with each other due to bivalent metal ions (mainly magnesium ions). Positively charged magnesium ions attract negative charges of phosphoric acid residues, ensuring the stability of the ribosomes. In double - stranded nucleic acids, the helices are connected to each other by means of nitrogenous bases complementarily interacting with each other by a hydrogen bond. However, the magnesium ions and nitrogen bases in nucleic acids could not be specifically located. It has been established that magnesium ions and flat nitrogenous bases can be located inside special polytopes of higher dimension. Here knowledge is needed of the higher dimension of phosphoric acid residues and sugar molecules. Such polytopes are polytopes with anti - parallel edges, i.e. cross - polytopes of higher dimension. Binding agents are located on the free coordinate planes of these polytopes in the vicinity the center of the polytope.

In this case, the two - dimensional coordinate plane on the boundary of the polytope should contain the objects to be joined. In the case of magnesium ions, there are four specific coordinate planes inside the 5 – cross - polytope, in which an ion can accommodate, combining negative charges. In the case of nitrogenous bases, the existence of 12 coordinate planes inside a cross - polytope of dimension 13, in which flat nitrogenous bases can be located, connecting the helix of nucleic acids, was discovered. Exactly as much as there are options for combining nitrogenous bases. It is given that each coordinate plane of these 12 planes has a specific environment of atoms. It should be assumed that only one of the 12 possible compounds of nitrogenous bases is placed in each of these planes. It is surprising that the existence of higher - dimensional polytopes with anti - parallel edges is possible only in the case of the opposite direction of interacting helices, and this is exactly what nature provides in the double - helix DNA and in the RNA segments with self - inversion of the helix in the opposite direction.

To build the polytopic of hereditary information, we need to supplement Figure 2, chapter 2 with edges that create a closed and convex figure. In this case, it is not necessary to connect the vertices symmetrically relative to the center of Figure 2, chapiter 2 with the edges. Each vertex will be connected by an edge to the other vertices. In this case, the polytope will be a cross - polytope and its dimension is equal to half the number of vertices. Indeed, according to (Zhizhin, 2019 a), there is a relation between f_i the number of dimension elements i in the cross - polytope and the dimension d of the cross - polytope itself,

$$f_i(d) = 2^{1+i} C_d^{d-1-i}. \tag{2}$$

At the vertex $i = 0$, therefore, according to (2) $f_0 = 2C_d^{d-1} = 2d, d = f_0 / 2 = 13$.

Thus, the polytope of the hereditary information has dimension 13. When portraying this polytope, we will use the technique that is used in portraying cross - polytopes. We distribute all 26 vertices on the circle so that the vertices opposite in Figure 2, chapter 2 remain opposite and there is no edge between them (Figure 3, chapter 2).

The edges corresponding to the chemical bonds are indicated in this figure by thick black solid lines. Anti-parallel edges highlighting coordinate planes with vertices corresponding to nitrogenous bases are indicated by bluey lines. The remaining edges are indicated by thin dash-dotted black lines. We emphasize that there is a chemical bond between the nitrogenous bases, but it is not covalent. Therefore, this connection is not indicated by an edge.

When the polytope of hereditary information moves along the common axis of the double helix of nucleic acids in accordance with the sequence of bases in DNA, the connection between the vertices $F_i^{/}, F_i^{//}$ is carried out by one of the twelve base pairs, which occupies one of the 12 coordinate planes of the polytope.

By relation (2), one can determine the number of elements of different dimensions in the polytope of hereditary information.

The number of edges is $f_1(13) = 2^2 C_{13}^{11} = 312$,
The number of triangles is $f_2(13) = 2^3 C_{13}^{10} = 2288$,
The number of tetrahedrons is $f_3(13) = 2^4 C_{13}^{9} = 10560$,
The number of fourth - dimensional faces (simplexes) is $f_4(13) = 2^5 C_{13}^{8} = 41184$,
The number of fives - dimensional faces (simplexes) is $f_5(13) = 2^6 C_{13}^{7} = 109824$,
The number of six - dimensional faces (simplexes) is $f_6(13) = 2^7 C_{13}^{6} = 219648$,

The number of seven - dimensional faces (simplexes) is $f_7(13) = 2^8 C_{13}^5 = 164736$,
The number of eight – dimensional faces (simplexes) is $f_8(13) = 2^9 C_{13}^4 = 183040$,
The number of nine - dimensional faces (simplexes) is $f_9(13) = 2^{10} C_{13}^3 = 292864$,
The number of ten - dimensional faces (simplexes) is $f_{10}(13) = 2^{11} C_{13}^2 = 159744$,
The number of eleven - dimensional faces (simplexes) is $f_{11}(13) = 2^{12} C_{13}^1 = 26624$,
The number of twelve - dimensional faces (simplexes) is $f_{12}(13) = 2^{13} = 4096$.

The obtained numbers determine the structure of the polytope of hereditary information.

The monograph (Zhizhin, 2019 a) introduced the concept of the incidence coefficients of elements of lower dimension with respect to elements of the higher dimension and elements of higher dimension with respect to elements of the lower dimension. The first characterizes the number of elements of a certain higher dimension to which the given element of a lower dimension belongs. The second characterizes the number of elements of a given lower dimension that are included in a particular element of a higher dimension. Here we must remember that the vertices of geometric elements of various dimensions are atoms, molecules or functional groups. Therefore, the incidence of geometric elements to a friend means contact between particles of the matter, including living matter. The contact between particles of matter can be interpreted as the transfer of information on material structures, including biological structures. It was proved that the sum of incidents retains its value when changes the direction of the relationship between the elements (the law of conservation of incidents) (Zhizhin, 2019 a). For the polytope of hereditary information this value is $1.78\ 10^9$. A significant value of the total incidence stream in the polytope of hereditary information indicates an intensive exchange of information between elements of the polytope of hereditary information. This may explain for the recently discovered epigenetic principle of the transmission of hereditary information without changing the sequence of genes in DNA and RNA molecules (Hawkes, et al, 2016; Lindquist, et al., 2016; Mancuso, 2017).

CONCLUSION

The studies carried out show that with a large dimension of interacting biomolecules, the condition for their correspondence (complementarity) can only be the condition that ensures the possibility of a chemical reaction between molecules. Moreover, since the molecules have a higher dimension, i.e. are representable as polytopes of higher dimension, then only the condition of correspondence to the geometry of polytopes of higher dimension can be a correspondence condition. In the examples considered, this correspondence condition is associated with a mirror image of the geometric elements of one polytope in relation to the geometric elements of the polytope interacting with it. So, when considering the formation of a chain of polytopes along a line (the interaction of amino acid molecules), the bound molecules turned out to be mirror images of each other relative to the line along which they line up. During the interaction of nucleic acids, the interacting sugar molecules are also connected by mirror reflection, forming polytopes with antiparallel edges, i.e. high-dimensional cross-polytopes.

As part of the definition of functional dimension, in which all functional groups of monosaccharide molecules are taken as vertices of the corresponding geometric figure, analytical expressions are obtained for calculating the dimension of polytopes of monosaccharide molecules depending on the number of chiral carbon atoms. They refer to both linear molecules and molecules with a closed carbon cycle (furanoses and pyranoses). It has been shown that the cyclization of monosaccharides leads to an increase in their dimension. The geometric shape of monosaccharide molecules has the form of a polytope of the simplex type of higher dimension. Based on the type of polytope of higher dimension corresponding to the pyranose monosaccharide molecule and calculations, a three - dimensional approximate image of the pyranose monosaccharide molecule was obtained. It has the form of a pentagonal prism connected by a carbon cycle to a pyramid, which has a pentagonal and hexagonal base. This image is fundamentally different from the commonly used in the literature image of the pyranose molecule in the form of an "armchair". The resulting simplified three - dimensional image of the pyranose monosaccharide molecule was used to analyze the possibility of the formation of disaccharides and monosaccharide chains.

It has been found that the statement about the linearity of the connection of β - anomers of monosaccharides, widespread in the literature, is not correct since a deviation from linearity is observed in the chain of monosaccharides. An even greater deviation from linearity in the monosaccharide chain is observed when α - anomers are combined. Here, the possibility of the formation of spirals with simultaneous rotation in opposite directions was discovered.

An analysis of the conditions for the connection of molecules of the highest dimension (dimension of more than three) with each other in three servants important for biochemistry: compounds of amino acid molecules (dimension 4), compounds of monosaccharides (dimension 15), nucleic acid compounds (dimension 12). It is obvious that in none of these examples the accepted state of complementarity, which is in accordance with the surfaces of the interacting molecules, makes no sense, since it is impossible to indicate that it is the surface of the molecule of the highest dimension, i.e. it is impossible to indicate what is the correspondence of the alleged surfaces (there is no No hollow, no protrusion). It can be argued that when combining molecules of the highest dimension, the opposite of electrical charges play a significant role. Ultimately, the attraction of electrical charges provides the compounds of the molecules. For example, when combining amino acid molecules, one of the molecules gives a proton (+H), and the other gives the hydrocele group (- OH). As a result, a neutral water molecule is distinguished, and the remaining parts due to the particles remaining with the opposite charge are attracted to each other.

In addition, in the geometry of interacting molecules, a mirror arrangement of valence ties to each other at the junction of the molecules is observed. It also leads to the opposite location of charges with opposite signs.

In the interaction of sugar molecules (glucose), the proton of one molecule (+H) is also combined with a hydrocele group (- OH) another molecule with the release of a water molecule. The remaining particles with opposite charges are attracted to each other by forming a compound or β - anomer or α -anomer. In the case of the formation of α - anomer, the opposite of the charge of particles leads to the simultaneous rotation of molecules in opposite directions in perpendicular planes. In the case of the formation of a β - anomer connection due to the connection, the anti-parallelism of the rib pair occurs.

The interaction of nucleic acids occurs using nitrogenous bases, which are connected to each other with a hydrogen connection. In this regard, as you know, the compound is provided by the opposite of charged with electronic charges. This is complementarity, and not that certain nucleotides are connected.

Thus, the choice of geometric elements connected by a mirror image is determined by the specific shape of the interacting molecules in a higher-dimensional space. Only when molecules of the simplest form interact, the common element of the interacting molecules is a point.

REFERENCES

Corey, R. B., & Pauling, L. (1953). Fundamental dimensions of polypeptid chains. *Proc R Soc Lord B Biol Sci,* 141(902), 10-20.

Crick, F. H. C. (1959). The present position of the coding problem. *Brookhaven Symposia in Biology, 12*, 35–39. PMID:13812855

Frank - Kamenetskiy, M. (1986). DNA structure: A simple solution to the stability of the double helix? *Nature, 324*, 325.

Frank - Kamenetskiy, M. (1988). H – form DNA and the hairpin – triplex model. *Nature, 333*, 214.

Gamow, G. (1954). Possible relation between deoxyribonucleic acid and protein structures. *Nature, 173*(4398), 318. doi:10.1038/173318a0

Gratias, D., & Cahn, J. W. (1986). Periodic and qusiperiodic crystals. *Scripta Metallurgica, 20*(9), 1193–1197. doi:10.1016/0036-9748(86)90030-X

Hawkes, E. J., Hennelly, S. P., Novikova, I. V., Irwin, J. A., Dean, C., & Sanbonmatsu, K. Y. (2016). COOLAIR antisense RNAs from evolutionarily conserved elaborate secondary structures. *Cell Reports, 16*(12), 3087–3096. doi:10.1016/j.celrep.2016.08.045 PMID:27653675

Koolman, J., & Roehm, K. (2013). *Color Atlas of Biochemistry*. Thieme.

Lehninger, A. L. (1985). *Principes of Biochemistry*. Worth Publishers, Inc.

Lindquist, S. (2016). Luminidependens (LD) is an arabidopsis protein with prion behavior. *Proceedings of the National Academy of Sciences of the United States of America, 113*(21), 6065–6070. doi:10.1073/pnas.1604478113 PMID:27114519

Mancuso, S. (2017). *PLANT REVOLUTION. Le piante hanno già inventato il nostro futuro*. GUIUTI EDITORE.

Metzler, D. E. (1980). *Biochemistry. The Chemical Reactions of Living Cells*. Academic Press.

Oswald, A. T., MacLeod, C. M., & McCarty, M. (1944). Studies on the Chemical Nature of the Substance Inducing Transformation of Pneumococcal Types: Induction of Transformation by a Desoxyribonucleic Acid Fraction Isolated from Pneumococcus Type III. *The Journal of Experimental Medicine, 79*(2), 137–158. doi:10.1084/jem.79.2.137 PMID:19871359

Pauling, L. (1987). So-called Icosahedral Decagonal Quasicrystals are Twins of an 820-atom Cubic Cristal. *Physical Review Letters, 58*(4), 365–368. doi:10.1103/PhysRevLett.58.365 PMID:10034915

Pauling, L., & Corey, R. B. (1950). Two Hydrogen-bonded Spiral Configurations of the Polypeptide Chain. *Journal of the American Chemical Society, 72*(11), 5349. doi:10.1021/ja01167a545

Pauling, L., & Corey, R. B. (1953). Compound Helical Configurations of the Polypeptide Chain:Structure of Proteins of the α - keratin Type. *Nature, 171*(4341), 59–61. doi:10.1038/171059a0 PMID:13025480

Pauling, L., Corey, R. B., & Branson, H. R. (1951). The Structure of Proteins: Two Hydrogen-bonded Helical Configurations of the Polypeptide Chain. *Proceedings of the National Academy of Sciences of the United States of America*, *37*(4), 205–211. doi:10.1073/pnas.37.4.205 PMID:14816373

Poincare, A. (1895). Analysis situs. *J. de e` Ecole Polytechnique, 1,* 1 – 121.

Shechtman, D., Blech, I., Gratias, D., & Cahn, J. W. (1984). Metallic phase with longerange orientational order and no translational symmetry. *Physical Review Letters*, *53*(20), 1951–1953. doi:10.1103/PhysRevLett.53.1951

Spirin, A. S. (2019). *Molecular biology. Structure of the ribosome and protein biosynthesis*. Knowledge Lab.

Spirin, A. S., & Gavrilova, L. P. (1971). *The Ribosome*. Science.

Watson, J. D. (1976). *Molecular Biology of the Gene*. Benjamin.

Watson, J. D., & Crick, F. H. C. (1953a). Molecular structure of nucleic acids. *Nature*, *171*(4356), 738–740. doi:10.1038/171737a0 PMID:13054693

Watson, J. D., & Crick, F. H. C. (1953b). Genetical implications of the structure of deoxyribose nucleic acid. *Nature*, *171*(4361), 964–967. doi:10.1038/171964b0 PMID:13063483

Zhizhin, G. V. (2013 a). Relations for the number of faces of different dimensions in the tower of n - dimensional convex polytopes. Materials 9th All-Russian Scientific School "Mathematical research in the natural sciences." 24-32. Geological institute KSC RAS, Apatity.

Zhizhin, G. V. (2013 b). Images of convex regular and semiregular n-dimensional polytopes. Materials 9th All-Russian Scientific School "Mathematical research in the natural sciences." 32-42. Geological institute KSC RAS, Apatity.

Zhizhin, G. V. (2014). *World – 4D*. Polytechnic Service.

Zhizhin, G. V. (2016). The structures, topological and functional dimension of biomolecules. *International Journal of Chemoinformatics and Chemical Engineering*, *5*(2), 44–58. doi:10.4018/IJCCE.2016070104

Zhizhin, G. V. (2018 a). *Chemical Compound Structures and the Higher Dimension of Molecules: Emerging Research and Opportunities*. IGI Global. doi:10.4018/978-1-5225-4108-0

Zhizhin, G. V. (2018b). The Metric Structure of Nucleic Acids and the Higher Dimension of Their Constituents. *International Journal of Chemoinformatics and Chemical Engineering*, *7*(2), 1–15. doi:10.4018/IJCCE.2018070101

Zhizhin, G. V. (2019 a). *The Geometry of Higher-Dimensional Polytopes*. IGI Global. doi:10.4018/978-1-5225-6968-8

Zhizhin, G. V. (2019b). The structure of polytope of hereditary information. J. *International Journal of Chemoinformatics and Chemical Engineering*, *8*(2), 7–22. doi:10.4018/IJCCE.2019070102

Zhizhin, G. V. (2019 c). *Attractors and higher dimensions in population and molecular biology: emerging research and opportunities*. IGI Global. doi:10.4018/978-1-5225-9651-6

Zhizhin, G. V. (2019d). The Polytope of Hereditary Information: Structure, Location, Signification. *Biochemistry and Modern Applications.*, *2*, 56–62. doi:10.33805/2638-7735.124

Zhizhin, G. V. (2020a). Dimension and Life. *Acta Scientific Biotechnology*, *1*(11), 35–40.

Zhizhin, G.V. (2020 b). Higher Dimensions in Linear, Helicoidal and Folder Structures of Protein Molecules. *Acta Scientific Biotechnology,* 1(12), 02-09.

Zhizhin, G. V. (2020c). Hidden Nucleic Acid Bond Order? *Acta Scientific Biotechnology*, *1*(4), 34–36.

Zhizhin, G. V. (2020d). Interaction of Nucleic Acids: Hidden Order of Interaction. *International Journal of Applied Research in Bioinformatics*, *11*(2), 1–15. doi:10.4018/IJARB.2021070101

Zhizhin, G. V. (2020e). Spirals of Polysaccharides Simultaneously Rotating in Opposite Directions. *Acta Scientific Biotechnology*, *1*(7), 13–16.

Zhizhin, G. V. (2021 a). *Normal partitions and hierarchical fillings of n - dimensional spaces*. IGI Global. doi:10.4018/978-1-7998-6768-5

Zhizhin, G. V. (2021b). Spatial Models of Sugars and Their Components. *International Journal of Applied Research in Bioinformatics*, *11*(2), 9–23. doi:10.4018/IJARB.2021070102

Zhizhin, G. V. (2021 c). The Structure of DNA Taking into Account the Higher Dimension of Its Components. In *Encyclopedia of Information Science and Technology* (5th ed., pp. 730–747). Ed. Mehdi Khosrow- Pour. doi:10.4018/978-1-7998-3479-3.ch050

Zhizhin, G. V. (2021d). Interaction of Nucleic Acids: Hidden Order of Interaction. *International Journal of Applied Research in Bioinformatics*, *11*(2), 1–8. doi:10.4018/IJARB.2021070101

Zhizhin, G. V. (2021e). Hidden Order of Interacting Nucleic Acids. *Journal of Nature Science and Sustainable Technology*, *15*(4), 1–9.

Zhizhin, G. V. (2022 a). *The Classes of Polytopes Higher Dimensions in Physics, Chemistry, and Biology*. IGI Global.

Zhizhin, G. V. (2022b). The Properties of Biomaterials Hidden in the Spaces of Higher Dimension. *Annals of Advanced Biomedical Sciences.*, *5*(2), 1–9. doi:10.23880/aabsc-16000189

Zhizhin, G. V. (2023 a). *Non-Euclidean Geometry in Materials of Living and Non-Living Matter in the Space of the Highest Dimension*. Nova Science Publisher.

Zhizhin, G. V. (2023b). To the Issue of the Complementarity of Biomolecules in the Space of the Highest Dimension. *Journal of Clinical Oncology and Research*, *3*, 1–15.

Zhizhin, G. V., & Diudea, M. V. (2016). Space of Nanoworld. In M. V. Putz & C. M. Marius (Eds.), *Sustainable nanosystems, development, properties, and applications* (pp. 221–245). IGI Global.

Chapter 12
Emergence of Life on Earth in a Space Higher Dimension

ABSTRACT

A critical analysis of existing scientific ideas about the origin of life on the land plug was carried out. A new idea is put forward about the origin of life on planet Earth in a flow mound, consisting of rocks of the earth`s crust, above a hydrothermal vent at the bottom of the ocean, where the necessary conditions for the synthesis and evolution of RNA, DNA, and the formation of cells of living organisms are observed. It is proved that water molecules, as the basis of living organisms and the source of the origin of life, have the highest dimension. The properties of chains of biomolecules are considered, geometrically investigating their structure in a space of higher dimension. It is shown that their movements and form are subject only to the conditions of their possible existence in given circumstances and are not subject to the imperatives set in advance from general considerations ("chiral purity"). We can say that the molecules of the chains of living matter prove by their existence the principle of the phenomenon of life.

INTRODUCTION

Thinking people have always been interested in the question of when and how life could arise on planet Earth. This request is also important for other planets, especially since humanity is looking for possible other planets on which life is possible. Over the course of history, many different hypotheses have been put forward in this regard. However, many hypotheses have been rejected over time as impossible, after corresponding advances in science have dismissed these hypotheses as impossible. It is necessary to find an environment in which life could arise, in addition, once a

DOI: 10.4018/979-8-3693-0588-1.ch012

living thing has arisen, it must be able to develop, participate in the evolution process, rejecting bad options and consolidating the best options. In this case, we are talking about molecules, i.e. molecules necessary for life - biomolecules. The idea that the ancient world was built using ribonucleic acid (RNA) molecules has now begun to seem convincing. It is necessary to find a place on planet Earth where ribonucleic acid molecules of sufficient length could originate and where they could further develop, turn into deoxyribonucleic acid (DNA) molecules, sugar molecules, and other molecules necessary for life. And how did the transition from biomolecules to the form of life, which we are now witnessing - to the cellular form of life, come about.

In addition, at present, the question of the dimension of molecules used by living organisms is becoming more and more clear. Since 2016, the author has been publishing articles in various reputable journals proving that the dimension of biomolecules is higher than three, i.e. that biomolecules have a higher dimension. It is proved that this fact is essential. It was found that a number of properties of substances are hidden in spaces of higher dimension. In order to use and know these properties, it is necessary to be able to use the space of higher dimension. This raises the question of where did the highest dimension of biomolecules come from? Apparently, the highest dimension of biomolecules arose together with the appearance of biomolecules on Earth in the process of the development of the planet itself. It should be noted here that the highest dimension is observed not only in biomolecules, but also in mineral molecules (Zhizhin, 2018, 2019, 2020, 2021, 2023). Maybe there is some connection between the dimensions of biomolecules and the dimensions of minerals? We will try in this chapter to answer these questions, or at least some of them.

THE FORMATION OF ORGANIC MATTER ON PLANET EARTH

One of the greatest problems of natural science is the problem of the emergence of life on planet Earth. It has long been worried about the human mind. Matter, i.e. The objective reality, which we observe and experimentally study, is in constant movement. Moreover, the movement of matter cannot be considered only as the movement of material bodies in space. Such mechanical movements are the simplest (lower) form of movement. Along with this lower form of movement, there are other more complex forms of movement that arise as new qualities in the process of the development of matter. These new qualities of material objects really appear again at this stage of development, they did not exist before. These new qualities are more complex, more diverse compared to existing earlier. In the event of new qualities, the previous forms of movement do not disappear, they, as it were, fading into the background. Moreover, new forms of movement cannot be reduced to the previous forms of movement.

According to the isotopic analysis of the most ancient breeds of the Earth, our planet has existed for about 5 MRD years. For a very long period after its formation, the earth was lifeless. At this time, she was characterized by only one inorganic form of movement of matter. In the absence of life, these forms of movement were associated with the movement of the earth's crust and its rocks (Vernadsky, 1925; Fersmann, T.1 - T.4, 1934). In a four -volume essay, A. Fersman describes in detail the migration of chemical elements of the earth's crust. An important part of this description is presented by studies on carbon migration. At the stage preceding the emergence of life on Earth, there was a long process of formation and gradual complication of organic substances. The simplest starting carbon compounds in this respect are hydrocarbons. An obligatory condition for the abiogenic formation of hydrocarbons on a particular space object is the abundance of hydrogen on it in a free or bound state. But since hydrogen is the main element of our Galaxy, it makes up more than 90% of its total matter, the occurrence of hydrocarbons is extremely common on various bodies of our stellar world. Hydrocarbons have indeed been found in almost all accessible celestial bodies. The data of planetary cosmogony show that the abiogenic formation of hydrocarbons on a large scale occurred in the very process of the emergence of the Earth as a planet. It also occurred later during the formation of the earth's crust. In the modern era of the existence of the Earth, the vast majority of organic compounds are synthesized on the basis of a new form of movement - biologically. Therefore, now the abiogenic way of formation of hydrocarbons has faded into the background. The atmosphere of the Earth, into which gaseous hydrocarbons were released from the earth's crust, was fundamentally different from the modern one. She was restorative (Galimov, 2004, 2009). The atmosphere of the lifeless Earth mainly consisted of hydrogen compounds: water vapor, ammonia, carbon disulfide, etc. S. Miller and H. Urey (Miller, 1953; Miller, Urey, 1959) tried to experimentally reproduce such an atmosphere. He passed silent discharges through a mixture of water vapor, ammonia, methane, and hydrogen. At the same time, he formed amino acids - the basic building blocks from which protein molecules are built. Similar results were obtained when the same gas mixture was exposed to short-wave ultraviolet rays, which penetrated deeply into the earth's primary atmosphere, due to the absence of oxygen at that time, which is formed as a result of the vital activity of living organisms (Oro, 1961; Bernal, 1967).

Organic substances were formed in the indicated way, first in the earth's atmosphere, and then passed into its hydrosphere, where their transformations and interactions did not stop. The experiments of a number of researchers (Akabori, 1957; Wilson, et al., 1961; Schramm and Pohlmann, 1962) demonstrated the full possibility of far-reaching polymerization reactions of amino acids under these conditions. In this case, high-molecular substances such as proteins or nucleic acids are formed. However, the orders of arrangement of monomers in these macromolecular substances observed in experiments were random and did not correspond to the orders that are present in real proteins and nucleic acids at the present time, i.e., achieved in the process of evolution and complication.

It is currently believed that in the process of development of the organic world, ribonucleic acids (RNA) first arose as simpler (single-stranded). DNA arose much later, in parallel with the complication of functions and with the formation of a narrower purposefulness of the properties of organisms (Belozersky, 1957). For a long time, only genetic functions were recognized for nucleic acids, including RNA. The main genetic function of complementary replication with the participation of an enzyme catalyzing polymerization is undoubtedly the prerogative of DNA, but it is also inherent in RNA. The function of RNA replication on an RNA template was demonstrated using the replicative cycle of a number of RNA viruses as an example (Baltimore et al., 1966; Spiegelman and Haruna, 1966). Such RNA replication, independent of DNA and forbidden in most cases in the cycles of normal cells of modern organisms, can be considered as a relic that has been preserved in the modern world thanks to viruses, as relic genetic agents that take replication and translation out of control of cellular DNA (Spirin, 2019). At present, it appears that relict RNA replication can also be detected in normal cells (Ahlquist, 2002). Another genetic function - protein coding is also inherent in RNA, mRNA codes proteins along with DNA. It was fundamentally important to establish the fact that RNA nucleotide chains, like the polypeptide chains of proteins, are able to fold into compact structures (Spirin et al., 1959; Spirin, 1960; Kiselev et al., 1961), and that folding leads to specific conformations of molecules RNA (Vasiliev et al., 1978, 1986). In the development of ideas about the function of RNA, the discovery of non-coding RNA, which primarily includes ribosomal RNA, was decisive. It was for high polymer noncoding RNAs that it was shown that their specific compact folding determines the conformation of the ribosome. The ability of RNA to form compact spatial structures provides the basis for specific interactions with other molecules, both macromolecules and ligands. The final recognition of RNA's widest ability to recognize other molecules and interact with them in a very specific way came thanks to aptamers - small-sized synthetic RNA obtained by selection from many variants of nucleotide sequences using procedures, the so-called "endless evolution" (evolution in a test tube) (Ellington and Shostak 1990; Turk and Gold 1990). It turned out that it is possible to select and multiply RNAs that have the ability to selectively bind almost any kind of molecules, from low molecular weight organic compounds to individual peptides and proteins (Gold et al., 1995; Puglis, Williamson, 1999). The catalytic function of RNA is based on the ability of RNA to specific molecular recognition. However, throughout the entire previous history of biochemistry, it has been argued that biochemical catalysis is the prerogative of exclusively proteins - enzymes. Therefore, all theories of the origin of life were forced to proceed from the primacy of proteins as macromolecules absolutely necessary for the emergence of biochemical metabolism.

The discovery of the catalytic function of RNA overturned all previous ideas about the exclusive role of proteins not only in the emergence of life, but also in understanding the very phenomenon of life. By analogy with proteins - enzymes - enzymes (Dixon, Webb, 1979), catalytic RNAs began to be called ribozymes. It turned out to be possible to create artificial ribozymes with a wider range of catalyzed reactions (Check and Golden, 1999). The discovery of catalytically active RNA ribozymes also played a decisive role in the development of the hypothesis that RNA, as an independent polymer, could reproduce and function in ancient times on Earth and in other space objects before the appearance of cellular life forms. it can be assumed that RNA molecules could do without not only DNA as a genetic material, but also without proteins to catalyze important synthetic and metabolic reactions.

The idea of an ancient protein-free RNA world as a possible precursor to modern life on Earth was finally formulated in 1986 (Gilbert, 1986). At present, this hypothesis is generally accepted (Cech, Gesteland, Atkins, 1993, 1999, 2006). However, the question of the origin of such a world on Earth is one of the most difficult in terms of the origin of life.

THE EVOLUTION OF ORGANIC MATTER ON PLANET EARTH

It can be assumed that primary oligoribonucleotides arose from abiogenically formed monoribonucleotides by polymerization on the surfaces of clays and minerals. In any case, the appearance of oligoribonucleotides should be the starting point for the emergence of the RNA world. For further development, it was necessary that the abiogenic synthesis of oligoribonucleotides, based on rare random events, be supplemented with a permanent mechanism that could generate variants of these oligomers and extend them. The elongation of short oligoribonucleotides into polyribonucleotides seems to be an absolutely necessary condition for the formation of compactly folded structures with specific ligand recognition properties and catalytic activities. The generation of variants in a population of abiogenic oligo- and polyribonucleotides is required in order to give a chance for the random occurrence of the desired functional, including catalytic activities. The possibility of such processes was confirmed experimentally. In 1991-1993, Chetverin A.B. and Chetverina H.V. with employees developed a molecular RNA cloning method (Chetverin, et al, 1991; Chetverina, Chetverin, 1993; Chetverina, Chetverin et al., 1999). These experiments showed that RNA molecules are able to form colonies on wet solid media, if they are provided on these media conditions for replication. In this case, conditions are created for very fast evolution: RNA colonies were not fenced off from the external environment and could easily exchange molecules, i.e. their genetic material.

Experiments showed that RNA chains in solution exchanged their parts in sequences. The exchange could be carried out both between different molecules and within the same molecule and could occur anywhere in the chains. This discovery, obviously, is directly related to the problem of the emergence of the ancient world of RNA. The appearance of sufficiently long polyribonucleides and the generation of variants during random cis- and trans-rearrangements should have led to the random appearance of ribozymes, and the appearance in the RNA population of a ribozyme catalyzing the process of complementary RNA replication should have been critical.

The emergence and existence of the RNA world on Earth naturally required an aquatic environment with metal salts dissolved in it.

In what form could the aquatic environment exist on the ancient Earth? Previously, it was believed that life originated in the seas and oceans and then came to Earth. However, most geologists and paleontologists now believe that seas and oceans apparently did not exist on the ancient Earth. Maybe these were shallow reservoirs ("Darwin's ponds and puddles")? But it is difficult to imagine that in a puddle there could be metal ions necessary for the creation of molecules of living organisms. In addition, in order for the selection of the best RNA molecules to arise in this aquatic environment, it is necessary to separate this aquatic environment from the surrounding space. Apparently, for the first time the idea of isolation from the environment as a prerequisite for the reproducing system of macromolecules to evolve was put forward by Oparin in 1924 (Oparin, 1924). As the most primitive method of primary compartmentalization, which could be carried out in the presence of abiogenically occurring polypeptides and proteins, Oparin proposed coacervation - the formation of droplets of a concentrated sol in a dilute medium with dissolved molecules. This idea was later supported by J.W.S. Haldane (Haldane, 2015). However, according to the hypothesis, the ancient world of RNA was single-polymeric, and due to the lack of polypeptides, polysaccharides, and other polymers capable of cocarceration, the compartmentalization of RNA ensembles in the form of coarcicular droplets seems unlikely (Spirin, 2019). Nevertheless, in order for each RNA ensemble to exist as a system, inherit acquired traits that are useful for the entire system, and evolve, its nucleotisynthetic ribozymes and synthesis products must be somehow connected in space. Therefore, in most theories of the origin of life, the emergence of some kind of limiting membranes is postulated as a necessary condition for the beginning of evolution. However, this condition contradicts another condition for the growth of the system, its openness for possible exchange with the environment.

Life began in water and continues in water. Water is matter and the matrix of life. Water is not only outside of us, but also inside us. Every living organism is 80% water. Recent studies (Qian, et al., 2010; Russel, et al., 2005; Simakov, et al., 2020) show that as a result of the interaction of water with pyrite FS_2, there is a possibility of the appearance of protocells LUCA in water layers on the surface of minerals, including on the surface of nanodiamonds, which can form under terrestrial conditions at moderate temperatures and pressures (Simakov, 2018).

The theory of the origin of life is formulated in the context of gravitational differentiation of water worlds massive enough to hold carbon dioxide and nitrogen atmosphere. On the early earth differentiation into various spheres - core, mantle, asthenosphere, lithosphere, hydrosphere and atmosphere - were in many ways the answer to radiogenic and gravitational heat production in the interior. The effect of this density differentiation was to collect the electrons in the nucleus in natural iron (Fe0 in full composition). However, the formation of the early core still remained. relatively electron-rich upper mantle (Fe II-bearing minerals), compared to exhaled and accumulated oxidized volatile substances (H_2 O, CO_2, N_2, NO, SO_2, P_4 O_{10}), which made up the early hydrosphere and atmosphere. The atoms that make up these oxidized molecules are forced to divide electrons covalently, which makes some of them potential electrons. Hydrothermal H_2 acted as a carrier gas for electrons generated from reduced iron in the crust. Recovery (hydrogenation), especially marine CO_2, NO (or NO_3^-) and Fe III on alkaline submarine hydrothermal vents, led to the emergence of metabolic pathways, first in inorganic compartments, then in prokaryotic cells that proliferate, led to the emergence of metabolic pathways, primarily inorganic compartments, then in prokaryotic cells that proliferate exponentially to produce an early biosphere, a last sphere for differentiating (Takai, et al., 2004). On a larger scale of the solar system volatiles differentiate from the sun outward in response to solar wind, radiation, gravitational field reduction and condensation and sublimation. But again, the same tension between waters containing CO_2, and hydrothermal H_2 will tend to bring life to other rocky terrestrial globes such as Mars and possibly Europa and Enceladus (Russell and Hall, 1999; Vance et al., 2007). Without this geochemical there will be no potential and there will be no life, because life is not the convergence of a myriad of organic molecules, but the resolution of the redox potential and pH voltage in an aqueous medium in which simple organic molecules synthesized from CO_2 and other minor substances before polymerization, interact and even hydrolyze. One-digit sinks from early cells will be acetate or acetic acid and its derivatives. Outside the solar system applies the same logic. Only those extra solar planets with rocky interiors and enclosing surface or groundwater containing dissolved carbon dioxide (metered with nitric oxide) will have a concomitant biosphere, one that would be forced to emerge in order to resolve the same geochemical and electrochemical stresses.

We tend to be overly confused about the complexity of life and our place in life. It was generally believed that the origin of life must be rare, or even a unique event that ended, as we liked to think, in ourselves. But science constantly destroys ideas about our place in the universe, while at the same time, revealing more of their wonders. We ourselves it would seem an uncredible combination of very complex interacting cells. However, in a broader sense, we pay for the planet to a more probable state, i.e. we waste energy like we don't tomorrow and produce mountains of waste, performing entropy imperative (Martyushev, Seleznev, 2006; Russell, 2007). The best way to understand the mystery of the origin of life is to try to unravel from what led to the existence of our ancestors. Maybe we can get in further than Pooh-Bach's "protoplasmic primordial atomic globule" (Gilbert and Sullivan, 1885). Here we will develop the idea that our ancestors were housed in inorganic compartments consisting of silica, silicates, carbonates, and hydroxides, with added metal sulfides with catalytic potential: we evolved from hydrogen and shaping fluid containing hydrogen sulfide exiting through cracks in serpentinization of oceanic crust into the earliest carbon dioxide ocean (Russell, et al.,1989). A hydrothermal mound grew out of the interaction of two fluids over a hydrothermal vent. This mound is retarded much further than communication, except for the partial penetration of hydrogen ions, carbon dioxide, phosphates, and nitrates from the ocean. We believe that it was in such an environment that hydrogenation began, which led to the emergence of life.

Our exploration of this idea will lead us to the conclusion that life must be as common as hydrothermal convection cells on icy, wet rocky planets lit by stars large enough to hold an atmosphere, or at least a carbon dioxide ocean, as in our solar system. All biomolecules ultimately evolved from a protocell that emerged over 4.5 billion years ago.

In the work (Qian et al., 2010) the results of an experimental study of the hydration of rock elements are presented in which the sulfidation of magnetite to form pyrite under hydrothermal conditions () vapor saturated pressure, a process associated with gold depositions in a number of ore deposits. The formation of pyrite marcasite was studied as a function of reaction time, temperature, PH, sulfide concentration, solid-weight-to fluid- volume ratio, and geometric surface area of magnetite in polytetrafluoroethylene – lined autoclave (PTFE) and a titanium and stainless steel flow – through cell.

Thus, the rock mound above the hydrothermal vent at the bottom of the ocean is the place where everything necessary for the formation and evolution of RNA is performed. Firstly, this mound contains all the metal ions necessary for the formation of RNA. Secondly, the mound has cracks, which ensure the existence in them of RNA colonies of different lengths, connected by a single space, i. e. the kurgan provides the necessary compartmentalization. Thirdly, the mound above the hydrothermal vent is flowing, i. e. it ensures the exchange of the RNA colony with the environment, i.e. ensures the evolution of RNA. Fourthly, all the necessary chemical reactions can and do take place in the mound to create RNA molecules and, subsequently, DNA molecules. Consequently, the difficulties in all previous theories of the origin of life on Earth are removed. There is no need to make all sorts of often conflicting assumptions.

DIMENSION OF BINARY COMPOUNDS OF ELEMENTS OF THE EARTH'S CRUST

Since the origin of life took place in a mound of rocks at the bottom of the ocean, we should consider the chemical compounds that this mound could consist of. A.E. Fersman, in his four-volume geochemistry (Fersman, 1934), investigating the geochemistry of the earth's crust, paid much attention primarily to binary compounds of chemical elements. He drew attention to the fact that many different binary compounds have the same type of structure. Among these structures, he singled out three structures: adamantane, wurtzite and fluorite. The structure of adamantane was already considered in the 10th chapter of this monograph and it was shown that the dimension of the compound of this structure is 9. Therefore, the dimension of nanodiamonds that can be formed in the earth's crust (according to the studies of S.K. Simakov (Simakov,)) is also equal to 9. In addition, as shown in the monograph by G.V. Zhizhin (Zhizhin, 2018), many more binary compounds with the participation of transition and non-transition elements have the adamantane structure. The structure of adamantane with the participation of transition elements has binary compounds. The structure of adamantane with the participation of only non-transitional elements has binary compounds.

The author's works (Zhizhin, 2019 a, b, c; 2018; 2020 a, b, c, d) prove that various biomolecules in nature have the highest dimension. This article proves that the water molecules themselves, from which biomolecules originated in contact with minerals, have the highest dimension. Thus, the highest dimension of biomolecules is a consequence of the highest dimension of water molecules. Water has the ability to adjust its structure to the structure of substances that are in it or border it. As an illustration of this property, the process of restructuring of water to match the structure of cyanide compounds dissolved in water is considered.

THE STRUCTURE OF WATER

The water molecule has a peculiar shape. Oxygen in the second energy level *p* orbital has one quantum cell with a pair of electrons with opposite spins and two quantum cells with one electron in each of them. These two electrons combine with two electrons of the hydrogen atoms to form a covalent chemical bond. In addition, the oxygen atom has one more pair of electrons at the second energy level in the *s* orbital. The hydrogen atoms, which donated their two electrons to form a bond, remain sufficiently distant from the oxygen atom and have a positive charge +. Unshared electron pairs of the oxygen atom of the outer energy level naturally have a negative charge -. Conventionally, the shape of a water molecule can be represented as shown in Figure 1.

Figure 1. The shape of a water molecule

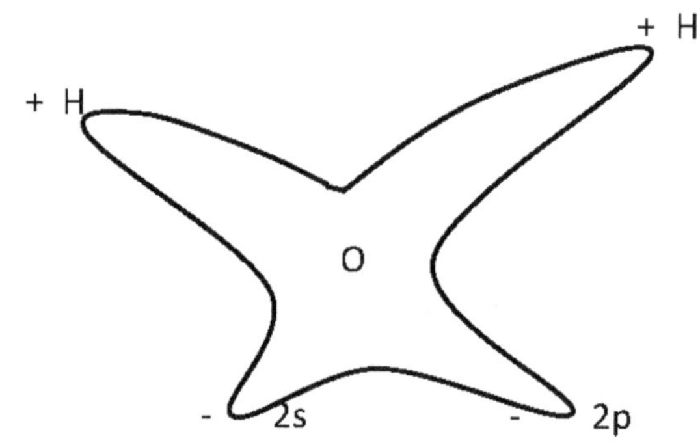

It is obvious that positively charged protons of other water molecules can join the orbitals with negative charges of the second energy level of the oxygen atom. The result is a tetrahedral structure with a center (Figure 2).

The tetrahedron with a center on Figure 2 has 5 vertices $a_0 \div a_4$, 10 edges

$$a_0a_1, a_0a_2, a_0a_3, a_0a_4, a_1a_2, a_3a_1, a_1a_4, a_4a_2, a_2a_3, a_4a_3,$$

10 two - dimensional faces

$$a_3a_4a_1, a_1a_3a_2, a_1a_4a_2, a_2a_3a_4, a_1a_0a_3, a_1a_0a_2, a_0a_4a_1, a_0a_4a_2, a_0a_3a_2, a_0a_3a_4,$$

Figure 2. The tetrahedral structure with a center of molecules water with joined two protons

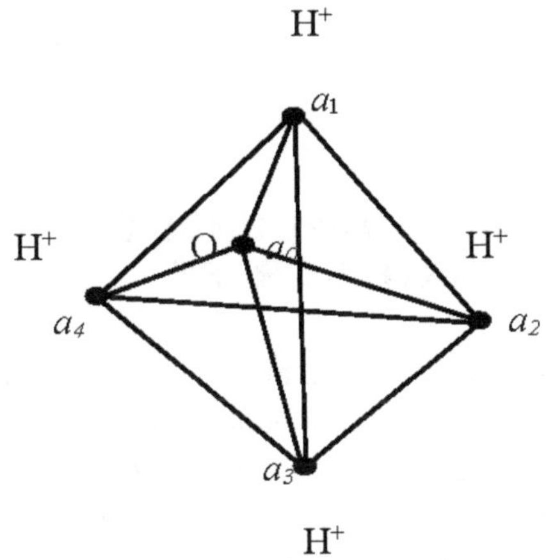

5 three - dimensional faces

$$a_3a_4a_1a_2, a_0a_1a_3a_2, a_1a_4a_2a_0, a_3a_4a_1a_0, a_3a_4a_0a_2.$$

To determine the dimension of a tetrahedron with a center, it is sufficient to use the Euler - Poincaré equation (Poincaré, 1895)

$$\sum_{i=0}^{n-1}(-1)^i f_i(P) = 1 + (-1)^{n-1}. \quad (1)$$

There n is dimension of a polytope P,f_i is the number of elements with dimension i in the polytope P.

In this case $f_0$5, f_1=10, f_2=10, f_3=5.

Substituting the obtained values into equation (1), we obtain, i. e. Euler - Poincaré's equation for a tetrahedron with a center holds for $n = 4$. This proves that a tetrahedron with a center has dimension 4.

The existence of a tetrahedral distribution of hydrogen atoms in the network of hydrogen bonds of water molecules is proved by the methods of computer experiment (Eisenberg, Kautsman, 1975; Lyashchenko, Dunyashev, 2003; Rahman, Stillinger, 1973). However, the possibility of the existence of non - tetrahedral dimers is noted. This is quite natural since hydrogen bonds do not differ in strength and chaotic rearrangements of the network of hydrogen bonds are possible in water. However, the tetrahedral distribution of hydrogen atoms in the structure of water is predominant. It should be remembered that oxygen atoms are in the center of the tetrahedron, and the tetrahedron with the center has dimension 4 (this is not mentioned in all works on the structure of water). Note that this is significant if we remember the statement that life originated in water.

DIMENSION OF BINARY COMPOUNDS OF ELEMENTS OF THE EARTH'S CRUST

Since the origin of life took place in a mound of rocks at the bottom of the ocean, we should consider the chemical compounds that this mound could consist of. A.E. Fersman, in his four-volume geochemistry (Fersman, 1934), investigating the geochemistry of the earth's crust, paid much attention primarily to binary compounds of chemical elements. He drew attention to the fact that many different binary compounds have the same type of structure. Among these structures, he singled out three structures: adamantane, wurtzite and fluorite. The structure of adamantane was already considered in the 10th chapter of this monograph and it was shown that the dimension of the compound of this structure is 9. Therefore, the dimension of nanodiamonds that can be formed in the earth's crust (according to the studies of S.K. Simakov (Simakov,)) is also equal to 9. In addition, as shown in the monograph by G.V. Zhizhin (Zhizhin, 2018), many more binary compounds with the participation of transition and non-transition elements have the adamantane structure. The structure of adamantane with the participation of transition elements has binary compounds Ag_2O, Cu_2O, $CuCl$. The structure of adamantane with the participation of only non-transitional elements has binary compounds Pb_4O_6, AS_4O_6, Sb_4, O_6, P_4O_{10}, P_4O_4, $SiCl_4$, BeF_2, HgS.

Therefore, all listed binary compounds have dimension 9.

The wurtzite structure with the participation of transition elements (Zhizhin, 2018) has compounds ZnO, CdO, ZnS. Compounds have the structure of wurtzite with the participation of only non-transition elements BeO, AlN. The dimension of all these wurtzite and fluorite structures, as shown (Zhizhin, 2018), is 4. Let us determine the dimension of the wurtzite structure using the binary compound as an example ZnS.

In the structure of wurtzite every atom of one component has a tetrahedral environment of the atoms of the other component. This results in the arrangement of tetrahedrons with center so that vertex one tetrahedron is the center of another tetrahedron (Figure 3).

Figure 3. The tetrahedral coordination atoms in wurtzite

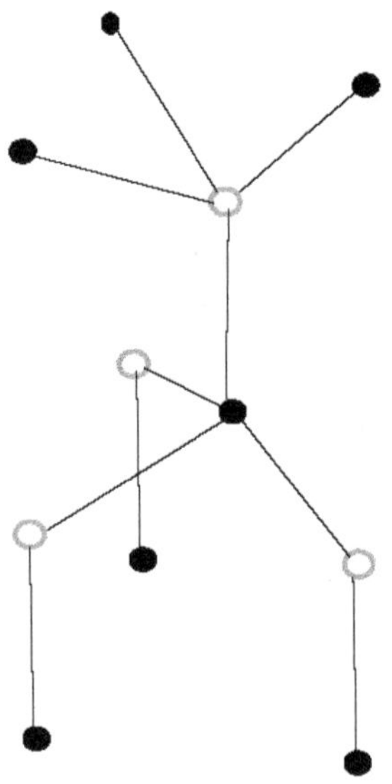

A white circle is atom of one component. A black circle is atom of other component

If to carry construction of atoms in Figure 3 on this principle, we obtain a spatial lattice, the unit cell the lattice is a convex shape it is shown in Figure 4.

This figure is the unit cell structure of the wurtzite. In Figure 4 solid lines represent chemical bonds of the atoms, and the dotted lines are only geometric sense outlining contours of the figure. We define the dimension of this figure by the Euler-Poincare equation (12.1). The number of vertices of this figure is equal to 14, i.e. f_0=14. The number of edges is equal to 29, i.e. f_1=29. The number of two-dimensional faces is the sum of the number of triangles (8) and number of quadrangles (13), i.e. f_2=21. The number of three-dimensional faces is equal to 6. This figures are *abcghkon, gceruo, cefump, cdfulm, bcdklu,* and all shape on Figure 12.4 without inner partitions, i.e. f_3=6. Substituting these values f_i, (i=0,1,2,3) in the Euler-Poincare equation (12.1), we find that it is satisfied for n = 4

$$14 - 29 + 21 - 6 = 0.$$

Thus, the dimension of polytope on Figure 4 is equal to 4, i.e. the unit cell structure of the wurtzite has dimension 4.

On example of compound $MnCl_2$ we look at the structure of fluorite. Isolate magnesium atoms lying at the centers of the cube faces (), and chlorine atoms (), forming a smaller cube inside the bigger cube, which are located at the vertices of magnesium atoms (Figure 5).

Figure 4. The unit cell of the wurtzite

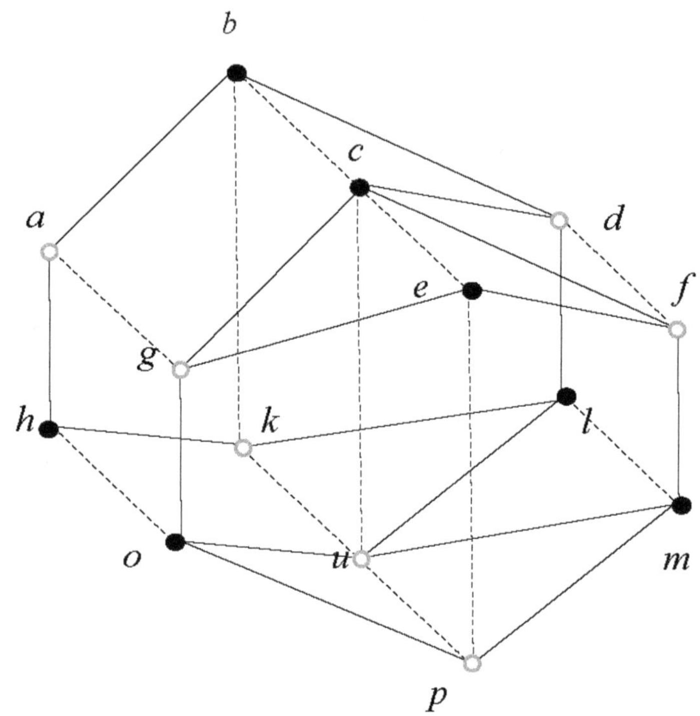

From Figure 5 it follows that the number of vertices is 14, i.e. f_0=14, the number of edges is 36, i.e. f_1=36, the number of flat faces is sum from number of triangles (24) and number of rectangles (6) (smaller cube faces), i.e. f_2=30. The number three - dimension shape to sum up from smaller cube (1), pyramids on its faces (6) and figure (22) without inner parts (1), i. e. f_3=8. Substituting these values f_i, (i=0,1,2,3) in the equation (2), we find that it is satisfied for $n = 4$

$$14 - 36 + 30 - 8 = 0.$$

Thus, the dimension of polytope on Figure 5 is equal to 4, i. e. the unit cell structure of the fluorite has dimension 4.

The given examples show that the dimension of many compounds of the elements of the earth's crust is more than three. The participation of several elements in a compound naturally only increases their dimension (Zhizhin, 2019, 2021).

Figure 5. The unit cell of the fluorite - magnesium atom, - chlorine atom

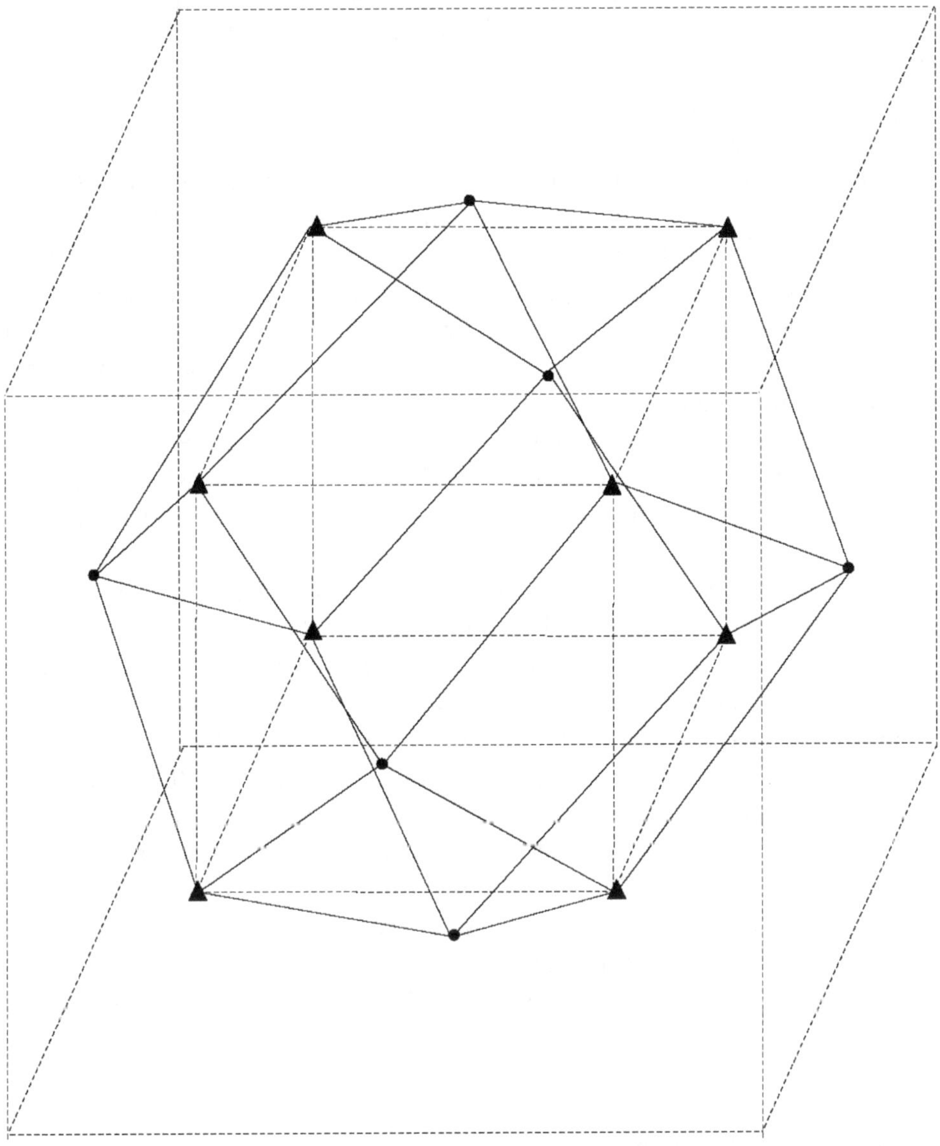

THE DIMENSIONAL OF BIOMOLECULES

In biomolecules, you can find many examples of compounds in the form of a tetrahedron with a center, that is, similar to the structure of a water molecule (for example CH_4, NH_4, PO_4, etc.). All of them have a dimension of 4. The most important representatives of biomolecules are molecules of carbohydrates, proteins, lipids, nucleic acids. When depicting biomolecules, schematic planar images of Fisher and Haworth (Metzler, 1980; Lehninger, 1982; Koolman, Roehm, 2013) are usually used. But they do not reflect the real spatial structure of biomolecules. In 2016, the author proposed spatial images of biomolecules in the form of polytopes (Zhizhin, 2016 a), the dimension of which is determined by the Euler - Poincaré equation (12.1). Since the concept of a functional group is widely used in organic compounds, when determining the structure of biomolecules and their dimensions, not only individual atoms, but also functional groups can be located at the vertices of the corresponding polytope. This is how the concept of the functional dimension of a biomolecule was introduced. As a result, the functional dimensions of many biomolecules were determined. In particular (Zhizhin, 2018, 2019 a, b), the functional dimension of the simplest carbohydrate monosaccharide aldose (with three carbon atoms) turned out to be 5, as well as the dimension of the tartaric acid molecule. The functional dimension of the D - ribose molecule is 12. The functional dimension of the α – D - glucose molecule is 15. The functional dimension of any amino acid is 4, as is the dimension of phosphoric acid. Biomolecules connecting in series with each other form chains: polysaccharides, polypeptides, nucleic acids. Geometrically, they represent chains of polytopes of the highest dimension. Additional bonds between sections of the chain or between different polymer chains with the help of hydrogen bonds or metal ions (iron, calcium, magnesium, zinc) lead to the formation of polytopes of even greater dimension. Such conformations of biomolecules play a very important role in living organisms, organizing the processes necessary for the vital activity of organisms. It has just been established (Zhizhin, 2019 a, b, c; 2020 a, b) that the bond of nucleic acids by means of a hydrogen bond of nitrogenous bases leads to the formation of a polytope of hereditary information of dimension 13, characterized by a powerful flow of information exchange between the elements of this polytope. This makes it possible to explain the existence of the so - called epigenetic process of the transfer of hereditary information, which is not associated with a change in the sequence of genes.

THE MEMORY OF WATER

There is no doubt about the existence of a certain relaxation time required for the water to return to its original state after the cessation of external influences, for example, a magnetic or electric field (Klassen, 1973; Martynova, et. al., 1969; Tatarsky, 1989; Lalalyants, 1990). Let us try to explain the existence of the memory of water about the chemical compounds dissolved in it, which are essential for the life of biological organisms. Let such compounds be cyanide compounds. It is well known that hydrocyanic acid *HCN* and its salts are strong poisons. The toxic effect of these compounds is that the ions CN^- formed in their solution inhibit the reduction of oxygen by the most important respiratory enzyme, cytochrodoxidation, in the tissue cells of living organisms (Lehninger, 1982; Ershov, 1989). Ions CN^- react with the oxidized form of cytochrome oxidase and form a complex compound with a trivalent iron atom of enzyme $[Fe(CN)_6]^{3-}$. As a result of the overlapping of the orbitals of the complexing agent Fe^{3+} by the ligands CN^- and the hybridization of the orbitals, the ion $[Fe(CN)_6]^{3-}$ has octahedral coordination (Figure 6). Geometrically, this ion is an octahedron with a center.

Figure 6. Geometric structure of the ion $[Fe(CN)_6]^{3-}$

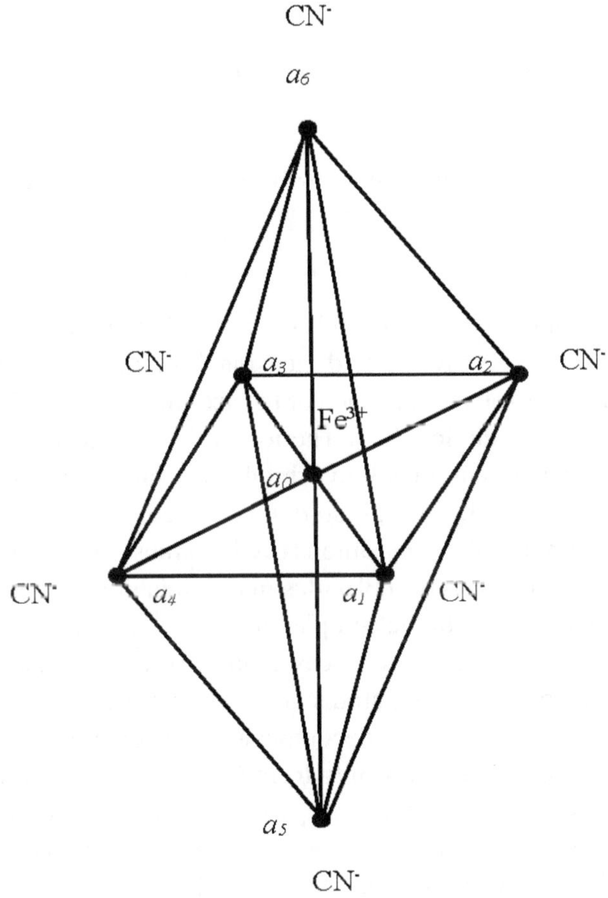

The functional dimension of this ion is 4. The octahedron with a center on Figure 12.6 has 7 vertices $a_0 \div a_6$, (f_0=7); 18 edges

$a_0a_1, a_0a_2, a_0a_3, a_0a_4, a_0a_5, a_0a_6, a_1a_2, a_2a_3, a_3a_4, a_1a_4, a_6a_1, a_6a_2, a_6a_3, a_6a_4, a_1a_5, a_5a_2, a_5a_3, a_4a_5$ (f_1=10);

20 two - dimensional faces

$a_0a_6a_1, a_0a_6a_2, a_0a_3a_6, a_0a_6a_4, a_1a_0a_5, a_5a_0a_2, a_3a_0a_5, a_0a_4a_5, a_1a_0a_2,$
$a_2a_3a_0, a_4a_0a_3, a_1a_0a_4, a_5a_2a_1, a_3a_5a_2, a_5a_3a_4, a_5a_1a_4, a_6a_2a_1, a_6a_3a_2, a_6a_3a_4, a_6a_1a_4,$
(f_2=20);

9 three - dimensional faces

$a_0a_4a_1a_5, a_0a_1a_5a_2, a_3a_5a_2a_0, a_3a_4a_5a_0, a_1a_4a_0a_6, a_0a_1a_2a_6, a_3a_6a_0a_2, a_0a_4a_3a_6, a_1a_4a_2a_5a_3a_6,$
(f_3=9).

Substituting the obtained numbers $f_i(i=0 \div 3)$ into equation (12.1), we obtain,

$7 - 18 + 20 - 9 = 0,$

i.e. the Euler - Poincaré equation holds for $n = 4$. This proves that the octahedron is centered, that is, ion $[Fe(CN)_6]^{3-}$has dimension 4. In this compound, each pair of electrons 2s of the nitrogen atom in each of the six ions CN^- occupies one of the six vacant quantum cells of the ion Fe^{3+}. The ion Fe^{3+} has two vacant quantum cells in orbit 3d, one vacant quantum cell in orbit 4s and three vacant quantum cells in orbit 4p. Hybridization of the orbitals sp^3d^2 takes place. It has been experimentally established (Kostromina, 1990) that anions CN^- bind most strongly with cations Fe^{3+} in comparison with other ions or molecules in the stoichiometric series in terms of the strength of interaction with the complexing agent. The strong bond of ions CN^- with cations Fe^{3+} disrupts the redox processes in cytochrome oxidase, consisting in the transfer of electrons to copper atoms and then to oxygen. As a result, the ions H^+ formed in the body during the decomposition of proteins, carbohydrates and amino acids do not react with oxygen to form water. The concentration of ions H^+ in the body increases sharply, which leads to poisoning of the body. Obviously, in this way, the degree of poisoning will be determined by the concentration of ions CN^- in the solution. They are formed during the dissociation of hydrocyanic acid or its salts. Moreover, hydrocyanic acid dissociates weakly, and the dissociation of salts goes almost to the end. The dissociation of hydrocyanic acid as a result of the formation of ions H^+ slows down the dissociation of water. An increase in the concentration of hydrocyanic acid leads to a decrease in the degree of dissociation of both hydrocyanic acid and water. Indeed, the concentration of ions and molecules in solution is determined by the relations

$$\left[H^+\right]=c_{H_2O}\alpha_{H_2O}+c_{HCN}\alpha_{HCN},\left[OH^-\right]=c_{H_2O}\alpha_{H_2O},$$
$$\left[CN^-\right]=c_{HCN}\alpha_{HCN},\left[H_2O\right]=c_{H_2O}(1-\alpha_{H_2O}),\left[HCN\right]=c_{HCN}(1-\alpha_{HCN}). \tag{2}$$

There are α_{H_2O} – degree of dissociation of water, α_{HCN}-*degree* of dissociation of hydrocyanic acid, c_{H_2O} – concentration of water before dissociation, c_{HCN}-concentration of hydrocyanic acid before dissociation. Substituting relations (12.2) into the equations for the dissociation constants

$$K_{H_2O}=\frac{\left[H^+\right]\left[OH^-\right]}{\left[H_2O\right]},K_{HCN}=\frac{\left[H^+\right]\left[CN^-\right]}{\left[HCN\right]}, \tag{3}$$

taking into account that the degrees of dissociation are small compared to unity, we obtain

$$K_{H_2O}=c_{H_2O}\alpha_{H_2O}^2+c_{HCN}\alpha_{HCN}\alpha_{H_2O},$$
$$K_{HCN}=c_{H_2O}\alpha_{H_2O}\alpha_{HCN}+c_{HCN}\alpha_{HCN}^2. \tag{4}$$

Solving system (12.4) with respect to $\alpha_{HCN,}$ α_{H_2O}, taking into account that the mass fractions of water x_{H_2O} and acid x_{HCN} satisfy the conditions $x_{H_2O}+x_{HCN}=1$, $x_{HCN}<<1$, we obtain

$$\alpha_{H_2O}=\frac{K_{H_2O}}{\sqrt{\rho(K_{H_2O}/\mu_{H_2O}+x_{HCN}K_{HCN}/\mu_{HCN})}},\frac{\alpha_{HCN}}{\alpha_{H_2O}}=\frac{K_{HCN}}{K_{H_2O}}. \tag{5}$$

There are ρ- *s*olution density, μ- *m*olecular weight.

Relations (2), (5) make it possible to calculate the degree of dissociation and ion concentration for a given initial concentration of hydrocyanic acid. In particular, taking the values of the dissociation constants $K_{H_2O}=10^{-14}$ mol/l, K_{HCN}=4.79•10^{-10}

mol/l at very low concentrations of hydrocyanic acid (x_{HCN}→0), we find $\alpha_{H_2O} = 10^{-8}$, α_{HCN}=4.79•10-4. At a lethal concentration of hydrocyanic acid (Ershov, 1989) x_{HCN}=1.4•10^{-6} these values are equal $\alpha_{H_2O} = 10^{-10}$, α_{HCN}=4.6•10^{-4}. Thus, the dissociation of water in solutions of hydrocyanic acid and its salts is insignificant, and the degree of dissociation of hydrocyanic acid weakly depends on its concentration. From (2), (5) it follows that despite a decrease in the degree of dissociation of hydrocyanic acid with an increase in its concentration, the concentration of ions CN^- increases proportionally x_{HCN}.

When any substances are dissolved, the solvent molecules are strongly influenced by the molecules of the dissolved substance. In this case, the solvent molecules are modified. It was shown (Semenchenko, 1932) that the water molecules surrounding the ions of the dissolved substance are under the influence of forces equivalent to a pressure of 50,000 atmospheres. In the process of hydration, molecules of hydrated water (Zundel, 1972) pass into an excited state with vibrations of nuclei relative to newly acquired positions that differ from unexcited positions. The nature of the fluctuations is determined by the nature of the hydrated ion. This is how the nature of the hydrated ion is transferred to the water molecules in its environment. Due to the excited vibrations, water molecules become more deformable and polarizable. It is these properties that put the ion CN^- on the first place in the spectrochemical series of the activities of ions and the forces of their interaction with the complexing agent. The greater the polarizability of the particle, the stronger its bond with the complexing agent. The greater the deformability of the molecule, the easier it is for the molecule to "fit" into the particular relief of the enzyme (in this case, cytochrome oxidase). The network of hydrogen bonds in water has the ability to transfer perturbations external to the network to the entire volume of the solution, while maintaining the stability of the structure of the H – network (Rodnikova, 1993). The phenomenon of negative solvation is associated with this property (Samoilov, 1957), i. e. an increase in the mobility of solvent molecules near ions. All this indicates that the excitation created by ions CN^- with an increase in concentration is transferred to almost all water molecules in solution. After dilution of this solution, the acquired properties of water are retained for some time, which is necessary for the relaxation of the H - bond network. The characteristic time for a change in the properties of water in the absence of external influences on it is, according to experiments (Miller, 1969), several tens of minutes. While the characteristic time for the absorption of ions and water molecules into the tissue cells of living organisms is fractions of a second (Ivanov, 1993). It follows from this that water molecules, having changed under the influence of ions CN^-, even after dilution of water, can retain their activity and, instead of ions CN^-, take their place as ligands around the complexing agent Fe^{3+} in the cytochrome oxidase enzyme, firmly binding with it. In this case, a pair of electrons 2s of the oxygen atoms in the water molecule is transferred to the vacant orbitals 3d, 4s, 4p of the iron atom of the complexing agent. Here, too, an octahedron with a center of dimension 4 is formed around the iron atom, and the violation of redox processes in cytochrome oxidase is preserved.

CHELATED COMPOUNDS

One of the main factors determining the affinity of molecules for metal ions is the chelate effect, i.e. a clearly expressed ability of molecules to bind metal ions in the presence of groups capable of complex formation in the molecules. Nature has successfully used the chelate effect to create such important metal-containing molecules as porphins, chlorophyll, calcium-binding proteins, ribosomes, and amino acids (Metzler, 1980). During the formation of life on planet Earth, metal ions were formed during the interaction of metals in the earth's crust with water, leading to their spread over the surface of the planet (Fersman, 1923; Vernadsky, 1925). Most metal ions form a hydrate shell. These include not only iron ions, but also manganese, cobalt, nickel, zinc, and magnesium. These are all elements that play an important role in living organisms. As a rule, there are six water molecules in the hydration shell of metal ions and an octahedral structure with a center is formed. The dimension of such a structure, as has just been proved, is equal to 4. However, the hydrate shell may not always be correct. In these cases, the dimension of the hydration shell, together with the metal ion, can be greater than four. During the synthesis of organic compounds in nature, instead of water molecules, organic molecules can be located in the metal hydrate shell. An important role in nature is played by complex compounds around iron ions (the previously discussed Fe - porphyrins), which have a higher dimension. The chlorophyll molecule has a structure similar Fe - porphyrin. The only metal ion in chlorophyll is the magnesium ion. This structure also has a higher dimension.

DIMENSION AND GEN

We simply understand that a gene is a certain segment of DNA responsible for the inheritance of a certain trait. However, this is not quite true. We know that DNA is located in chromosomes, but in addition to DNA, there are also proteins in chromosomes. This is not an inert part of the chromosomes since proteins with metal ions form enzymes. It is enzymes that organize all chemical processes in the body. In particular, they can cut DNA, highlighting certain sections from them. In addition, there are areas in DNA that are not responsible for the transmission of hereditary information – introns (Guttman, et al., 2002).

It has recently been experimentally established that protein histone modifications affect the active inheritance of traits, and this effect is not associated with the sequence of nitrogenous bases in nucleic acids (epigenetics) (Lindquist, et al., 2016; Sanbonmatsu, et al.,2016; Mancuso, et. al., 2014; Mancuso, et. al., 2017). It follows from this that the genes are more complex formations than just a segment of the sequence of nucleotides in DNA. Timofeeff - Ressovsky wrote about this in his theory of the target (Timofeeff, Zimmer, 1947). In particular Timofeeff - Ressovsky argued that the gene is a physics - chemical unit, i. e. a large molecule, micelle, or, in the broadest sense, a crystalline structure. Obviously, the dimension of such a unit will be much more than three, since even its constituent parts - nucleotides and proteins have a higher dimension. Timofeeff - Ressovsky studied the effect of irradiation on the processes of heredity. It should be noted here that these processes require additional analysis. The fact is that since a gene has a higher dimension, gene mutations under the influence of irradiation are the result of the movement of irradiating particles in a space of higher dimension. But in chapter 3 of this work, it is shown that even the inertial motions of particles in the space of higher dimension do not obey the laws of classical mechanics due to the non - Euclidean nature of the space of higher dimension (Zhizhin, 2021 h).

THE "PHENOMEN OF LIFE" IN THE BEHAVIOR OF THE MOLECULES OF LIVING MATTER

At the end of the 20th century, under the influence mainly of the works of L. Pasteur (Valeri-Rado, 1950; Pasteur, 1960), an idea was formed about the special properties of living matter. For example, in the work of N.F. Reimers (1994) we read: "L. Pasteur's law of chiral purity: living matter consists of chiral pure structures (chiral purity is the presence of only objects that are incompatible with their mirror image). Living proteins are built only from "left" amino acids (polarizing light to the left), nucleic acids are composed exclusively of "right" sugars, etc. Substances of non-biogenic origin are chiral symmetrical - "left" and "right" molecules in them are equal in number. Chiral purity determines the specificity of the living, its irreducibility to the inanimate."

Similar views are repeated in the works of other authors (for example, Voitkevich, 1996). These ideas are based on a deep (unconscious) belief that the dimension of molecules cannot be more than three. However, in numerous works of the author, it has recently been proven that the dimension of biomolecules is more than three (Zhizhin, 2016, 2018, 2019 b, 2020 d, 2022). In this regard, it is necessary to clarify what the phrase about the polarization of light to the right or to the left means, and how this is related to the structure of the corresponding molecules. To do this, one should refer to the initial experiments of L. Pasteur, taking into account the highest dimensionality of molecules. In addition, since the movement along the chain of biomolecules is usually accompanied by a rotation of the molecules at a certain angle, the possibility of rotation of the molecules in these movements should be considered, choosing to study the chain of amino acid molecules, the chain of sugar molecules and the chain of nucleic acid molecules, taking with account they dimension. This study is the subject of this work.

Spatial images of the molecules wine acid

The molecules of the wine acid are in two forms (*L* and *D*). Spatial images of these forms are presented in Figure 7 and Figure 8. (Zhizhin, 2018).

Figure 7. Spatial structure of the D – wine acid

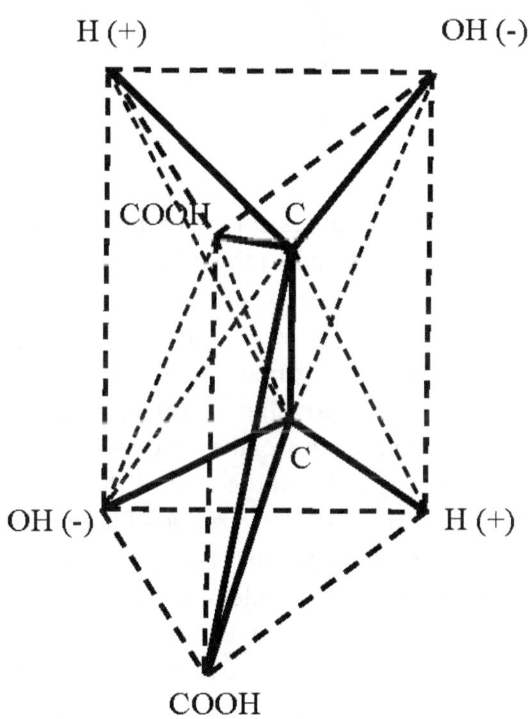

Figure 8. Spatial structure of the L – wine acid

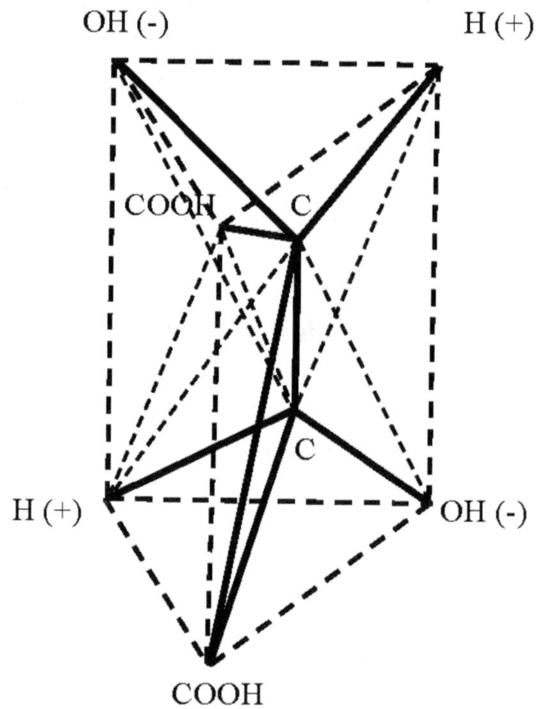

The solid black lines in these figures correspond to the valence bonds in these molecules, the dotted lines correspond to the edges of the polytope, outlining its shape. Figures 1 and 2 differ from each other only by the opposite arrangement of ions *H* (+) and *OH* (-) (as well as in the flat Fisher projections). Therefore, they are incompatible with each other. The dimension of these polytopes is 5 (Zhizhin, 2018). The images obtained make it possible to explain the main property of tartaric acid - rotation of the plane of polarization of the incident light in different directions: in the case of the *D* form to the right, in the case of the *L* form to the left. It is a known device for rotating the plane of polarization of light, having the appearance of two folded triangular prisms, the boundary between which serves to reflect light (Wood, 1936). Can be said that the molecule of tartaric acid is a natural device for rotating the plane of light polarization. Two carbon atoms into structure molecules play the role of the reflecting partition in the molecule. The rotation occurs in the forms *D* and *L* in different directions because of the opposite arrangement of the charges of the hydrogen ions *H* (+) and the hydroxyl group *OH* (-) in these forms. Thus, the reason for the different rotation of the plane of polarization of light lies not in the different forms of the crystals of *D* - tartaric acid and *L*-tartaric acid, as

Pasteur suggested, but in different forms of molecules, clearly visible in the image in the space $5D$.

It should be noted that a characteristic feature of tartaric acid molecules, which leads to rotation of the light polarization plane, is characteristic of organic compounds. Of course, organic compounds are present in living matter, but organic compounds themselves are not living matter. Therefore, it is incorrect to say that the rotation of the plane of polarization is an innate property of living matter.

States of Amino Acid Molecules in Protein Structures

The simplest structure of a protein is a chain of amino acid molecules. Amino acid molecules exist in two enantiomorphic species L and D (Figure 2 a, b).

The dimension of these forms is 4 (Zhizhin, 2016). The works of L. Pauling et al. (Pouling, Corey, Branson, 1951; Pouling, Corey, 1950, 1953; Corey, Pouling, 1953) were widely used to describe the chains of amino acid molecules. However, these works do not take into account the higher dimension of amino acid molecules, which leads to contradictions, and the methods developed in these works cannot be taken into account (Zhizhin, 2016, 2020 d, 2022, 2023 a, b, c). Then the polypeptide chain of two molecules of L - amino acids, taking into account the four - dimensionality of amino acid molecules, has the form shown in Figure 4.

In the case of a D - amino acid, such a polypeptide chain is shown in Figure 5.

From Figure 4 and Figure 5 it can be seen that in order for the corresponding vertices of tetrahedrons with the center (i.e., specific atoms or functional groups) could lie on a system of parallel lines (i.e., form a linear polypeptide chain), it is necessary that the amino acid molecules in the chain alternate with their mirror image relative to the edge in the tetrahedron connecting the bond centers of the polypeptide chain (i.e., relative to the segment $CO - NH$). The chains in Figure 4 and Figure 5 differ by the mirror reflection of the molecules relative to the perpendicular to the segment $CO - NH$ (L - and D - molecules). In Figures 9.4 and 9.5, you can see the translational symmetry of the circuits. Moreover, the elementary translational element in this case is a group of two linked amino acid molecules, one of which is a mirror image of another amino acid molecule relative to the segment. From these two linked amino acid molecules, you can create a convex polytope by connecting the edges of each vertex of any of the molecules with all other vertices in the group. This is how a simplex polytope of dimension 9 is formed, since the dimension of a simplex polytope is one less than the number of vertices in a simplex (Zhizhin, 2019), and the number of vertices in two tetrahedra with a center is 10. Therefore, a polypeptide linear amino acid chain has translation symmetry, in which an elementary translation element is a simplex polytope of dimension 9.

The polypeptide chain can change its direction (Spirin, 2019). This, in particular, is required in the formation of globular proteins. In this case, loops of the polypeptide chain are formed. Since the polypeptide chain, as we have seen, is a chain of polytopes of higher dimension, it is necessary to verify the possibility of such turns in the chain of polytopes of higher dimension, which are a chain of amino acid molecules. Figure 7 shows a possible implementation of such a rotation. It is interesting to note that when the polypeptide chain is rotated, the enantiomorphic shape of the amino acid molecules changes.

This follows from the geometric analysis of the rotation presented in Figure 7. If the upper part of the polypeptide chain in Figure 9.7 is considered as the initial one, then it is a sequence of D -molecules of the amino acid from left to right. When the chain is rotated, a polypeptide chain is formed, going from right to left (right turn), while the amino acid molecules look like L -molecules. If the lower part of the polypeptide chain in Figure 7 is considered the initial one, then the chain is a sequence of L - molecules going from left to right. When turning in this case (left turn), the molecules take the form of D - molecules going in sequence from right to left. The change in the shape of amino acid molecules during the rotation of the polypeptide chain can be considered, in a sense, the justification for the existence of enantiomorphic forms of amino acid molecules.

Changes in the enantiomorphic forms of amino acid molecules during chain rotations have not previously been paid attention to. Thus, statements about the chiral purity of proteins in living organisms, which are characterized by the formation of globules with a turn of amino acid chains by 180 degrees, are without foundation.

States of the Saccharide Molecules in the Monosaccharide Chains

As shown in the works (Zhizhin, 2019 b, 2020 c, e, 2021 a), all monosaccharides with a closed cycle have the highest dimensions (monosaccharides furanoses have a dimension of 12, monosaccharides pyranoses have a dimension of 15). For example, a molecule of α – D - glucose monosaccharide has the form of a polytope in Figure 4, chapter 3.

Based on this image, you can get a simplified three - dimensional image, taking into account all the features of the image of dimension 15 (Zhizhin, 2022). So, two images of the α – D - glucose β - anomer and α - anomer (Figure 7 and Figure 8, chapter 3) were obtained.

Monosaccharide molecules combine with each other thanks to the combination of two hydroxyl groups with the release of a water molecule. The remaining oxygen atom connects the remaining part of the monosaccharides. The most common chains of α -D - glucose monosaccharide residues. Here, residual α - D - glucose molecules can be joined through an oxygen atom, connecting carbon atoms 1 - 1, 1 - 4, 4 - 4, 4 - 1. In some cases, the carbon atom $C_{(2)}$ in the cycle will be involved in the compound. Branching chains of glucose molecules occurs at the sixth carbon atom in the functional group CH_2OH. Monosaccharide chains have different forms depending on the type of these molecules, i. e. depending on whether these molecules are *β*- anomers or α - anomers. Using the obtained simplified three - dimensional models of α - *D* - glucose molecules, we consider in more detail these compounds in a metric image, taking into account the angles between the valence bonds. Let two *β*- anomers of the α - *D* - glucose molecule join together, linking the carbon atom $C_{(1)}$ of the first α - *D* - molecule to the carbon atom $C_{(4)}$ of the second α - *D* - molecule (lactose). Based on the geometric image in Figure 3.11, you can imagine a top view of this connection.

It is easy to see that the same attachment of the third α - *D* - glucose molecule leads to a linear arrangement of molecules when viewed from above on this compound. Linearity will not change with the following similar connections. However, given the three - dimensionality of the image, you should look at this connection from a different view. Figure 11 a) shows an image of the junction of two α - *D* - glucose molecules when viewed from the front. The black segments in Figure 11 a) are the traces of the intersection of the upper bases of the prism with a plane passing through the valence bonds of the oxygen atom and carbon atoms. According to the obtained solution, the angle between the base of the prism and the valence bond is 36 degrees. both the first α -*D* - glucose molecule and the second α -*D* - glucose molecule. The valence angle at the oxygen atom in compounds with two carbon atoms in the chain of α -*D* - glucose molecules, as you know, is slightly larger than the normal valence angle and is about 112 degrees. Thus, the angle 1O2 in Figure 11 a) is 176 degrees. This means that the base of the second α -*D* - glucose molecule has a slope to the base of the first α - *D* - glucose molecule other than zero. More precisely, this slope is 4 degrees, therefore the second α - *D* - glucose molecule, when viewed in full view, turns to the left relative to the first α - *D* - glucose molecule. It is clear that such an addition of a third α - *D* - glucose molecule will increase the angle of inclination of this molecule relative to the first molecule. Thus, the statement about the linearity of the chain of α - *D* - glucose molecules when connecting them in the case of anomers is not entirely accurate. One can say that this statement is somewhat one - sided.

Let two *α* - anomers of the *α* - *D* - glucose molecule join together, linking the carbon atom $C_{(1)}$ of the first *α* -*D* - molecule to the carbon atom $C_{(4)}$ of the second *α* - *D* - molecule (maltose). Based on the geometric image in Figure 11, you can imagine a top view of this connection (Figure 3.12).

It is easy to see that the same attachment of the third *α* - *D* - glucose molecule leads to a nonlinear arrangement of molecules when viewed from above on this compound. Nonlinearity will save with the following similar connections. Thus, the chain of molecules turns to the right. However, given the three - dimensionality of the image, you should look at this connection from a different view. Figure 3.12 a) shows an image of the junction of two α - D - glucose molecules when viewed from the front. The black segments in Figure 12 a) are the traces of the intersection of the upper bases of the prism with a plane passing through the valence bonds of the oxygen atom and carbon atoms. According to the obtained solution, the angle between the base of the prism and the valence bond is 36 degrees both the first α - D - glucose molecule and the second α - D - glucose molecule. Thus, In Figure.12 a) the angle 1H $C_{(1)}$ is 36 degrees and the angle $C_{(1)}OC_{(4)}$ is 112 degrees. Since the angle $C_{(4)}O2$ is 36 degrees too so the angle between bases 1H and O2 is 14 degrees. This means that the base of the second α - D - glucose molecule has a consequent slope to the base of the first α - D - glucose molecule to the left. It is clear that such an addition of a third α - D - glucose molecule will increase the angle of inclination of this molecule relative to the first molecule. In the chains of β- anomers, one of the projections is linearly transmitted along the chain, and the projection perpendicular to it along the chain along a curved line. In chains of α - anomers (spirals), it is observed that when moving along a chain of projections in mutually perpendicular planes, the pyranose molecules rotate in opposite directions (Figure 13). Thus, chains of α - anomers of α - D - glucose molecules rotate simultaneously in opposite directions in perpendicular planes.

Formation of "Right" and "Left" Nucleic Acids

Nucleic acid is a chain of D – ribose (α - D - ribose or 2 – deoxy – D – ribose. Further analysis refers to the case α - D – ribose. In the case 2 – deoxy – D – ribose of an analysis like to, it is given in (Zhizhin, 2023 a) molecules between which are the phosphoric acid residues that connect these molecules. Let us construct this circuit, taking into account the concept of repulsion of electron pairs (Gillespie, 1972). According to this concept, tetrahedral coordination of atoms and functional groups in these compounds is carried out around the phosphorus atom in the phosphoric acid residue, and in the carbon atoms in the D - ribose molecule also. It is this concept that leads to the higher dimensionality of the remainder of phosphoric acid and the molecule of D - ribose. When a phosphoric acid residue is added to the D - ribose molecule from the hydroxyl group of the phosphoric acid residue and the hydroxyl group of the D - ribose molecule, a water molecule that leaves the compounds forms and an oxygen atom that binds the phosphoric acid residue and the D - ribose molecule. The concept of repulsion of electron pairs also operates here, according to which two valence electrons of the oxygen atom maximally repel each other, forming electron pairs with electrons of the phosphorus atom and the carbon atom.

Therefore, taking into account that tetrahedral coordination is carried out around the atoms of phosphorus and carbon already, it can assume that the valence bonds of the oxygen atom are located linearly with respect to the oxygen atom. When the second *D* - ribose molecule is added to the phosphoric acid residue, the second hydroxyl group of the phosphoric acid residue with the compound CH_2OH of this *D* - ribose molecule also forms molecules of water that leaves the compound and an oxygen atom connecting the phosphorus atom and the functional group CH_2 of the *D* - ribose molecule. By virtue of the concept of repulsion of electron pairs, the linear arrangement of the valence bonds of the oxygen atom can be considered here also. In addition, as the first step within the framework of the notions of the functional dimension of molecules (Zhizhin, 2016), can shall assume the linear arrangement of the valence bonds also in the neighborhood of the functional group CH_2 of the *D* - ribose molecule. One will take into account this when constructing a chain of *D* - ribose molecules and phosphoric acid residues. To construct the sequence of *D* - ribose molecules and the phosphoric acid residue in the chain can will use a simplified image of the *D* - ribose molecule of dimension 12. Let us leave only the outer contour and valence bonds necessary for interaction with the environment from the *D* - ribose molecule. At the same time, one will not depict the hydrogen atoms inside the contour of carbon atoms. We approximately to consider that the lengths of chemical bonds carbon - carbon, carbon - oxygen close to each other. The simplified images (Zhizhin, 2021 b) of the *D* – ribose molecules of the enantiomer form *A* and of the enantiomer form *B* under these conditions presented in Figure 11 and Figure 13.

The absence of edges and atoms in Figures 14, 15 does not mean that they really are not. This is only the conventionality of the image, the full image these molecules of dimension 12 are in (Zhizhin, 2021 b).

It should be noted that under the assumed projection conditions, the *D* - ribose of molecules on a two - dimensional plane in which a closed carbon ring is located has a number of properties that are convenient for further constructions. The enclosed carbon ring is a regular pentagon with an internal angle at the vertices $\pi - \dfrac{2\pi}{5} = 108^o$. The angles in the three trapezoids formed by the valence bonds and the outer contour, because of the symmetry and parallelism of the bases, are 54 degrees and 126 degrees. In two obtuse - angled isosceles triangles formed by valence bonds and the outer contour, the angles are 126 degrees and 27 degrees.

Now, in a projection onto a two - dimensional plane, one shall depict a sequence of the simplified images of *D* - ribose molecules and phosphoric acid residues, taking into account the separation of water molecules at the junction of phosphoric acid residues and *D* - ribose molecules and the linear arrangement of the valence bonds in vicinity of these compounds. In Figure 15 is presented the sequence for *D* – ribose molecules in form *A*.

In Figure 16 is presented the sequence structure of the nucleic acid molecule with for *D* – ribose molecules in form *B*.

It follows from Figures 16, Figure 15 that the *D* - ribose molecules in both chains are not linearly distributed.

The projection of the nucleic acid molecules depicted in Figures 16 and 17 can be described mathematically. Let us turn to a detailed picture of the connection of two *D* – ribose molecules of the form *B* in Figure 16 (Figure 19). In Chapter 4 of this monograph, a system of equations describing the geometry of Figure 19 is given, and the corresponding analytical solution is obtained. Thus, the representation of phosphoric acid residues and D – ribose molecules composing the chains of nucleic acids, it is used in the form of polytopes of higher dimension. The geometrical cause of formation of different form by molecules nucleic acids (right and left spirals with different number of D – ribose molecules in period, including closed chains) has been determined. It is connected with the mutual arrangement of polytopes of polytopes of phosphoric acid residues and D – ribose molecules. Substituting the known effective values of the lengths of chemical bonds (carbon – carbon, oxygen – oxygen, phosphorus – oxygen) into the structure of polytopes, the values of the characteristic geometric parameters of nucleic acid were calculated: their effective diameter and period. It turned out that the calculated values of these parameters are in good agreement with their values, determined earlier experimentally (Watson, Crick, 1953 a, b; Frank – Kamenetskiy, 1988, 2004, 2010; Ha, et al., 2005).

CONCLUSION

A variety of theoretical ideas about the possible mechanisms of the origin of life on planet Earth, put forward at different times, starting with the so-called "Darwin's ponds", are critically examined. Taking into account modern views on the existence of the "RNA World" on Earth in antiquity, as well as recently discovered gas emissions from the surface of the earth's crust at the bottom of the oceans (Russell et al.,1989, 1999, 2005, 2007), it was concluded that mounds above the vent of hydrothermal emissions, consisting of rocks of the earth's crust, are the most possible site for the origin of life on Earth. Firstly, this mound contains all the metal ions necessary for the formation of RNA. Secondly, the mound has cracks, which ensure the existence in them of RNA colonies of different lengths, connected by a single space, i.e. the kurgan provides the necessary compartmentalization. Thirdly, the mound above the hydrothermal vent is flowing, i.e. it ensures the exchange of the RNA colony with the environment, i.e. ensures the evolution of RNA. Fourthly, all the necessary chemical reactions can and do take place in the mound to create RNA molecules and, subsequently, DNA molecules. Consequently, the difficulties in all previous theories of the origin of life on Earth are removed. There is no need to make all sorts of often conflicting assumptions.

Data are given on the dimensionality of various binary compounds that are widely distributed in the materials of the earth's crust (Fersmann, 1934), studied by the author over a number of years, indicating a higher dimensionality of these compounds. The data on the highest dimension of biomolecules (of various carbohydrates, proteins, nucleic acids and other biomolecules) are also given, which were determined in particular in this monograph, as well as in previous works of the author. The properties of water are noted to rebuild its structure, taking into account the complementary interaction of water with solid surfaces that are in contact with it, as well as the property of the memory of water about the chemical compounds in it. It is concluded that the higher dimension of biomolecules is a consequence of the conditions for the origin of life in the mound above the source of emissions through the earth's crust at the bottom of the oceans, where water was in contact with ore bodies and acquired their properties, i.e. higher dimension, and as a result, the biomolecules in the water also acquired a higher dimension.

A geometric analysis of the behavior of molecules of living matter (amino acids, sugars, nucleic acids) in a higher-dimensional space has shown that their movements and shapes of molecules obey only the conditions of their possible existence, and do not obey the imperatives set in advance by someone. Thus, amino acids, contrary to the principles of chiral purity, can simultaneously coexist with both dextrorotatory (polarized light) and levorotatory molecules, forming, for example, the most important globular proteins for life. Sugar molecules can form spirals simultaneously rotating both to the right and to the left in different planes of multidimensional space. Nucleic acids form spirals rotating both to the right and to the left. The statements of some authors about the existence of only one of the types of biomolecules are erroneous, based on the observed property of tartaric acid to rotate polarized light in one direction. This property is associated not with some obscure ideas about living things, but with the geometric characteristics of the tartaric acid molecule. All properties of the molecules of living matter obey only the previously defined general principle of the phenomenon of life (Korogodin, 2010-2012).

REFERENCES

Ahliquist, P. (2002). RNA -dependent RNA polymerases viruses and RNA silencing. *Science*, *296*(5571), 1270–1273. doi:10.1126/science.1069132 PMID:12016304

Akabori, Sh. (1957). *On the origin of preproteins. In book. Origin of life on Earth.* Science.

Baltimore, D., Girard, M., & Darnell, J. E. (1966). Aspects of the synthesis of poliovirus RNA and the formation of virus particles. *Virology, 29*(2), 179–189. doi:10.1016/0042-6822(66)90024-9 PMID:4287327

Belozersky, A. N. (1957). On the species specificity of nucleic acids. In the book. In A. I. Oparin (Ed.), *The Emergence of Life on Earth* (pp. 198–205). Publishing House of the Academy of Sciences of the USSR.

Bernal, J. D. (1967). *Origin of Life*. World Publishing Co.

Cech, T. R., Gesteland, R. F., & Atkins, J. F. (Eds.). (1993, 1999, 2006). RNA World. Cold Spring Harbor Press. New York.

Cech, T. R., & Golden, B. L. (1999). Building a catalytic active site using only RNA. In R. F. Gesteland, T. R. Ceck, & J. F. Atkins (Eds.), *The RNA World* (2nd ed., pp. 321–347). Cold Spring Harbor Laboratory Press.

Chetverina, H. V., & Chetverin, A. B. (1993). Cloning of RNA molecules in vitro. *Nucleic Acids Research, 21*(10), 2349–2353. doi:10.1093/nar/21.10.2349 PMID:7685078

Chetverina, H. V., Demidenko, A. A., Ugarov, V. I., & Chetverin, A. B. (1999). Spontaneas rearrangements in RNA sequence. *FEBS Letters, 450*(1-2), 89–94. doi:10.1016/S0014-5793(99)00469-X PMID:10350063

Chetvernin, A. B., Chetverina, H. V., & Munishin, A. V. (1991). On the nature of Spontaneous RNA sythesis by Qβ replcase. *Journal of Molecular Biology, 222*(1), 3–9. doi:10.1016/0022-2836(91)90729-P PMID:1719219

Corey, R. B., & Pauling, L. (1953). Fundamental dimensions of polypeptid chains. *Proc R Soc Lord B Biol Sci,* 141(902), 10-20.

Dixon, M., & Webb, E. C. (1979). *Ensymes*. Longman Group Ltd.

Eisenberg, D., & Kautsman, V. (1975). *Structure and properties of water*. Gidrometyoizdat.

Ellington, A., & Szostak, J. (1990). In vitro selection of RNA molecules that bind specific ligands. *Nature, 346*(6287), 818–822. doi:10.1038/346818a0 PMID:1697402

Ershov, Y. A. (1989). *Mechanisms of toxic action of inorganic compounds*. Medicine.

Euclid. (2012). *Beginnings*. URSS.

Fersmann, A. E. (1923). *Chemical elements of the Earth and Space.*

Fersmann, A. E. (1937). Geochemistry. V. 1 - 4. Leningrad: ONTI.

Frank - Kamenetskiy, M. (1988). H – form DNA and the hairpin – triplex model. *Nature, 333*, 214.

Frank – Kamenetskiy, M.D. (2004). *The DNA Age.* Moscow, Publishing house KDU.

Frank – Kamenetskiy. M.D. (2010). Queen of a living cell. From the structure of DNA to the biotechnological revolution. Moscow: Act – Press.

Galimov, E. M. (2004). Phenomena of life: Equilibrium and non- linearity. Origins of life and evolution of the Biosphere. *Origin Life. Biosphere.*, *6*, 599–613.

Galimov, E. M. (2009). Conditions for the Emergence of Life on Earth. *Biosphere*, *1*(1), 39–47.

Gilbert, W. (1986). Origin of life: The RNA world. *Nature*, *319*(6055), 618. doi:10.1038/319618a0

Gillespie, R. J. (1972). *Molecular Geometry.* Van Nostrand Reinold Company.

Gold, L., Polisky, B., Uhlenbeck, O., & Yaris, M. (1995). Diversity of oligonucleotide functions. *Annual Review of Biochemistry*, *64*(1), 763–797. doi:10.1146/annurev.bi.64.070195.003555 PMID:7574500

Ha, S. C., Lowenhaupt, R., Rich, Y.-G., Kim, K. K., & Kim, K. K. (2005). Cristal structure of a junction between B – DNA and Z – DNA revels two extruded bases. *Nature*, *437*(7062), 1183–1186. doi:10.1038/nature04088 PMID:16237447

Haldane, J. B. S. (2015). *Selected Genetic Papers. (Rout liege Revivals)* (K. R. Dronamraju, Ed.).

Ivanov, K. P. (1993). *Fundamentals of the body's energy. Biological oxidation and oxygen supply.* St. Petersburg. *Science.*

Kisselev, N. A., Gavrilova, L. P., & Spirin, A. S. (1961). Configurations of high-polymer ribonucleic acid macromolecules as revealed by electron microscopy. *Journal of Molecular Biology*, *3*(6), 778–783. doi:10.1016/S0022-2836(61)80083-1 PMID:14456472

Klassen, V. I. (1973). *Water and magnet.* Science.

Koolman, J., & Roehm, K.-H. (2013). *Color Atlas of Biochemistry.* Georg Thieme Verlag KG.

Korogodin, V.I. (2012). *Phenomenon of life. Selected works* (in 2 volumes). Managing editors S.G. Inge-Vechtomov, A.F. Tsybin. Moscow: Nauka.

Kostromina, N. A. (1990). *Chemistry of coordination compounds*. Chemistry.

Lalalyants, I. (1990). "Closure" in Paris. *Science and Life, 1*, 86–89.

Lehninger, A. L. (1982). *Principles of Biochemistry* (Vol. 1-3). Worth Publisher, Inc.

Lindquist, S. (2016). Luminidependens (LD) is an arabidopsis protein with prion behavior. *Proceedings of the National Academy of Sciences of the United States of America, 113*(21), 6065–6070. doi:10.1073/pnas.1604478113 PMID:27114519

Lyashchenko, A. K., & Dunyashev, V. S. (2003). Complementary organization of the water structure. *Journal of Structural Chemistry, 44*(5), 909–915. doi:10.1023/B:JORY.0000029822.94836.67

Mancuso, S. (2017). *PLANT REVOLUTION. Le piante hanno già inventato il nostro futuro*. GUIUTI EDITORE.

Mancuso, S. (2014). Experience teaches plants to learn faster and forget slower in environments where it matters. *Oecologia, 175*(1), 63–72. doi:10.1007/s00442-013-2873-7 PMID:24390479

Martynova, O. I., Gusev, B. T., & Leont'ev, E. A. (1969). To the question of the mechanism of the influence of the magnetic field on aqueous solutions of salts. *Успехи физических наук, 98*(1), 195–199. doi:10.3367/UFNr.0098.196905n.0195

Martynshev, L. (2006). Maximum entropy production principle in physics, chemistry, and biology. *Physics Reports, 426*(1), 1–45. doi:10.1016/j.physrep.2005.12.001

Metzler, D. E. (1980). *Biochemistry. The Chemical Reactions of Living Cells* (Vol. 1-3). Academic Press.

Miller, E. V. (1969). On the spontaneous decrease in the density of water condensed from steam. *Reports of the USSR Academy of Sciences, 184*(1), 136–141.

Miller, S. L. (1953). The production of amine acids under possible primitive earth conditions. *Science, 117*(3046), 528–529. doi:10.1126/science.117.3046.528 PMID:13056598

Miller, S. L., & Urey, H. C. (1959). Organic compound synthesis on the primitive Earth. *Science, 130*(3370), 245–259. doi:10.1126/science.130.3370.245 PMID:13668555

Nelson, D. L., & Cox, M. M. (2022). Lehninger Principles of Biochemistry. (Vol. 1 – 3). W.H. Freeman. Macmillan Learning, New York.

Oro, J. (1961). Mechanism of synthesis of adenine from hydrogen cyanide under plausible primitive Earth conditions. *Nature*, *191*(4794), 1193–1194. doi:10.1038/1911193a0 PMID:13731264

Paster, L. (1960). *Selected words*. Academy of Sciences of the USSR.

Pauling, L., & Corey, R. B. (1950). Two Hydrogen-bonded Spiral Configurations of the Polypeptide Chain. *Journal of the American Chemical Society*, *72*(11), 5349. doi:10.1021/ja01167a545

Pauling, L., & Corey, R. B. (1953). Compound Helical Configurations of the Polypeptide Chain:Structure of Proteins of the α - keratin Type. *Nature*, *171*(4341), 59–61. doi:10.1038/171059a0 PMID:13025480

Pauling, L., Corey, R. B., & Branson, H. R. (1951). The Structure of Proteins: Two Hydrogen-bonded Helical Configurations of the Polypeptide Chain. *Proceedings of the National Academy of Sciences of the United States of America*, *37*(4), 205–211. doi:10.1073/pnas.37.4.205 PMID:14816373

Poincaré, A. (1895). Analysis situs. *J. de é Ecole Polytechnique, 1*, 1-121.

Puglisi, J., & Williamson, J. R. (1999). RNA interaction with small ligands and peptides. In R. F. Gesteland, T. R. Cech, & J. F. Atkins (Eds.), *The RNA World* (2nd ed., pp. 403–425). Cold Spring Harbor Laboratory Press.

Qian, G., Brugger, J., Chen, G., & Skinner, W. M. (2010). An experimental study of the mechanism of the replacement of magnetite by pyrite up to 300 ^{0} C. *Geochimica et Cosmochimica Acta*, *74*(19), 5610–5630. doi:10.1016/j.gca.2010.06.035

Rahman, A., & Stillinger, F. H. (1973). Hydrogen-Boden patterns in liquid water. *Journal of the American Chemical Society*, *95*(24), 7943–7948. doi:10.1021/ja00805a003

Rodnikova, M. N. (1993). Features of solvents with a spatial network of H-bonds. *Journal of Physical Chemistry*, *67*(2), 275–280.

Russel, M. G., Hall, A. J., Boyce, A. J., & Fallick, A. E. (2005). On hydrothermal convection systems and the emergence of life. *Economic Geology and the Bulletin of the Society of Economic Geologists*, *100*, 419–438.

Russell, M. J. (2007). The alkaline solution to the emergence of life: Energy, entropy and early evolution. *Acta Biotheoretica*, *55*(2), 133–179. doi:10.1007/s10441-007-9018-5 PMID:17704896

Russell, M. J., & Hall, A. J. (1999). On the inévitable émergence of life on Mars. In: Hiscox, J.A. (Ed.) *The Search for Life on Mars, Proceedings of the 1st UK Conference*. British Interplanetary Society.

Russell, M. J., Hall, A. J., & Turner, D. (1989). In vitro growth of iron sulfide chimneys: Possible culture chambers for origin -of-life experiments. *Terra Nova*, *1*(3), 238–241. doi:10.1111/j.1365-3121.1989.tb00364.x

Samoilov, O. Y. (1957). *Structure of aqueous solutions and hydration of ions"*. Science.

Schramm, G., Grötsch, H., & Pollmann, W. (1962). Nicht-enzymatische von Polysacchariden, Nucleosiden und Nucleinsäuren und die Entstehung selbst -vermehrungsfähiger systeme. *Angewandte Chemie*, *74*(2), 53–59. doi:10.1002/ange.19620740202

Semenchenko, V. N. (1932). *What are the solutions?* Science.

Simakov, S. (2018). Nano- and micro-sized diamond genesis in nature: An overview. *Geoscience Frontiers*, *9*(6), 1849–1858. doi:10.1016/j.gsf.2017.10.006

Simakov S., Barone G., Melnik N.N., Scribano V. (2021). Sicilian serpentinite xenoliths containing abiotic organics with nanodiamond clusters as key model for prebiotic processes. *Geoscience Frontiers*, *12*(1), 255 – 263. . doi:10.1016/j.gsf.2020.04.008

Spiegelman, S., & Haruna, I. (1966). A rational for an analysis of RNA replication. *Proceedings of the National Academy of Sciences of the United States of America*, *55*(6), 1539–1554. doi:10.1073/pnas.55.6.1539 PMID:5227673

Spirin, A. C. (1960). On macromolecular structure of native high-polymer ribonucleic acid in solution. *Journal of Molecular Biology*, *2*(6), 436–446. doi:10.1016/S0022-2836(60)80054-X

Spirin, A.C. (2019). *Molecular Biology. Structure of the ribosomes and protein biosynthesis*. Moscow: High school.

Spirin, A. S., Gavrilova, A. P., Bresler, S. E., & Mosovetsky, M. I. (1959). Study of the macromolecular structure of infectious ribonucleic acid from tobacco mosaic virus. *Biochemistry (Moscow)*, *24*, 938–947. PMID:13833381

Takai, K., Gamo, T., Tsunogai, U., Nakayama, N., Nealson, K. H., & Horikoshi, K. (2004, August). Geochemical and microbiological evidence for a hydrogen-based, hyper-thermophilic subsurface lithoautotrophic microbioal ecosystem (HeperSLIHE) beneath an active deep-sea hydrothermal field. *Extremophiles*, *8*(4), 269–282. doi:10.1007/s00792-004-0386-3 PMID:15309563

Tatarsky, F. (1989). Struggle for "strange" water. *Chemistry and Life*, *12*, 57–59.

Timofeeff – Ressovsky, N.W., & Zimmer, K.G. (1947). *BIOPHYSIK. Das Trefferprinzip in der Biologie*. Leipzig: Arbeitsgemeinschaft medizinischer Verlage G.m.b.H.

Tuerk, C., & Gold, L. (1990). Systematic evolution of ligands by exponential enrichment. *Science*, *249*(4968), 505–510. doi:10.1126/science.2200121 PMID:2200121

Valeri – Rado. R. (1950). Life of Paster. Publishing Hause of Forging Literature: Moscow.

Vance, S., Harnmeijer, J., Kimura, J., Hussmann, H., Demartin, B., & Braun, J. M. (2007). Hydrothermal systems in small ocean planets. *Astrobiology*, *7*(6), 987–1005. doi:10.1089/ast.2007.0075 PMID:18163874

Vasiliev, V. D., Selivanova, O. M., & Koteliasky, V. E. (1978). Specific self-packing of the ribosomal 16S RNA. *FEBS Letters*, *95*(2), 273–276. doi:10.1016/0014-5793(78)81009-6 PMID:363457

Vasiliev, V. D., Serdyuk, I. N., Gudkov, A. T., & Spirin, A. S. (1986). Self-organization of ribosomal RNA. In: Structure, Function, and Genetics of Ribosomes (eds. Hardesty, B and Kramer, G.), 128 – 142. Springer-Verlag, New-York. doi:10.1007/978-1-4612-4884-2_8

Vernadsky, V.I. (1925). The history of the minerals of the earth's crust. *Scientific chemical and technical publishing house, 1*(1), 1 - 208.

Voitkevich, G. V. (1996). Birth of the Earth: Reader for secondary schools. Phoenix: Rostov-on-Don.

Watson, J. D., & Crick, F. H. C. (1953a). Molecular structure of nucleic acids. *Nature*, *171*(4356), 738–740. doi:10.1038/171737a0 PMID:13054693

Watson, J. D., & Crick, F. H. C. (1953b). Genetical implications of the structure of deoxyribose nucleic acid. *Nature*, *171*(4361), 964–967. doi:10.1038/171964b0 PMID:13063483

Wilson, A., Kaplan, N. O., & Ciotti, M. M. (1961). Evolution and differentiation of dehydrogenasis. *Annals of the New York Academy of Sciences*, *94*(3), 701–722. doi:10.1111/j.1749-6632.1961.tb35567.x PMID:14037006

Wood, R. (1936). *Research in physical optics*. Wentworth Press.

Zhizhin, G. V. (2016). The structure, topological and functional dimension of biomolecules. *J. International Journal of Chemoinformatics and Chemical Engineering*, *5*(2), 44–58. doi:10.4018/IJCCE.2016070104

Zhizhin, G. V. (2018). *Chemical Compound Structures and the Higher dimension of Molecules: Emerging Research and Opportunities*. IGI Global. doi:10.4018/978-1-5225-4108-0

Zhizhin, G. V. (2019a). The polytope of hereditary information the structure, location, signification. *Biochemistry and Modern Applications*, *2*, 56–62. doi:10.33805/2638-7735.124

Zhizhin, G. V. (2019 b). *Attractors and higher dimensions in population and molecular biology: emerging research and opportunities*. IGI Global. doi:10.4018/978-1-5225-9651-6

Zhizhin, G. V. (2019c). The structure of polytope of hereditary information. *International Journal of Chemoinformatics and Chemical Engineering*, *2*(2), 7–22. doi:10.4018/IJCCE.2019070102

Zhizhin, G. V. (2020a). Hidden Nucleic Acid Bond Order? *Acta Scientific Biotechnology*, *1*(4), 34–36.

Zhizhin, G. V. (2020b). Interaction of Nucleic Acids: Hidden Order of Interaction. *International Journal of Applied Research in Bioinformatics*, *11*(2), 1–15. doi:10.4018/IJARB.2021070101

Zhizhin, G. V. (2020c). Spirals of Polysaccharides Simultaneously Rotating in Opposite Directions. *Acta Scientific Biotechnology*, *1*(7), 13–16.

Zhizhin, G. V. (2020d). The Dimension and Life. *Acta Scientific Biotechnology*, *1*(11), 35–40.

Zhizhin, G. V. (2020e). Enantiomorphism of Tartaric Acid as a Property of Five-Dimensional Polytopes. *Acta Scientific Biotechnology*, *1*(12), 16–19.

Zhizhin, G. V. (2021a). Spatial Models of Sugars and their Compounds. *International Journal of Applied Research in Bioinformatics*, *11*(2), 9–23. doi:10.4018/IJARB.2021070102

Zhizhin, G. V. (2021 b). The Structures of DNA Taking into Account the Higher Dimensional of its Components. In Kosrupohr (Ed.) Encyclopedia of Organizational Knowledge, Administration, and Technologies. Hershey, PA: IGI Global.

Zhizhin, G. V. (2022). *The Classes of Polytopes Higher Dimensions in Physics, Chemistry and Biology*. IGI Global.

Zhizhin, G. V. (2023 a). *Non-Euclidean geometry in Materials of Living and Non-Living Matter in the space of the Higher Dimension*. Nova Science Publisher.

Zhizhin, G.V. (2023 b). *Adamantane as a Means of Higher Dimension, Preventing the Formation of the Beta- Amyloid,* 6(1), 1 – 9.

Zhizhin, G. V. (2023c). To the Issue of the Complementary of Biomolecules in the Space of the Highest Dimension. *Journal of Clinical Oncology and Research*, *2*(1), 1–15.

Zundel, G. (1972). Hydrotation and intermolecular interactions. Mir: Moscow.

Conclusion

Thus, being in the positions of an ancient person, we believe that more than three are difficult, and we do not want to see and understand that everything around us and inside us develops in the dimension of more than three. It is proved that the Schrödinger equation has a solution in a space of higher dimension. Consequently, the discrete nature of matter exists in a space of higher dimension. The study showed that the motion of an electron around a nucleus in a space of higher dimension leads to the appearance of quantum numbers with a total number equal to the dimension of space. On a level with the principal quantum number, characterizing the average distance of the electron from the nucleus, and the orbital quantum number, characterizing the shape of the electron orbitals, a series of magnetic quantum numbers appear in a space of higher dimension. Its correspondent the possible values of the projections of the angular momentum of an electron on the coordinate axes of space. The number of these magnetic quantum numbers is equal to the dimension of the space minus two. Taking into account the principal quartic number and orbital quantum number, the total number of quantum numbers characterizing the motion of an electron around the nucleus is equal to the dimension of space. For example, in three - dimensional space the atom has besides a spin quantum number yet of three more quantum numbers. The main quantum number n, which determines the average distance of the electron from the nucleus. Orbital quantum number l, characterizing the momentum of an electron. The magnetic quantum number m_l, characterizing the possible value of the projection of the angular momentum of the electron on the axis z in a magnetic field. The main quantum number can take the values 1, 2, 3, ... The orbital quantum number can take the values 0, 1, 2, ...

Magnetic quantum numbers can take the values 0, ±1, ±2, ±3,... .

The values of the orbital quantum number l correspond to the form of electron orbital. Taking into account the large dimensionality of the neighborhood of the nucleus in the atom, one should expect an increase in the number of quantum cells in orbitals *p, d, f* while maintaining their shape. For example, not 6 but 8 electrons (for each quantum cell with a corresponding pair of electrons per axis) can be on the orbital *p*.

On the other hand, the analysis of real forms of molecules shows that their dimensionality, determined by the shape of the corresponding convex body, whose vertices are atoms, is often higher than three.

Undoubtedly, the possible loss of the number of quantum cells in orbitals *p, d, f* should not lead to a change in the number of electrons in the atomic structure in neutral atoms. The number of electrons under these conditions is of course equal to the charge of the nucleus. But the possible increase in the number of vacant quantum cells contributes to an increase in the chemical activity of atoms, allowing the creation of complex chemical compounds. This can be seen, for example, in biology.

The creation of vacant orbitals in higher-dimensional space can just serve as the basis for the electrophilicity of elements, which is characteristic of biochemical reactions.

In the framework of this monograph, the polytopes of higher dimension are the mathematical elements that model material bodies in nanoworld, including in the nanoworld of living matter. The geometric elements in this case are the faces of different dimensions of polytopes. Of interest is the question of transferring information from elements of lower dimension to elements of higher dimension and vice versa from elements of higher dimension to elements of lower dimension. This monograph establishes the existence of integral equality in relation to the issue of the transmission of information by elements of lower and higher dimensions that describe natural objects of nanoworld. This integral equality is called the law of conservation of incidents. It shows that the sum of incidents of elements of lower dimension with respect to elements of higher dimension is equal to the sum of incidents of elements of higher dimension with respect to elements of lower dimension.

A method has been developed for obtaining simplified images of polytopes of higher dimension in a space of lower dimension. This method is used to obtain three - dimensional images of saccharide molecules: pyranoses (the monosaccharide molecules in which the rings of carbon atoms containing one oxygen atom and five carbon atoms) and furanoses (the monosaccharide molecules in which the rings of carbon atoms containing one oxygen atom and four carbon atoms). It is proved that the furanose molecules are a convex three - dimensional body, and the pyranose molecules consist of two convex bodies, in general, forming a non - convex body. This led to a fundamental difference in the structure of polysaccharide chains of pyranose and furanose molecules. If in the furanose spirals each projection of the furanose molecule onto the coordinate planes preserves the general direction of rotation when moving along the spiral, then a complex picture is observed in the chains of pyranose molecules. In the chains of β- anomers, one of the projections is linearly transmitted along the chain, and the projection perpendicular to it along the chain along a curved line. In chains of α - anomers (spirals), it is observed that when moving along a chain of projections in mutually perpendicular planes, the pyranose molecules rotate in opposite directions.

The currently known forms of the helical structure of nucleic acid molecules *A, B*, *Z*, and nucleic acid molecules in the form of a closed ring were experimentally found. In this monograph is proved that the main reason for the formation of nucleic acid molecules of different shapes is the different geometric realization of the chemical bond between the phosphoric acid residue and the *D* - ribose molecule, as objects of higher dimensionality. In other words, the residues of phosphoric acid can geometrically act as a bridge, connecting the molecules of *D* - ribose with each other. It is shown that the angle γ between the projections of the valence bonds of the phosphorus atom in the phosphoric acid residue on the plane of the carbon ring in the *D* - ribose molecule is a parameter characterizing the various geometric forms of chemical bonding of phosphoric acid residues with *D* - ribose molecules. The angle γ characterizes the orientation of the phosphoric acid residue in the three - dimensional space. This parameter can therefore be regarded as an internal degree of freedom of the structure of the nucleic acid molecule.

A mathematical model describing the geometric features of the chemical bond of these objects is constructed. Calculations on this model showed that the characteristics of known nucleic acid structures, determined experimentally, and calculated from the constructed model, are in good agreement with each other, while the model does not contain any empirical parameters. Calculations show that known nucleic acid structures are particular cases in the model and it can be considered that known examples of nucleic acid structure forms are experimental confirmation of the correctness of the mathematical model. However, according to the mathematical model, its solutions are many other nucleic acid structures that have not been experimentally discovered so far. One can hope that their experimental discovery is ahead.

The existence of two enantiomeric forms of nucleic acid molecules, connected with the existence of two enantiomeric forms of *D* - ribose molecules, is established.

The resulting variety of nucleic acid molecule structures is represented in the form of analytical dependencies, graphs and tables.

The higher dimensionality of the constituent nucleic acid molecules, which allows to describe mathematically the structure of nucleic acid, requires reconsidering the issues of tight packing of nucleic acid molecules in cells, viruses and bacteria, provided that the nucleic acid chains necessary for the preservation and transfer of genetic information are complementary.

The monograph discusses various interactions of nucleic acids in the process of transmission of hereditary information. They all result from hydrogen bonding between nitrogenous bases covalently attached to five-carbon sugar molecules in nucleic acids. In double-stranded DNA, the connection between the helices is carried out mainly with the help of the canonical pairs of nitrogenous bases adenine - thymine, guanine - cytosine. The elements in these pairs can be interchanged. In double-stranded RNA, thymine is replaced by uracil in canonical pairs of nitrogenous bases. This interaction of nitrogenous bases (and, consequently, the interaction of nucleic acids) is provided by the chemical structure of nitrogenous bases. In all interactions, a nitrogen atom is involved, which is included in all nitrogenous bases. With the help of correspondence (complementarity) of nitrogenous bases in interaction pairs, the information embedded in the original DNA is transmitted during the transition from the original DNA to subsequent nucleic acids mRNA, rRNA, tRNA, mRNA up to the formation of a protein in ribosomes according to the given information in the original DNA. However, in addition to canonical pairs of nitrogenous bases, there are non-canonical interactions of nitrogenous bases and, consequently, non-canonical pairs of nitrogenous bases, the transmission of information through which differs from the information transmitted by canonical pairs of nitrogenous bases.

This monograph lists both canonical and non-canonical base pairs. It is significant that in one part of the pairs of nitrogenous bases, their bond is carried out using two parallel hydrogen chemical bonds, in the other part of the pairs of nitrogenous bases, their bond is carried out using three parallel hydrogen chemical bonds. However, in non-canonical pairs of nitrogen bases there are pairs in which the bond is carried out by two non-parallel hydrogen bonds, and there are even pairs of nitrogen bases with single hydrogen bonds. Thus, under the action of non-canonical pairs of nitrogenous bases, there is no unambiguous coding of the information contained in the original DNA. In addition, non-canonical pairs of nitrogenous bases make it possible to assert the possibility of the arrangement of nitrogenous bases in a pair in different planes. Indeed, if nitrogenous bases are considered flat, then in canonical pairs of nitrogenous bases, in which the bond is carried out by two and three parallel hydrogen bonds, the planes of nitrogenous bases must coincide, forming a certain single plane of the pair. In non-canonical pairs, there is no reason to assert this, especially for pairs in which the bond is carried out by a single hydrogen bond.

It is important that the interaction of single-stranded nucleic acids in double-stranded nucleic acids occurs between helices directed towards each other. This complicates the process of reading information from one helix to another. It is shown that the features of such a reading are described by the geometry of cross-polytopes (a special kind of higher-dimensional polytopes). The characteristics of such polytopes are given. In particular, for the first time, simplified images of double-stranded nucleic acids were constructed, in which five-carbon sugar molecules are presented as three-dimensional polyhedron that reflect all the main features of these molecules in full, i. e. in the form of polytopes of dimension 12. Certain vertices of these three-dimensional polyhedron are the points of attachment of nitrogenous bases to them. It is essential that, as follows from the constructed images, the points of connection of nitrogenous bases to sugar molecules are located in a space of higher dimension. The location of nitrogenous bases, taking into account their geometry, must be studied separately. This will be done in the next chapter.

The interactions of nucleic acids in ribosomes, which carry out the creation of protein molecules according to the program specified in the original DNA, are quite complex. Descriptions of the processes of these interactions, despite the results already available, require further development, taking into account the higher dimension of the chemical compounds of nucleic acids. The exceptional role of magnesium ions in these processes, which ensure the stability of ribosomes at a sufficient concentration of these ions in ribosomes, can be noted. It can be assumed that this stability is affected by the possibility that magnesium ions have vacant electron orbitals in higher-dimensional space.

The representations of the sugar molecule and the residue of phosphoric acid in the form of polytopes of higher dimension are used. Based on these ideas and their simplified three - dimensional images, a three-dimensional image of nucleic acids is constructed. The geometry of the neighborhood of the compound of two nucleic acid helices with nitrogen bases has been investigated in detail. It is proved that this neighborhood is a polytope with anti-parallel edges of dimension 13 (13 – cross - polytope). This polytope is called the polytope of hereditary information. The geometry of the polytope of hereditary information is investigated. It is shown that in the flat coordinate planes of the polytope of hereditary information there are located flat complementary hydrogen compounds of the nitrogenous bases of two nucleic acid helices. It turned out that the number of these coordinate planes

(12) is exactly as many as there are various options for hydrogen compounds of nitrogenous bases. An image of the polytope of hereditary information with a specific indication of its coordinate planes is constructed. It is shown that not all 12 coordinate planes can have flat nitrogenous bases. Only five coordinate planes can have nitrogenous bases. Considering that there are 4 canonical base pairs, and there are 5 planes in which each pair can be located, the total number of base pairs location options is 20.

It is essential that if a nucleic acid molecule contains several helices of nucleic acids, then the polytope of the hereditary information of the nucleic acid consists of several polytopes of the hereditary information of pairs of chemically linked single-stranded nucleic acids. Combining polytopes of hereditary information of pairs of nucleic acids linked by a hydrogen bond into one polytope contradicts the geometry of interacting nucleic acids. The law of conservation of incidence is only valid within a pair of chemically linked nucleic acids.

From experimental studies of the chemistry of nucleic acids, it follows that methylation of nucleic acids to a certain extent contributes to the protection of nucleic acids from various damaging factors. However, their role in protecting nucleic acids is not always understood. On the other hand, it follows from studies of viral infections that the concentration of methyl groups in the DNA of viruses is significant. Apparently, this allows the DNA of viruses to suppress the activity of the DNA of living cells and force them to work for the production of viruses in viral infections. Impact investigated of nucleic acid methylation on the geometric characteristics of nucleic acids. Such methylation is characteristic of viral nucleic acids when infected with viruses of organisms. It was shown that methylation of nitrogenous bases leads to an increase in their dimension, creating a higher dimension of the bases. This, in accordance with law for the conservation of incidents, sharply increases the intensity of information transfer between elements of different dimensions of nucleic acids. At the same time, the intensity of transmission of hereditary information in the polytopic of hereditary information increases significantly. It has been established that the intensity of information transfer in the methylated polytope of hereditary information increases by more than 2000 times in comparison with the unmethylated polytope of hereditary information. It should be assumed that it is this property that activates viral nucleic acids when they infect organisms, weakening organisms, including due to the use of methyl groups of organism cells by viruses. To counteract this activation of viral nucleic acids it is necessary to increase the concentration of methyl groups in the nitrogenous bases of the nucleic acids of the organism.

This will enhance the transfer of information in the nucleic acids of the organism, activating them in the process of confrontation with the nucleic acids of viruses. It should be emphasized that the dimension of the methylated polytope of hereditary information is 18, on 5 units higher than the dimension of unmethylated polytope of hereditary information. This indicates that the interaction of nucleic acids during infection of the organism with viruses occurs in space of a very high dimension.

The use of methylation for the experimental study of possible practical counteraction to viral infections will require further development of the methylation technique, measurement of the methylation degree, and determination of ways to change this degree. These issues are currently practically not investigated and have not been considered in this work.

An analysis of the scientific literature on the studies of the three -fluid genetic code showed that despite the many serious scientific works performed by outstanding researchers in many countries of the world, the proposed genetic code has many short comings, and it is universal to seize it. Experimental studies in recent years have found many deviations from the proposed genetic code. This applies to animals and protozoa, bacteria and fungi, algae and plants. The triplets encoding the amino acid residues of many of them differ from the triplets recommended by the triplet genetic code. More examples of deviations from the universal genetic code are expected to be discovered in the future. The heterogeneous degeneracy of the genetic code is amazing. Some amino acid residues do not have degeneracy, and some amino acid residues have degeneracy equal to 6. There is no explanation for this situation within the framework of the universal genetic code. No relationship was found between the degree of degeneracy and the chemical structure of amino acid residues. The reason for many contradictions in the genetic code is the original position of its triplet nature. This position was based on an arithmetic discrepancy between the number of types of nucleotides and the number of types of amino acid residues. To coordinate these numbers, triplet codons were introduced, and this required even more complicating the genetic code. Determining the content of triplets required complex experiments related to changing the number of nucleotides in the nucleotide chain using chemical reagents. Synthetic statistical heteropolyribonucleotides of various compositions were used in the studies. Statistical methods cannot lead to unambiguous correct results.

These studies relied on old geometric ideas about the three-dimensional shape of biomolecules. Although almost 10 years ago it was proved that biomolecules have a higher dimension. This completely changes the idea of the interaction of molecules. In this regard, this chapter proposes a new concept of the genetic code, taking into account the higher dimension of nucleic acids and their components. Within the framework of this approach, it was proved that the area in which the transmission of hereditary information occurs is a polytope of dimension 13. This area has 12 two-dimensional flat coordinate planes, in which pairs of nitrogenous bases connecting nucleotides could be located. The geometrical analysis of polytopes of hereditary information showed that five pairs of nitrogenous bases can be located only in five of these 12 coordinate planes, since other coordinate planes are filled with elements of the polytope. Considering that any of the four pairs of nitrogenous bases can, in principle, be located in each of these planes, then 20 variants of coordinate planes arise with the corresponding variants of pairs of nitrogenous bases. Thus, the number of these variants of coordinate planes is exactly equal to the number of types of amino acids. Each amino acid corresponds to one of the coordinate planes with the corresponding pair of nitrogenous bases. The genetic code within the framework of these ideas consists in establishing a correspondence between the selected amino acid of a certain coordinate plane and a certain pair of nitrogenous bases. Establishing this correspondence should be the task of new experimental studies. Within the framework of these representations, it is not necessary to introduce triplets and the complexities and uncertainties associated with them.

It is shown that the widespread quasi - plane Pauling models do not reflect and even contradict the spatial structure of polypeptide amino acid molecules, taking into account their highest dimension. This applies to all types of polypeptide chain conformation. In this regard, new models of linear polypeptide chains were construct and it was found that these chains have translational symmetry in the space of higher dimensions. The elementary element of the translational symmetry of these chains is the 9 - dimensional polyhedron, which consists of two mirror - opposite amino acid molecules. Helical conformations of polypeptide chains are also composed of higher - dimensional elements. The rotation of monomeric amino acid units in a helical conformation is determined only by the alternation of active centers in an asymmetric amino acid molecule. The study of the amino acid β - structures showed that they also have translational symmetry in the space of the highest dimension. It was found that the β - structures can have various forms, including not containing amino acid molecules in either parallel or antiparallel positions. Moreover, β - structures have elements of translation symmetry with dimension 23, consisting of two topologically equivalent polyhedrons of dimension 5 and two topologically equivalent polyhedrons of dimension 4. The resulting polyhedrons in high dimension, as translation elements of the construction of spaces protein molecules, should be further compared with the general theory of normal partitioning of n - dimensional spaces and corresponding high - dimensional parallelohedrons. The turn of the polypeptide chain, which is often found in globular proteins, has been studied in the higher dimensional space. It was found that the rotation of the polypeptide chain is accompanied by a change in the enantiomorphic form of the protein molecules.

The deposition of protein molecules in the brain with the formation of β - amyloid is the cause of many diseases of the nervous system. Until now, the possible structure of the β - amyloid complex and methods of treating these diseases have been little studied. In this work, with the aim of constructing mathematical models of the volumetric filling of space with protein molecules, this possibility is investigated without using simplified flat Pauling models. Analyzing the process of attachment of amino acid molecules under these conditions, it was found that the formed layers of molecules have the highest dimension both in antiparallel and parallel conformation. Based on the recently developed theory of polytopic prismahedrons, a mathematical model of the volumetric filling of space with amino acid molecules was built, which for the first time makes it possible to describe the structure of plaques of protein molecules in the brain. This model can serve as a mathematical model of the native conformation of a globular protein in the general case. Obviously, the specific type of the native structure of the globular protein depends on the type of protein molecules and the conditions for the formation of the globule.

It has been established that the structure of the main part of the adamantane molecule is much more complex than previously thought. The dimension of this structure is determined by reducing it to the well-known simplex polytope form. It is determined that the dimension of this polytope is 9. Adding various functional groups to the main part of the adamantane molecule only increases its dimension. Consequently, many drugs, which include adamantane molecules with different functional groups, have a higher dimension. Examples of calculating the dimensions of known medicines are given. It has been established that 10 carbon atoms in the main part of the adamantane molecule have 16 active (non-extinguished) valence bonds. At least 16 amino acid residues can join them. In this case, in addition, the amino acid residues, connecting with each other, form a chemically inert shell around the main part of the adamantane molecule. Thus, the main part of the adamantane molecule without functional groups added in various drugs can concentrate amino acid residues around itself, forming chemically inert particles, thereby preventing the creation of beta-amyloids, leading to the development of various prion diseases of nerve cells, such as Alzheimer's disease, Parkinson's disease, and other diseases.

The studies carried out show that with a large dimension of interacting biomolecules, the condition for their correspondence (complementarity) can only be the condition that ensures the possibility of a chemical reaction between molecules. Moreover, since the molecules have a higher dimension, i.e. are representable as polytopes of higher dimension, then only the condition of correspondence to the geometry of polytopes of higher dimension can be a correspondence condition. In the examples considered, this correspondence condition is associated with a mirror image of the geometric elements of one polytope in relation to the geometric elements of the polytope interacting with it. So, when considering the formation of a chain of polytopes along a line (the interaction of amino acid molecules), the bound molecules turned out to be mirror images of each other relative to the line along which they line up. During the interaction of nucleic acids, the interacting sugar molecules are also connected by mirror reflection, forming polytopes with antiparallel edges, i.e. high-dimensional cross-polytopes.

As part of the definition of functional dimension, in which all functional groups of monosaccharide molecules are taken as vertices of the corresponding geometric figure, analytical expressions are obtained for calculating the dimension of polytopes of monosaccharide molecules depending on the number of chiral carbon atoms. They refer to both linear molecules and molecules with a closed carbon cycle (furanoses and pyranoses). It has been shown that the cyclization of monosaccharides leads to an increase in their dimension. The geometric shape of monosaccharide molecules has the form of a polytope of the simplex type of higher dimension. Based on the type of polytope of higher dimension corresponding to the pyranose monosaccharide molecule and calculations, a three - dimensional approximate image of the pyranose monosaccharide molecule was obtained. It has the form of a pentagonal prism connected by a carbon cycle to a pyramid, which has a pentagonal and hexagonal base. This image is fundamentally different from the commonly used in the literature image of the pyranose molecule in the form of an "armchair". The resulting simplified three - dimensional image of the pyranose monosaccharide molecule was used to analyze the possibility of the formation of disaccharides and monosaccharide chains.

It has been found that the statement about the linearity of the connection of β - anomers of monosaccharides, widespread in the literature, is not correct since a deviation from linearity is observed in the chain of monosaccharides.

An even greater deviation from linearity in the monosaccharide chain is observed when α - anomers are combined. Here, the possibility of the formation of spirals with simultaneous rotation in opposite directions was discovered.

An analysis of the conditions for the connection of molecules of the highest dimension (dimension of more than three) with each other in three servants important for biochemistry: compounds of amino acid molecules (dimension 4), compounds of monosaccharides (dimension 15), nucleic acid compounds (dimension 12). It is obvious that in none of these examples the accepted state of complementarity, which is in accordance with the surfaces of the interacting molecules, makes no sense, since it is impossible to indicate that it is the surface of the molecule of the highest dimension, i.e. it is impossible to indicate what is the correspondence of the alleged surfaces (there is no hollow, no protrusion). It can be argued that when combining molecules of the highest dimension, the opposite of the electrical charges plays a significant role. Ultimately, the attraction of electrical charges provides the compounds of the molecules. For example, when combining amino acid molecules, one of the molecules gives a proton (+H), and the other gives the hydrocele group (- OH). As a result, a neutral water molecule is distinguished, and the remaining parts due to the particles remaining with the opposite charge are attracted to each other.

In addition, in the geometry of interacting molecules, a mirror arrangement of valence ties to each other at the junction of the molecules is observed. It also leads to the opposite location of charges with opposite signs.

In the interaction of sugar molecules (glucose), the proton of one molecule (+H) is also combined with a hydrocele group (- OH) another molecule with the release of a water molecule. The remaining particles with opposite charges are attracted to each other by forming a compound or β - anomer or α -anomer. In the case of the formation of α - anomer, the opposite of the charge of particles leads to the simultaneous rotation of molecules in opposite directions in perpendicular planes. In the case of the formation of a β - anomer connection due to the connection, the anti-parallelism of the rib pair occurs.

The interaction of nucleic acids occurs using nitrogenous bases, which are connected to each other with a hydrogen connection. In this regard, as you know, the compound is provided by the opposite of charged with electronic charges. This is complementarity, and not that certain nucleotides are connected.

A variety of theoretical ideas about the possible mechanisms of the origin of life on planet Earth, put forward at different times, starting with the so-called "Darwin's ponds", are critically examined. Taking into account modern views on the existence of the "RNA World" on Earth in antiquity, as well as recently discovered gas emissions from the surface of the earth's crust at the bottom of the oceans, it was concluded that mounds above the vent of hydrothermal emissions, consisting of rocks of the earth's crust, are the most possible site for the origin of life on Earth. Firstly, this mound contains all the metal ions necessary for the formation of RNA. Secondly, the mound has cracks, which ensure the existence in them of RNA colonies of different lengths, connected by a single space, i.e. the kurgan provides the necessary compartmentalization. Thirdly, the mound above the hydrothermal vent is flowing, i.e. it ensures the exchange of the RNA colony with the environment, i.e. ensures the evolution of RNA. Fourthly, all the necessary chemical reactions can and do take place in the mound to create RNA molecules and, subsequently, DNA molecules. Consequently, the difficulties in all previous theories of the origin of life on Earth are removed. There is no need to make all sorts of often conflicting assumptions.

Data are given on the dimensionality of various binary compounds that are widely distributed in the materials of the earth's crust, studied by the author over a number of years, indicating a higher dimensionality of these compounds. The data on the highest dimension of biomolecules (of various carbohydrates, proteins, nucleic acids and other biomolecules) are also given, which were determined in particular in this monograph, as well as in previous works of the author. The properties of water are noted to rebuild its structure, taking into account the complementary interaction of water with solid surfaces that are in contact with it, as well as the property of the memory of water about the chemical compounds in it. It is concluded that the higher dimension of biomolecules is a consequence of the conditions for the origin of life in the mound above the source of emissions through the earth's crust at the bottom of the oceans, where water was in contact with ore bodies and acquired their properties, i.e. higher dimension, and as a result, the biomolecules in the water also acquired a higher dimension.

Gennadiy Vladimirovich Zhizhin
Russian Academy of Natural Sciences, Russia

Compilation of References

Agol, V. I., Bogdanov, A. A., & Gvozdev, V. A. (1990). *Molecular biology: Structure and biosynthesis of nucleic acids*. Higher School.

Ahliquist, P. (2002). RNA -dependent RNA polymerases viruses and RNA silencing. *Science, 296*(5571), 1270–1273. doi:10.1126/science.1069132 PMID:12016304

Akabori, Sh. (1957). *On the origin of preproteins. In book. Origin of life on Earth*. Science.

Albert, B. (2002). *Molecular Biology of the Cell* (4th ed.). Garland Science.

Babenko, Yu. I. (2015). Antispace. In I.N. Toganov, Yu.I. Babenko. Anti-time and anti - space. S.-Petersburg: TIN.

Baltimore, D., Girard, M., & Darnell, J. E. (1966). Aspects of the synthesis of poliovirus RNA and the formation of virus particles. *Virology, 29*(2), 179–189. doi:10.1016/0042-6822(66)90024-9 PMID:4287327

Belov, O.V. (2011). Modeling of excisional repair of damaged stems in bacterial cells. *Physics of elementary particles and the atomic nucleus, 8,* 237 - 248.

Belozersky, A. N. (1957). On the species specificity of nucleic acids. In the book. In A. I. Oparin (Ed.), *The Emergence of Life on Earth* (pp. 198–205). Publishing House of the Academy of Sciences of the USSR.

Belozersky, A. N., & Spirin, A. S. (1958). A correlation between the compositions of deoxyribonucleic and ribonucleic acids. *Nature, 182*(4628), 111–112. doi:10.1038/182111a0 PMID:13566202

Benzer, S. (1961). On the topography of the genetic fine structure. *Proceedings of the National Academy of Sciences of the United States of America, 47*(3), 403–415. doi:10.1073/pnas.47.3.403 PMID:16590840

Berg, J., Tymoczko, J., & Stryer, L. (2002). *Biochemistry*. W.H. Freeman & Co Ltd.

Bergman, G., Waugh, J. L. T., & Pauling, L. (1952). Crystal structure of the intermetallic compound Mg_{32} (Al, Zn) $_{49}$ and related phases. *Nature, 169*(4312), 1057–1058. doi:10.1038/1691057a0

Bergman, G., Waugh, J. L. T., & Pauling, L. (1957). The crystal structure of the metallic phase Mg_{32} (Al, Zn) $_{49.}$ *Acta Crystallographica, 10*(4), 254–259. doi:10.1107/S0365110X57000808

Berlin, A. P., Wolfson, S. A., & Enikolopyan, N. S. (1978). *Kinetics of polymerization processes*. Chemistry.

Bernal, J. D. (1967). *Origin of Life*. World Publishing Co.

Brachet, J. (1941 – 1942). La détection histochimique et la microdosage des acides pentosenucleiques. *Enzymologia, 10*, 87–96.

Brachet, J. (1941-1942). La detection histochimique et la microdosage des acides pentosenucleiques. *Enzymologia, 10*, 87–96.

Büchel, W. (1963). Warum hat ger Raume drei Dimensionen? *Physikalische Blätter, 19*(12), 547–549. doi:10.1002/phbl.19630191204

Butler, J. M. (2005). *Forensic DNA Typing*. Gardners Books.

Caly, L., Wagstaff, K. M., & Jans, D. A. (2012). Nuclear trafficking of proteins from RNA viruses: Potenial targetfor anti-viral? *Antiviral Research, 95*(3), 202–206. doi:10.1016/j.antiviral.2012.06.008 PMID:22750233

Cantor, C. R., & Schimmel, P. R. (1980). *Biophysical Chemistry. Part I. The Conformation of Biological Macromolecules*. W.H. Freeman & Co Ltd.

Carlin, S. (1971). *Fundamentals of the theory of random processes*. World.

Caspersson, T., Landeström – Hyden, H., Aquilonius, L. (. (1941). Cytoplasmsnukleotide in eiweissproduzierenden Drüsenzellen. *Chromosoma, 2*, 111–131. doi:10.1007/BF00325956

Cech, T. R., Gesteland, R. F., & Atkins, J. F. (Eds.). (1993, 1999, 2006). RNA World. Cold Spring Harbor Press. New York.

Cech, T. R., & Golden, B. L. (1999). Building a catalytic active site using only RNA. In R. F. Gesteland, T. R. Ceck, & J. F. Atkins (Eds.), *The RNA World* (2nd ed., pp. 321–347). Cold Spring Harbor Laboratory Press.

Chetverina, H. V., & Chetverin, A. B. (1993). Cloning of RNA molecules in vitro. *Nucleic Acids Research, 21*(10), 2349–2353. doi:10.1093/nar/21.10.2349 PMID:7685078

Chetverina, H. V., Demidenko, A. A., Ugarov, V. I., & Chetverin, A. B. (1999). Spontaneas rearrangements in RNA sequence. *FEBS Letters, 450*(1-2), 89–94. doi:10.1016/S0014-5793(99)00469-X PMID:10350063

Chetvernin, A. B., Chetverina, H. V., & Munishin, A. V. (1991). On the nature of Spontaneous RNA sythesis by Qβ replcase. *Journal of Molecular Biology, 222*(1), 3–9. doi:10.1016/0022-2836(91)90729-P PMID:1719219

Corey, R. B., & Pauling, L. (1953). Fundamental dimensions of polypeptid chains. *Proc R Soc Lord B Biol Sci, 141*(902), 10-20.

Crick, F. H. C. (1958). On the protein synthesis. *Symposia of the Society for Experimental Biology, 12*, 138–163. PMID:13580867

Crick, F. H. C. (1959). The present position of the coding problem. *Brookhaven Symposia in Biology, 12*, 35–39. PMID:13812855

Crick, F. H. C. (1966). The genetic code: Yesterday, today and tomorrow. *Cold Spring Harbor Symposia on Quantitative Biology, 31*(0), 1–9. doi:10.1101/SQB.1966.031.01.007 PMID:5237190

Crick, F. H. C., Barnet, L., & Brenner, S. (1961). General nature of the genetic code for proteins. *Nature, 253*(4809), 374–375. doi:10.1038/1921227a0 PMID:13882203

Crotty, S., Cameron, C., & Audino, R. (2002). Rubaviriu's antiviral mechanism of action: Lethal mutagenesis? *Journal of Molecular Medicine, 80*(2), 86–95. doi:10.1007/s00109-001-0308-0 PMID:11907645

Dahm, R. (2005). Friedrich Miescher and the discovery of DNA. *Developmental Biology, 278*(2), 274–288. doi:10.1016/j.ydbio.2004.11.028 PMID:15680349

De Clerey, E. (2004). Antiviral drugs in current clinical use. *Journal of Clinical Virology, 30*, 2. PMID:15125867

De Maio, D., & Coen, D.M. (2001). Replia Yion Strategies of DNA Viruses. *Fields Virology,* 1.

Dixon, M., & Webb, E. C. (1979). *Ensymes*. Longman Group Ltd.

Dong, L., Hu, S., & Gao, J. (2020). Discovering drugs to treat coronavirus disease 2019(COVID-19). *Drug Discoveries & Therapeutics, 14*(1), 58–60. doi:10.5582/ddt.2020.01012 PMID:32147628

Doty, P., Boedtker, H., Fresco, J. R., Haselkorn, R., & Litt, M. (1959). Secondary structure in ribosomes acids. *Proceedings of the National Academy of Sciences of the United States of America, 45*(4), 482–499. doi:10.1073/pnas.45.4.482 PMID:16590404

Efimov, N. V. (1945). *Higher geometry*. OGIZ.

Eisenberg, D., & Kautsman, V. (1975). *Structure and properties of water*. Gidrometyoizdat.

Ellington, A., & Szostak, J. (1990). In vitro selection of RNA molecules that bind specific ligands. *Nature, 346*(6287), 818–822. doi:10.1038/346818a0 PMID:1697402

Engelman, D. M., Moore, F. B., & Schoenborn, B. P. (1975). Neutron scattering measurements of separation and shape of proteins in 30S ribosomal subunits of *Escherichia coli S2 – S5, S5 – S8, S3 – S7. Proceedings of the National Academy of Sciences of the United States of America, 72*(10), 3888–3892. doi:10.1073/pnas.72.10.3888 PMID:1105567

Ershov, Y. A. (1989). *Mechanisms of toxic action of inorganic compounds*. Medicine.

Euclid. (2012). *Beginning*. URSS.

Euclid. (2012). *Beginnings*. URSS.

Fersmann, A. E. (1937). Geochemistry. V. 1 - 4. Leningrad: ONTI.

Fersmann, A. E. (1923). *Chemical elements of the Earth and Space*.

Frank - Kamenetskiy, M. (1986). DNA structure: A simple solution to the stability of the double helix? *Nature, 324*, 325.

Frank - Kamenetskiy, M. (1986). DNA structure: A simple solution to the stability of the double helix? *Nature, 324*.

Frank - Kamenetskiy, M. (1988). H – form DNA and the hairpin – triplex model. *Nature, 333*, 214.

Frank - Kamenetskiy, M. (1988). H – form DNA and the hairpin – triplex model. *Nature, 333*.

Frank – Kamenetskiy, M.D. (2004). *The DNA Age*. Moscow, Publishing house KDU.

Frank – Kamenetskiy. M.D. (2010). Queen of a living cell. From the structure of DNA to the biotechnological revolution. Moscow: Act – Press.

Freeman, I. M. (n.d.). Why is Space Three-Dimensional? *American Journal of Physics*, *37*, 1222–1224. doi:10.1119/1.1975283

Gale, M., Tan Seng-Lai, J. R., & Katze, M. G. (2000). Translational Control of Viral Gene Expression in Eukaryotes. *Microbiology and Molecular Biology Reviews*, *64*(2), 239–280. doi:10.1128/MMBR.64.2.239-280.2000 PMID:10839817

Galimov, E. M. (2004). Phenomena of life: Equilibrium and non- linearity. Origins of life and evolution of the Biosphere. *Origin Life. Biosphere.*, *6*, 599–613.

Galimov, E. M. (2009). Conditions for the Emergence of Life on Earth. *Biosphere*, *1*(1), 39–47.

Gamow, G. (1954). Possible relation between deoxyribonucleic acid and protein structures. *Nature*, *173*(4398), 318. doi:10.1038/173318a0

Gamow, G., Rich, A., & Ycas, M. (1956). The problem of information transfer from the nucleic acids to proteins. *Advances in Biological and Medical Physics*, *4*, 23–68. doi:10.1016/B978-1-4832-3110-5.50006-6 PMID:13354508

Gerald, K. (2013). *Cell and Molecular Biology. Concepts and Experiments*. John Wiley and Sons.

Gilbert, W. (1986). Origin of life: The RNA world. *Nature*, *319*(6055), 618. doi:10.1038/319618a0

Gillespie, D. N. (1977). Stochastic simulations of coupled chemical reactions. *Journal of Physical Chemistry*, *81*(25), 2340–2361. doi:10.1021/j100540a008

Gillespie, D. N. (2001). Approximate accelerated stochastic stimulation of chemically reacting systems. *The Journal of Chemical Physics*, *115*(4), 1716–1733. doi:10.1063/1.1378322

Gillespie, R. J. (1972). *Molecular Geometry*. Van Nostrand Reinhold Company.

Gillespie, R. J., & Hargittai, I. (1991). *The VSEPR Model of Molecular Geometry*. Allyn and Bacon.

Gold, L., Polisky, B., Uhlenbeck, O., & Yaris, M. (1995). Diversity of oligonucleotide functions. *Annual Review of Biochemistry*, *64*(1), 763–797. doi:10.1146/annurev.bi.64.070195.003555 PMID:7574500

Golubev, A. G. (2009). *Biology of life expectancy and aging*. N-L Publishing House.

Gongadze, G.M. (2001). $5S_p$ RNA and ribosome. *The successes of modern biochemistry, 51*, 162 - 165.

Gonzalez Canga, A., Sahagún Prieto, A. M., Diez Liébana, M. J., Fernández Martínez, N., Sierra Vega, M., & García Vieitez, J. J. (2008). The pharmacokinetics and interactions of ivermectin in humans-a mini-review. *The AAPS Journal*, *10*(1), 42–46. doi:10.1208/s12248-007-9000-9 PMID:18446504

Gratias, D., & Cahn, J. W. (1986). Periodic and qusiperiodic crystals. *Scripta Metallurgica*, *20*(9), 1193–1197. doi:10.1016/0036-9748(86)90030-X

Gray, H. B. (1965). *Electrons and Chemical Bonding. W.A. BENJAMIN*. INC.

Grunbaum, B. (1967). *Convex Polytopes*. Springer.

Grunberg-Manago, M., & Ochoa, S. (1955). Enzymatic synthesis and breakdown of polynucleotides: Polynucleotide phosphorylase. *Journal of the American Chemical Society*, *77*(11), 3165–3166. doi:10.1021/ja01616a093

Gubin, S. P. (2019). *Chemistry clusters. Basics of classification and structure*. Editorial URSS.

Gurevich, L., & Mostepanenko, V. (1971). On the Existence of Atoms in *n* - Dimensional Space. *Physics Letters. [Part A]*, *35*(3), 201–202. doi:10.1016/0375-9601(71)90148-4

Haldane, J. B. S. (2015). *Selected Genetic Papers. (Rout liege Revivals)* (K. R. Dronamraju, Ed.).

Hall, C. E., & Slayter, H. C. (1959). Electron microscopy of ribonucleoprotein particles from *E., coli. Journal of Molecular Biology*, *1*(4-5), 329–332. doi:10.1016/S0022-2836(59)80014-0

Harris, T. (1966). *Theory of branching random processes*. World.

Ha, S. C., Lowenhaupt, K., Rich, A., Kim, Y.-G., & Kim, K. K. (2005). Cristal structure of a junction between B – DNA and Z – DNA revels two extruded bases. *Nature*, *437*(7062), 1183–1186. doi:10.1038/nature04088 PMID:16237447

Hawkes, E. J., Hennelly, S. P., Novikova, I. V., Irwin, J. A., Dean, C., & Sanbonmatsu, K. Y. (2016). COOLAIR antisense RNAs from evolutionarily conserved elaborate secondary structures. *Cell Reports*, *16*(12), 3087–3096. doi:10.1016/j.celrep.2016.08.045 PMID:27653675

Hayor, F., Jayaprakash, C.A. (2006). A tutorial on cellular stochastisity and Gillespies algorithm.

Hershey, A., & Chase, M. (1952). Independent functions of viral protein and nucleic acid in growth of bacteriophage. *The Journal of General Physiology*, *36*(1), 39–56. doi:10.1085/jgp.36.1.39 PMID:12981234

Hilbert, D. (1901). Gesammelte Abhandlagen. *Archivf. Mathematics and Physics*, *2*, 263–272.

Hobson, E. W. (1931). *The theory of spherical and ellipsoidal harmonics*. At the University Press.

Hubsher, G., Haider, M., & Okun, M. S. (2012). Amantadine the journey from fighting flu to treating Park'inson disease. *Neurology*, *78*(14), 1096–1099. doi:10.1212/WNL.0b013e31824e8f0d PMID:22474298

Huxley, H. E., & Zubay, G. (1960). Electron microscope observations on the structure of microsomal particles from *Escherichia coll. Journal of Molecular Biology*, *2*(1), 10–18. doi:10.1016/S0022-2836(60)80003-4

Ivanov, K. P. (1993). *Fundamentals of the body's energy. Biological oxidation and oxygen supply*. St. Petersburg. *Science*.

Jacob, F., & Monod, J. (1961). Genetic regulatory mechanisms in the synthesis of proteins. *Molecular Biology*, *3*(3), 318–356. doi:10.1016/S0022-2836(61)80072-7 PMID:13718526

Jacob, F., & Monod, J. (1961a). On regulation of gene activity. *Cold Spring Harbor Symposia on Quantitative Biology*, *26*(0), 193–209. doi:10.1101/SQB.1961.026.01.024

Karapet'yants, M. X., & Drakin, S. I. (1994). *General and Inorganic Chemistry*. Chemistry.

Kisselev, N. A., Gavrilova, L. P., & Spirin, A. S. (1961). Configurations of high-polymer ribonucleic acid macromolecules as revealed by electron microscopy. *Journal of Molecular Biology*, *3*(6), 778–783. doi:10.1016/S0022-2836(61)80083-1 PMID:14456472

Klassen, V. I. (1973). *Water and magnet*. Science.

Koolman. J. & Roehm. K.- H. (2013). Color Atlas of Biochemistry. Georg Thieme Verlag KG.

Koolman, J., & Roehm, K. (2013). *Color Atlas of Biochemistry*. Thieme.

Korogodin, V.I. (2012). *Phenomenon of life. Selected works* (in 2 volumes). Managing editors S.G. Inge-Vechtomov, A.F. Tsybin. Moscow: Nauka.

Korostelev, A., Trakhanov, S., Laurberg, M., & Noller, H. F. (2006). Crystal structure of a 70S ribosome- tRNA complex reveals functional interaction and rearrangements. *Cell*, *126*(6), 1065–1077. doi:10.1016/j.cell.2006.08.032 PMID:16962654

Kosrapohr, (Ed). (2003). *Managing data mining technologies in organizations: techniques and applications*. Hershey, PA: Idea Group Publishing.

Kostromina, N. A. (1990). *Chemistry of coordination compounds*. Chemistry.

Lake, J. A. (1976). Ribosome structure is determined by electroscope of *Escherichia coli* small subunits, large subunits, and monomeric ribosomes. *Journal of Molecular Biology*, *105*(1), 131–159. doi:10.1016/0022-2836(76)90200-X PMID:792456

Lalalyants, I. (1990). "Closure" in Paris. *Science and Life*, *1*, 86–89.

Lam, V., Takechi, R., & Mamo, J. C. L. (2021, September 14). Synthesis of human amyloid restricted to liver results in an alzheimer disease – like neurodegenerative phenotype. *PLoS Biology*, *19*(9), e3001358. doi:10.1371/journal.pbio.3001358 PMID:34520451

Landa, S., & Machacek, V. (1933). Sur l'adamantane, nouvel hydrocarure exait du naphte. *Collection of Czechoslovak Chemical Contributions*, *5*, 1–5. doi:10.1135/cccc19330001

Landau, L. D., & Lifshitz, E. M. (1963). Quantum mechanics. Nonrelativistic theory. Moscow: Gos. Ed. Phys.- Math. Lit.

Lehninger, A. L. (1982). *Principes of Biochemistry*. Worth Publishers, Inc.

Lehninger, A. L. (1982). *Principles of Biochemistry* (Vol. 1-3). Worth Publisher, Inc.

Lehn, J. M. (1998). *Supramolecular chemistry. Concept and perspectives*. Science.

Lindquist, S. (2016). Luminidependens (LD) is an arabidopsis protein with prion behavior. *Proceedings of the National Academy of Sciences of the United States of America*, *113*(21), 6065–6070. doi:10.1073/pnas.1604478113 PMID:27114519

Lobachevsky, N.I. (1835). *Imaginary geometry*. Scientific notes of Kazan University.

Lobachevsky, N. I. (1829). *About the beginnings of geometry*. Kazan University Bulletin.

Lord, E. A., Mackay, A. L., & Ranganathan, S. (2006). *New Geometry for New Materials*. Cambridge University Press.

Lundberg, L., Pinkham, C., Baer, A., Amaya, M., Narayanan, A., Wagstaff, K. M., Jans, D. A., & Kehn-Hall, K. (2013). Nuclear import and export inhibitors alter capsid protein distribution in mammalian cells and reduce Venezuelan Equine Encephalitis Virus replication. *Antiviral Research*, *100*(3), 662–672. doi:10.1016/j.antiviral.2013.10.004 PMID:24161512

Lyashchenko, A. K., & Dunyashev, V. S. (2003). Complementary organization of the water structure. *Journal of Structural Chemistry*, *44*(5), 909–915. doi:10.1023/B:JORY.0000029822.94836.67

Lyles, D.S. (2000). *Cytopathogenesis and Inhibition of Host Gene Expression by RNA Viruses*.

Lyles, D. S. (2000, December). Cytopathogenesis and Inhibition of Host Gene Expression by RNA Viruses. *Microbiology and Molecular Biology Reviews*, *64*(4), 709–724. doi:10.1128/MMBR.64.4.709-724.2000

Mackay, A. (2001). On Complexity. *Crystallography Reports*, *46*(4), 524–526. doi:10.1134/1.1387117

Magerramov, A. M., Makhmudov, K. T., Kopylovich, M. N., & Pombeiro, A. J. L. (Eds.). (2016). *Non-covalent interactions in the synthesis and design of new compounds*. Wiley. doi:10.1002/9781119113874

Mancuso, S. (2014). Experience teaches plants to learn faster and forget slower in environments where it matters. *Oecologia*, *175*(1), 63–72. doi:10.1007/s00442-013-2873-7 PMID:24390479

Mancuso, S. (2017). *PLANT REVOLUTION. Le piante hanno già inventato il nostro futuro*. GUIUTI EDITORE.

Mancuso, S., & Viola, A. (2013). *Sensibilità e intelligenza del mondo vegetale*. GUIUTI EDITORE.

Martynova, O. I., Gusev, B. T., & Leont'ev, E. A. (1969). To the question of the mechanism of the influence of the magnetic field on aqueous solutions of salts. *Успехи физических наук*, *98*(1), 195–199. doi:10.3367/UFNr.0098.196905n.0195

Martynshev, L. (2006). Maximum entropy production principle in physics, chemistry, and biology. *Physics Reports*, *426*(1), 1–45. doi:10.1016/j.physrep.2005.12.001

Melekhova, O. P. (2010). *Free radical processes in epigenomic regulation of development*. Nauka.

Metzler, D. E. (1977). *BIOCHEMISTRY. The Chemical Reactions of Living Cells*. Academic Press.

Metzler, D. E. (1977). *The Chemical Reaction of Living Cells*. Academic Press.

Metzler, D. E. (1980). *Biochemistry. The Chemical Reactions in Living Cells* (Vol. 1-3). Academic Press.

Metzler, D. E. (1980). *Biochemistry. The Chemical Reactions of Living Cells* (Vol. 1-3). Academic Press.

Miller, E. V. (1969). On the spontaneous decrease in the density of water condensed from steam. *Reports of the USSR Academy of Sciences*, *184*(1), 136–141.

Miller, S. L. (1953). The production of amine acids under possible primitive earth conditions. *Science*, *117*(3046), 528–529. doi:10.1126/science.117.3046.528 PMID:13056598

Miller, S. L., & Urey, H. C. (1959). Organic compound synthesis on the primitive Earth. *Science*, *130*(3370), 245–259. doi:10.1126/science.130.3370.245 PMID:13668555

Milton, R. C., Milton, S. C., & Kent, S. B. (1992). Total chemical synthesis of a D – enzyme: The enantiomers of HIV – 1 proteus show demonstration of reciprocal chiral substrate specificity. *Science*, *256*(5062), 1445–1448. doi:10.1126/science.1604320 PMID:1604320

Nelson, D. L., & Cox, M. M. (2022). Lehninger Principles of Biochemistry. (Vol. 1 – 3). W.H. Freeman. Macmillan Learning, New York.

Nelson, D. L., & Cox, M. M. (2022). *Lehninger Principles of Biochemistry*. W.H. Freeman.

Nirenberg, M. W., Jones, O. W., Leder, P., Clarck, B. F. C., Sly, W. S., & Pestka, S. (1963). On the coding of genetic information. *Cold Spring Harbor Symposia on Quantitative Biology, 28*(0), 549–557. doi:10.1101/SQB.1963.028.01.074

Nirenberg, M. W., & Leder, P. (1964). RNA codeword and protein synthesis: The effect of trinucleotides upon the binding of sRNA to ribosomes. *Science, 145*, 549–557. doi:10.1126/science.145.3639.1399 PMID:14172630

Nirenberg, M. W., & Mattei, J. H. (1961). The dependence of cell-free protein synthesis in *E. coli* upon naturally occurring or synthetic polyribonucleotides. *Proceedings of the National Academy of Sciences of the United States of America, 47*(10), 1588–1602. doi:10.1073/pnas.47.10.1588 PMID:14479932

Nonomura, Y., Blobel, G., & Sabatini, D. (1971). Structure of liver ribosomes studied by negative staining. *Journal of Molecular Biology, 60*(2), 303–324. doi:10.1016/0022-2836(71)90296-8 PMID:4106575

Novikova, N. A. (2007). *Storage and sale of virus genetic information.* Lower Novgorod State University named N.I. Lobachevsky: Lower Novgorod.

Novikova, N.A., Novikov, V.V., Dobrotina, N.A., & Mazepa, V.N. (2002). *Virology.* – Lower.

Nussbaum, J. M., Seward, M. E., & Bloom, G. S. (2013). Alzheimer disease a tale of two prions. *Prion, 7*(1), 14–19. doi:10.4161/pri.22118 PMID:22965142

Oro, J. (1961). Mechanism of synthesis of adenine from hydrogen cyanide under plausible primitive Earth conditions. *Nature, 191*(4794), 1193–1194. doi:10.1038/1911193a0 PMID:13731264

Oswald, A. T., MacLeod, C. M., & McCarty, M. (1944). Studies on the Chemical Nature of the Substance Inducing Transformation of Pneumococcal Types: Induction of Transformation by a Desoxyribonucleic Acid Fraction Isolated from Pneumococcus Type III. *The Journal of Experimental Medicine, 79*(2), 137–158. doi:10.1084/jem.79.2.137 PMID:19871359

Paster, L. (1960). *Selected words.* Academy of Sciences of the USSR.

Pasteur, L. (1960). *Selected works.* Publishing House of the Academy of Sciences of the USSR.

Pauling, L. (1987). So-called Icosahedral Decagonal Quasicrystals are Twins of an 820-atom Cubic Cristal. *Physical Review Letters, 58*(4), 365–368. doi:10.1103/PhysRevLett.58.365 PMID:10034915

Pauling, L., & Corey, R. B. (1950). Two Hydrogen-bonded Spiral Configurations of the Polypeptide Chain. *Journal of the American Chemical Society, 72*(11), 5349. doi:10.1021/ja01167a545

Pauling, L., & Corey, R. B. (1953). Compound Helical Configurations of the Polypeptide Chain: Structure of Proteins of the α - keratin Type. *Nature, 171*(4341), 59–61. doi:10.1038/171059a0 PMID:13025480

Pauling, L., Corey, R. B., & Branson, H. R. (1951). The Structure of Proteins: Two Hydrogen-bonded Helical Configurations of the Polypeptide Chain. *Proceedings of the National Academy of Sciences of the United States of America, 37*(4), 205–211. doi:10.1073/pnas.37.4.205 PMID:14816373

Piller, C. (2022). Research backing experimental Alzheimer`s drug was first target of suspicion. *Science, 377*, 6604.

Poincare, A. (1895). Analysis situs. *J de é Ecole Polytechnique, 1*, 1 – 121.

Poincare, A. (1895). Analysis situs. *J. de é Ecole Polytechnique, 1*, 1 – 121.

Poincaré, A. (1895). Analysis situs. *J. de é Ecole Polytechnique, 1*, 1 – 121.

Poincaré, A. (1895). Analysis situs. *J. de é Ecole Polytechnique, 1*, 1-121.

Poincare, A. (1895). Analysis situs. *J. de e` Ecole Polytechnique, 1*, 1 – 121.

Poincare, A. (1895). Analysis situs. *J de e'Ecole Polytechnique., 1*, 1–121.

Poincare, A. (1895). Analysis situs. *J. de e'Ecole Polytechnique, 1*, 1–121.

Poincare, A. (1895). Analysis Situs. *Journal de Ecole Polytechnique, 1*, 1–121.

Ponnuswamy, P., & Gromiha, M. (1994). On the conformational stability of oligonucleotide duplexes and tRNA molecules. *Journal of Theoretical Biology, 169*(4), 419–432. doi:10.1006/jtbi.1994.1163 PMID:7526075

Puglisi, J., & Williamson, J. R. (1999). RNA interaction with small ligands and peptides. In R. F. Gesteland, T. R. Cech, & J. F. Atkins (Eds.), *The RNA World* (2nd ed., pp. 403–425). Cold Spring Harbor Laboratory Press.

Qian, G., Brugger, J., Chen, G., & Skinner, W. M. (2010). An experimental study of the mechanism of the replacement of magnetite by pyrite up to 300 ^{0} C. *Geochimica et Cosmochimica Acta, 74*(19), 5610–5630. doi:10.1016/j.gca.2010.06.035

Rabinovich, S., Baldini, J. T., & Bannister, R. (1969). The therapeutic efficacy of rimantadine HCl in naturally influenza A2 outbreak. *The American Journal of the Medical Sciences, 257*(5), 328–335. doi:10.1097/00000441-196905000-00005 PMID:4891873

Rahman, A., & Stillinger, F. H. (1973). Hydrogen-Boden patterns in liquid water. *Journal of the American Chemical Society, 95*(24), 7943–7948. doi:10.1021/ja00805a003

Ramanathan, A., Brett, R.G., Siu – Hong, Chan. (2016). mRNA capping: biological functions and applications. *Nucleic Acids Res., 19*, 7511 – 7526.

Rasonable, R. R. (2011). Antiviral Drugs for Viruses Otber Than Humanimmunodeficiency virus. *Mayo Clinic Proceedings, 86*(10), 1009–1026. doi:10.4065/mcp.2011.0309 PMID:21964179

Riemann, G. F. B. (1854). On the Hypotheses Which Lie at the Foundations of Geometry. Gottingen: Gottingen Observatory.

Riemann, B. (1854). *Ueber die Hypothesen, welche der Geometrie zu Grunde liegen*. University of Göttingen.

Rodger, J. A. (2015). Discovery of medical Big Data analytics: Improving the prediction of traumatic brain injury survival rates by data mining Patient Informatics Processing Software Hybrid Hadoop Hive. *Informatics in Medicine Unlocked*, *1*, 17–26. doi:10.1016/j.imu.2016.01.002

Rodnikova, M. N. (1993). Features of solvents with a spatial network of H-bonds. *Journal of Physical Chemistry*, *67*(2), 275–280.

Russell, M. J., & Hall, A. J. (1999). On the inévitable émergence of life on Mars. In: Hiscox, J.A. (Ed.) *The Search for Life on Mars, Proceedings of the 1st UK Conference*. British Interplanetary Society.

Russell, M. J. (2007). The alkaline solution to the emergence of life: Energy, entropy and early evolution. *Acta Biotheoretica*, *55*(2), 133–179. doi:10.1007/s10441-007-9018-5 PMID:17704896

Russell, M. J., Hall, A. J., & Turner, D. (1989). In vitro growth of iron sulfide chimneys: Possible culture chambers for origin -of-life experiments. *Terra Nova*, *1*(3), 238–241. doi:10.1111/j.1365-3121.1989.tb00364.x

Russel, M. G., Hall, A. J., Boyce, A. J., & Fallick, A. E. (2005). On hydrothermal convection systems and the emergence of life. *Economic Geology and the Bulletin of the Society of Economic Geologists*, *100*, 419–438.

Samoilov, O. Y. (1957). *Structure of aqueous solutions and hydration of ions"*. Science.

Samson, S. (1972). Complex cubic A_6B compounds. II. The crystal structure of Mg_6Pd. *Acta Crystallographica. Section B, Structural Crystallography and Crystal Chemistry*, *28*(3), 936–945. doi:10.1107/S0567740872003437

Schaich, K. M. (2005). In F. Shahidi (Ed.), *Lipid Oxidation: Theoretical Aspects. Bailey's Industrial Oil and Fat Products* (1st ed., pp. 269–355).

Schifitto, G., Navia, B.A., & Yiannoutsos, C.T. (2007). Memantine and HIV- associated conqnitive impairment o neuropsychological and proton magnetic resonance spectroscopy study. *AIDS journal, 21*(14), 1877-1886.

Schramm, G., Grötsch, H., & Pollmann, W. (1962). Nicht-enzymatische von Polysacchariden, Nucleosiden und Nucleinsäuren und die Entstehung selbst -vermehrungsfähiger systeme. *Angewandte Chemie*, *74*(2), 53–59. doi:10.1002/ange.19620740202

Schrodinger, (1928). *Abhandlungen zur Wellenmechanik*. Leipzig.

Semenchenko, V. N. (1932). *What are the solutions?* Science.

Semenov, N. N. (1934). *Chain reactions*. Goshimtekhizdat.

Shafranovsky, I. I. (1964). *Diamonds*. Science.

Shechtman, D., Blech, I., Gratias, D., & Cahn, J. W. (1984). Metallic phase with longer range orientational order and no translational symmetry. *Physical Review Letters*, *53*(20), 1951–1953. doi:10.1103/PhysRevLett.53.1951

Shevchenko, V., Zhizhin, G., & Mackay, A. (2013). On the structure of the quasicrystals in the high dimension space. In M. V. Diudea (Ed.), *Diamond and Related Nanostructures*. Springer. doi:10.1007/978-94-007-6371-5_17

Simakov S., Barone G., Melnik N.N., Scribano V. (2021). Sicilian serpentinite xenoliths containing abiotic organics with nanodiamond clusters as key model for prebiotic processes. *Geoscience Frontiers*, *12*(1), 255 – 263. . doi:10.1016/j.gsf.2020.04.008

Simakov, S. (2018). Nano- and micro-sized diamond genesis in nature: An overview. *Geoscience Frontiers*, *9*(6), 1849–1858. doi:10.1016/j.gsf.2017.10.006

Speyer, J. F., Lengyel, P., Basilio, C., Wahba, A. J., Gardner, R. S., & Ochoa, S. (1963). Synthetic polynucleotides and the amino acid code. *Cold Spring Harbor Symposia on Quantitative Biology*, *28*(0), 559–567. doi:10.1101/SQB.1963.028.01.075

Spiegelman, S., & Haruna, I. (1966). A rational for an analysis of RNA replication. *Proceedings of the National Academy of Sciences of the United States of America*, *55*(6), 1539–1554. doi:10.1073/pnas.55.6.1539 PMID:5227673

Spirin, A.C. (2019). *Molecular biology. Ribosome and protein biosynthesis. Moskow: Laboratory of knowledge.*

Spirin, A.C. (2019). *Molecular Biology. Structure of the ribosomes and protein biosynthesis.* Moscow: High school.

Spirin, A.S. (1986). *Molecular biology. Structure of the ribosome and protein biosynthesis.* Moscow: High school.

Spirin, A. C. (1960). On macromolecular structure of native high-polymer ribonucleic acid in solution. *Journal of Molecular Biology*, *2*(6), 436–446. doi:10.1016/S0022-2836(60)80054-X

Spirin, A. S. (1969). Informosomes. *European Journal of Biochemistry*, *10*, 20–35. doi:10.1111/j.1432-1033.1969.tb00651.x PMID:4899460

Spirin, A. S. (1996). Masked and translatable messenger ribonucleoproteins in higher eukaryotes. In J. W. B. Hershey, M. B. Mathews, & N. Sonenberg (Eds.), *Translational Control* (pp. 319–334). Cold Spring Harbor Laboratory Press.

Spirin, A. S. (2019). *Molecular biology. Structure of the ribosome and protein biosynthesis.* Knowledge Lab.

Spirin, A. S. (2019). *Molecular biology. The ribosomes and protein biosynthesis.* Knowledge Lab.

Spirin, A. S. (Ed.). (1990). *Molecular biology. The structure and biosynthesis of nucleic acids.* High School.

Spirin, A. S., Belozersky, A. N., Shugaeva, A. N., & Vanyushin, B. F. (1957). Study of the species specificity of nucleic acids in bacteria. *Biochemistry (Moscow)*, *22*, 744–754. PMID:13479533

Spirin, A. S., Gavrilova, A. P., Bresler, S. E., & Mosovetsky, M. I. (1959). Study of the macromolecular structure of infectious ribonucleic acid from tobacco mosaic virus. *Biochemistry (Moscow)*, *24*, 938–947. PMID:13833381

Spirin, A. S., & Gavrilova, L. P. (1971). *The Ribosome*. Science.

Spirin, A. S., Kiselev, N. A., Shakulov, R. S., & Bogdanov, A. A. (1963). Study of the structure of ribosomes. Reversible unfolding of ribosomal particles into ribonucleoprotein strands and a folding pattern. *Biochemistry*, *28*, 920–930. PMID:14092472

Spirin, A. S., & Nrer, M. (1965). Messenger RNA in early see – urchin embryos: Cytoplasmic particles. *Science*, *150*(3693), 214–217. doi:10.1126/science.150.3693.214 PMID:5837337

Steed, J. V., & Atwood, J. L. (2000). *Supramolecular chemistry*. John Wiley and Sons.

Stewart, M., & McLachlan, A. D. (1975). Fourteen actin-binding sites on tropomyosin? *Nature*, *257*(5524), 331–333. doi:10.1038/257331a0 PMID:1161036

Takai, K., Gamo, T., Tsunogai, U., Nakayama, N., Nealson, K. H., & Horikoshi, K. (2004, August). Geochemical and microbiological evidence for a hydrogen-based, hyper-thermophilic subsurface lithoautotrophic microbioal ecosystem (HeperSLIHE) beneath an active deep-sea hydrothermal field. *Extremophiles*, *8*(4), 269–282. doi:10.1007/s00792-004-0386-3 PMID:15309563

Tatarnikova, O.G., Orlov, M.A., Bobkova, N.V. (2015). Beta-amyloid and tau protein: structure, interaction and primary properties. *The successes of biological chemistry*, *55*, 351-390.

Tatarsky, F. (1989). Struggle for "strange" water. *Chemistry and Life*, *12*, 57–59.

Timofeeff – Ressovsky, N.W., & Zimmer, K.G. (1947). *BIOPHYSIK. Das Trefferprinzip in der Biologie.* Leipzig: Arbeitsgemeinschaft medizinischer Verlage G.m.b.H.

Tuerk, C., & Gold, L. (1990). Systematic evolution of ligands by exponential enrichment. *Science*, *249*(4968), 505–510. doi:10.1126/science.2200121 PMID:2200121

Valeri – Rado. R. (1950). Life of Paster. Publishing Hause of Forging Literature: Moscow.

Vance, S., Harnmeijer, J., Kimura, J., Hussmann, H., Demartin, B., & Braun, J. M. (2007). Hydrothermal systems in small ocean planets. *Astrobiology*, *7*(6), 987–1005. doi:10.1089/ast.2007.0075 PMID:18163874

Vasiliev, V. D., Serdyuk, I. N., Gudkov, A. T., & Spirin, A. S. (1986). Self-organization of ribosomal RNA. In: Structure, Function, and Genetics of Ribosomes (eds. Hardesty, B and Kramer, G.), 128 – 142. Springer-Verlag, New-York. doi:10.1007/978-1-4612-4884-2_8

Vasiliev, V. D. (1971). Electron microscopy study of 70 S ribosomes of *Escherichia coli. FEBS Letters*, *14*(4), 203–205. doi:10.1016/0014-5793(71)80617-8 PMID:11945758

Vasiliev, V. D., Selivanova, O. M., & Koteliasky, V. E. (1978). Specific self-packing of the ribosomal 16S RNA. *FEBS Letters*, *95*(2), 273–276. doi:10.1016/0014-5793(78)81009-6 PMID:363457

Vernadsky, V.I. (1925). The history of the minerals of the earth's crust. *Scientific chemical and technical publishing house*, *1*(1), 1 - 208.

Voitkevich, G. V. (1996). Birth of the Earth: Reader for secondary schools. Phoenix: Rostov-on-Don.

Watanabe, K., & Osawa, S. (1955). tRNA sequences and variations in the genetic code. In: tRNA: Structure, Biosynthesis, and Function (eds. Soell D. and RajBhandary, U.L.): 225 – 250. ASM Press, Washington, DC.

Watson, J. D. (1976). *Molecular Biology of the Gene*. Benjamin.

Watson, J. D., & Crick, F. H. C. (1953a). Molecular structure of nucleic acids. *Nature*, *171*(4356), 738–740. doi:10.1038/171737a0 PMID:13054693

Watson, J. D., & Crick, F. H. C. (1953b). Genetical implications of the structure of deoxyribose nucleic acid. *Nature*, *171*(4361), 964–967. doi:10.1038/171964b0 PMID:13063483

Wilson, A., Kaplan, N. O., & Ciotti, M. M. (1961). Evolution and differentiation of dehydrogenasis. *Annals of the New York Academy of Sciences*, *94*(3), 701–722. doi:10.1111/j.1749-6632.1961.tb35567.x PMID:14037006

Wlodamer, A., Miller, M., Jaskólski, M., Sathayanarayana, B. K., Baldin, E., Weber, I. T., Selk, L. M., Chawson, L., Schneider, J., & Kent, S. B. (1989). Conserved folding in retroviral proteases: Crystal structure of synthetic HIV – 1 protease. *Science*, *245*(4918), 616–621. doi:10.1126/science.2548279 PMID:2548279

Wood, R. (1936). *Research in physical optics*. Wentworth Press.

Wulan, W. N., Heydet, D., Walker, E. J., Gahan, M. E., & Ghildyal, R. (2015). Nucleocytoplasmic transport of nucleocapsid proteins of enveloped RNA viruses. *Frontiers in Microbiology*, *6*, 553. doi:10.3389/fmicb.2015.00553 PMID:26082769

Wurm, T., Chen, H., Hodgson, T., Britton, P., Brooks, G., & Hiscox, J. A. (2001). Localization to the nucleolus is a common feature of coronavirus nucleoproteins, and the protein may disrupt host cell division. *Journal of Virology*, *75*(19), 9345–9356. doi:10.1128/JVI.75.19.9345-9356.2001 PMID:11533198

Ycas, M. (1969). *The Biological Code*. North – Holland Publ. Co.

Zhizhin, G. V. (2012, October). *Hierarchical filling of spaces with polytopes.* Paper presented at "St. Petersburg Scientific Forum: Science and Human Progress". 7th St.- Petersburg meeting of Nobel Prize laureates, St. – Petersburg, Russia.

Zhizhin, G. V. (2013 a). Relations for the number of faces of different dimensions in the tower of n - dimensional convex polytopes. Materials 9th All-Russian Scientific School "Mathematical research in the natural sciences. Geological institute KSC RAS, Apatity.

Zhizhin, G. V. (2013 a). Relations for the number of faces of different dimensions in the tower of n - dimensional convex polytopes. Materials 9th All-Russian Scientific School "Mathematical research in the natural sciences." 24-32. Geological institute KSC RAS, Apatity.

Zhizhin, G. V. (2013 b). Images of convex regular and semiregular n-dimensional polytopes. Materials 9th All-Russian Scientific School "Mathematical research in the natural sciences." 32-42. Geological institute KSC RAS, Apatity.

Zhizhin, G. V. (2013). Relations for the numbers of faces of various dimensions in the tower of n - dimensional convex regular polytopes. *Proceedings of the 9th All - Russian School "Mathematical Studies in the Natural Sciences."* Apatity: Geological Institute, Russian Academy of Sciences.

Zhizhin, G. V. (2013b). Images of convex regular and semiregular n-dimensional polytopes. Materials 9th All-Russian Scientific School Mathematical research in the natural sciences. Geological institute KSC RAS, Apatity.

Zhizhin, G. V. (2015). November. *Polytopic prismahedrons – fundamental regions of the n -dimension nanostructures.* Paper presented at The International conference Nanoscience in Chemistry, Physics, Biology and Mathematics, Cluj – Napoca, Romania.

Zhizhin, G. V. (2015, November). *Polytopic prismahedrons – fundamental regions of the n-dimension nanostructures.* Paper presented at The International conference "Nanoscience in Chemistry, Physics, Biology and Mathematics," Cluj – Napoca, Romania.

Zhizhin, G. V. (2021 a). The Structures of DNA Taking into Account the Higher Dimensional of its Components. In Kosrupohr (Ed.) Encyclopedia of Organizational Knowledge, Administration, and Technologies. Hershey, PA: IGI Global.

Zhizhin, G. V. (2021 b). The Structures of DNA Taking into Account the Higher Dimensional of its Components. In Kosrupohr (Ed.) Encyclopedia of Organizational Knowledge, Administration, and Technologies. Hershey, PA: IGI Global.

Zhizhin, G. V. (2021). The Structures of DNA Taking into Account the Higher Dimensional of its Components. In Kosrupohr (Ed.) Encyclopedia of Organizational Knowledge, Administration, and Technologies. Hershey, PA: IGI Global.

Zhizhin, G. V., Khalaj, Z., & Diudea, M. V. M.V. (2016). Geometrical and topological dimensions of the diamond. Distance, symmetry and topology in carbon nanomaterials. Springer.

Zhizhin, G.V. (2020 b). Higher Dimensions in Linear, Helicoidal and Folder Structures of Protein Molecules. *Acta Scientific Biotechnology,* 1(12), 02-09.

Zhizhin, G.V. (2020 c). Higher Dimensions in Linear, Helicoidal and Folder Structures of Protein Molecules. *Acta Scientific Biotechnology, 1*(12), 02-09.

Zhizhin, G.V. (2023 b). *Adamantane as a Means of Higher Dimension, Preventing the Formation of the Beta- Amyloid,* 6(1), 1 – 9.

Zhizhin, G. V. (2014). On High Dimensionalities in Nature. *Biosphere., 6*(4), 313–318.

Zhizhin, G. V. (2014). On the higher dimension in nature. *Biosphere*, *6*(4), 313–318.

Zhizhin, G. V. (2014). *World - 4D*. Polytechnic Service.

Zhizhin, G. V. (2014a). On the higher dimensions in nature. *Biosphere*, *6*(4), 313–318.

Zhizhin, G. V. (2014a). The fractal nature of disproportionate phases. [th all-Russian Scientific School "*Mathematical research in the natural sciences*". Apatity: Geological institute KSC RAS.]. *Materials (Basel)*, 10.

Zhizhin, G. V. (2014b). Incommensurable and fluctuating structures in the terrestrial space. *Biosphere*, *3*, 211–221.

Zhizhin, G. V. (2014b). *World 4D*. Polytechnic Service.

Zhizhin, G. V. (2014c). *World – 4D*. Polytechnic Service.

Zhizhin, G. V. (2015a). The dimensions of supramolecular compounds. *Biosphere*, *7*(2), 149–154.

Zhizhin, G. V. (2015b). Dimensions of compounds in supramolecular chemistry. *Int. J. Chem. Modeling*, *7*(2), 215–238.

Zhizhin, G. V. (2016). The structure, topological and functional dimension of biomolecules. *J. International Journal of Chemoinformatics and Chemical Engineering*, *5*(2), 44–58. doi:10.4018/IJCCE.2016070104

Zhizhin, G. V. (2016a). From multidimensional phase spaces to multidimensionality of space in Nature. *Biosphere*, *8*(3), 258–267.

Zhizhin, G. V. (2017). The structure, topological and functional dimension of biomolecules. *J. Chemoinformatics and Chemical Engineering*, *5*(2), 44–58.

Zhizhin, G. V. (2018). *Chemical Compound Structures and the Higher Dimension of Molecules: Emerging Research and Opportunities*. IGI Global. doi:10.4018/978-1-5225-4108-0

Zhizhin, G. V. (2018b). The Metric Structure of Nucleic Acids and the Higher Dimension of Their Constituents. *International Journal of Chemoinformatics and Chemical Engineering*, *7*(2), 1–15. doi:10.4018/IJCCE.2018070101

Zhizhin, G. V. (2019 c). *Attractors and higher dimensions in population and molecular biology: emerging research and opportunities*. IGI Global.

Zhizhin, G. V. (2019 e). *Attractors and Higher Dimensions in Population and Molecular Biology: Emerging Research and Opportunities*. IGI Global. doi:10.4018/978-1-5225-9651-6

Zhizhin, G. V. (2019). Methylated Polytope of Hereditary Information. *Int. Journal of Applied Research on Public Health Management*, *6*(2), 49–60.

Zhizhin, G. V. (2019). *The Geometry of Higher – Dimensional Polytopes*. IGI Global. doi:10.4018/978-1-5225-6968-8

Zhizhin, G. V. (2019a). The Law of Conservation of Incidents in the Space of Nanoworld. *International Journal of Chemoinformatics and Chemical Engineering*, *8*(1), 25–46. doi:10.4018/IJCCE.2019010103

Zhizhin, G. V. (2019a). The Polytope of Hereditary Information the Structure, Location, Signification. *Biochemistry and Modern Applications*, *2*(1), 56–62. doi:10.33805/2638-7735.124

Zhizhin, G. V. (2019a). The structure of polytope of hereditary information. J. *Chemoinformatics and Chemical Engineering*, *8*(2), 7–22.

Zhizhin, G. V. (2019b). The Polytope of Hereditary Information: Structure, Location, Signification. *Biochemistry and Modern Applications.*, *2*, 56–62.

Zhizhin, G. V. (2019b). The Structure of the Polytope of Hereditary Information. *International Journal of Chemoinformatics and Chemical Engineering*, *8*(2), 7–22. doi:10.4018/IJCCE.2019070102

Zhizhin, G. V. (2019c). The Metric Structure of Nucleic Acids and Higher Dimension of Their Constituents. *J. Chemoinformatics and Chemical Engineering*, *7*(2), 1–15.

Zhizhin, G. V. (2019c). The metric structure of nucleic acids and the higher dimension of their constituents. *J. Chemoinformatics and Chemical Engineering*, *7*(2), 1–15.

Zhizhin, G. V. (2019d). Higher Dimension of Clusters of Intermetallic Compounds. *J. Applied Nanotechnology Research*, *4*(1), 8–25. doi:10.4018/IJANR.2019010102

Zhizhin, G. V. (2020). Hidden Nucleic Acids Bond Order? *Acta Scientific Biotechnology*, *1*(4), 34–36.

Zhizhin, G. V. (2020a). Dimension and Life. *Acta Scientific Biotechnology*, *1*(11), 35–40.

Zhizhin, G. V. (2020a). Hidden Nucleic Acid Bond Order? *Acta Scientific Biotechnology*, *1*(4), 34–36.

Zhizhin, G. V. (2020a). The Dimension and Life. *Acta Scientific Biotechnology*, *1*(11), 35–40.

Zhizhin, G. V. (2020b). Interaction of Nucleic Acids: Hidden Order of Interaction. *International Journal of Applied Research in Bioinformatics*, *11*(2), 1–15.

Zhizhin, G. V. (2020b). Interaction of Nucleic Acids: Hidden Order of Interaction. *International Journal of Applied Research in Bioinformatics*, *11*(2), 1–15. doi:10.4018/IJARB.2021070101

Zhizhin, G. V. (2020c). Spirals of Polysaccharides Simultaneously Rotating in Opposite Directions. *Acta Scientific Biotechnology*, *1*(7), 13–16.

Zhizhin, G. V. (2020e). Enantiomorphism of Tartaric Acid as a Property of Five-Dimensional Polytopes. *Acta Scientific Biotechnology*, *1*(12), 16–19.

Zhizhin, G. V. (2021 a). *Normal partitions and hierarchical fillings of n - dimensional spaces*. IGI Global. doi:10.4018/978-1-7998-6768-5

Zhizhin, G. V. (2021 a). The Structure of DNA Taking into Account the Higher Dimension of Its Components. In *Encyclopedia of Information Science and Technology* (5th ed., pp. 730–747). Ed. Mehdi Khosrow- Pour. doi:10.4018/978-1-7998-3479-3.ch050

Zhizhin, G. V. (2021 c). *Nanotechnology and Clusters in the Space of Higher Dimension: Emerging Research and Opportunities*. IGI Global. doi:10.4018/978-1-7998-3784-8

Zhizhin, G. V. (2021). Spatial Models of Sugars and Their Components. *International Journal of Applied Research in Bioinformatics, 11*(2), 9–23.

Zhizhin, G. V. (2021b). Hidden Order of Interacting Nucleic Acids. *Journal of Nature Science and Sustainable Technology, 15*(4), 1–5.

Zhizhin, G. V. (2021b). Spatial Models of Sugars and their Compounds. *International Journal of Applied Research in Bioinformatics, 11*(2), 9–23. doi:10.4018/IJARB.2021070102

Zhizhin, G. V. (2021c). Methylated Polytope of Hereditary Information. *International Journal of Applied Research on Public Health Management, 6*(2), 49–60. doi:10.4018/IJARPHM.2021070104

Zhizhin, G. V. (2022 b). *Non - Euclidean Geometry in Materials of Living and Non – Living Matter in the Space of the Highest Dimension*. Nova Science Publisher.

Zhizhin, G. V. (2022 c). *The Classes of Polytopes Higher Dimensions in Physics, Chemistry, and Biology. Emerging Research and Opportunities*. IGI Global.

Zhizhin, G. V. (2022). *The Classes of Polytopes Higher Dimensions in Physics, Chemistry and Biology*. IGI Global.

Zhizhin, G. V. (2022). *The Classes of Polytopes Higher Dimensions in Physics, Chemistry, and Biology*. IGI Global.

Zhizhin, G. V. (2022a). On the Possible Spatial Structures of the β – Amyloid. Towards the theory of the native structure of proteins. *Int. J. of Applied Research on Public Health Management, 20*(Jan).

Zhizhin, G. V. (2022a). Structure of Garnet. *J. of Nature Science and Sustainable Technology., 16*(2), 1–9.

Zhizhin, G. V. (2022b). Structure of Minerals of the Garnet Family in the Higher-Dimensional Space. *J. of Nature Science and Sustainable Technology., 16*(4), 1–9.

Zhizhin, G. V. (2022b). The Properties of Biomaterials Hidden in the Spaces of Higher Dimension. *Annals of Advanced Biomedical Sciences, 5*(2), 1–9. doi:10.23880/aabsc-16000189

Zhizhin, G. V. (2023 a). *Non-Euclidean geometry in Materials of Living and Non-Living Matter in the space of the Higher Dimension*. Nova Science Publisher.

Zhizhin, G. V. (2023 a). *Non-Euclidean Geometry in Materials of Living and Non-Living Matter in the Space of the Highest Dimension*. Nova Science Publisher.

Zhizhin, G. V. (2023). Adamantane as a Means of High Dimension, Preventing the Formation of the Beta – Amyloid. *Annals of Advanced Biomedical Sciences*, *6*(1), 1–4.

Zhizhin, G. V. (2023). *Non – Euclidean Geometry in Materials of Living and Non-Living Matter in the Space of the Highest Dimension*. Nova Science Publisher.

Zhizhin, G. V. (2023b). To the Issue of the Complementarity of Biomolecules in the Space of the Highest Dimension. *Journal of Clinical Oncology and Research*, *3*, 1–15.

Zhizhin, G. V. (2023c). To the Issue of the Complementary of Biomolecules in the Space of the Highest Dimension. *Journal of Clinical Oncology and Research*, *2*(1), 1–15.

Zhizhin, G. V., & Diudea, M. V. (2016). Space of nanoworld. In M. V. Putz & C. M. Marius (Eds.), *Sustainable nanosystems, development, properties, and applications* (pp. 221–245). IGI Global.

Zhizhin, G. V., & Diudea, M. V. (2016). Space of Nanoworld. In M. V. Putz & C. M. Marius (Eds.), *Sustainable nanosystems, development, properties, and applications* (pp. 221–245). IGI Global.

Zhizhin, G. V., & Diudea, M. V. (2016). Space of Nanoworld. In M. V. Putz & M. C. Mirica (Eds.), *Sustainable Nanosystems, Development, Properties, and Applications* (pp. 214–236). IGI Global.

Zhizhin, G. V., Khalaj, Z., & Diudea, M. V. (2016). Geometrical and topology dimensions of the diamond. In A. R. Ashrafi & M. V. Diudea (Eds.), *Distance, symmetry and topology in carbon nanomaterials* (pp. 167–188). Springer. doi:10.1007/978-3-319-31584-3_12

Zlydnikov, D. M., Kubar, O. I., Kovaleva, T. P., & Kamorin, L. E. (1969). Study of Rimantadine in USSR: A Review of the literature. *Clinical Infectious Diseases*, *3*(3), 408–421. doi:10.1093/clinids/3.3.408 PMID:7025146

Zundel, G. (1972). Hydrotation and intermolecular interactions. Mir: Moscow.

About the Author

Gennadiy Vladimirovich Zhizhin was born February 11, 1942. Zhizhin evacuated from Leningrad during the blockade and has Russian citizenship.

Index

O

P

Q

R

S

T

V

W

www.ingramcontent.com/pod-product-compliance
Lightning Source LLC
LaVergne TN
LVHW081339300926
849161LV00008B/2022
* 9 7 9 8 3 6 9 3 0 5 8 8 1 *